APPLIED FRESHWATER BIOLOGY

John S. Richardson, Ph.D.

ISBN-13: 978-1-60427-169-0
e-ISBN: 978-1-60427-857-6

Printed and bound in the U.S.A. Printed on acid-free paper.

10 9 8 7 6 5 4 3 2 1

Library of Congress Cataloging-in-Publication Data
Names: Richardson, John S., 1956– author.
Title: Applied freshwater biology / John Richardson.
Description: Plantation, FL : J. Ross Publishing, [2024] | Includes bibliographical references
    and index. | Summary: "In this comprehensive book, Richardson lays out the origins and
    nature of the most prominent environmental stressors to freshwater systems. The first two
    chapters provide a review of freshwater biology and hydrology. Each of the next 12 focuses
    on a particular class of stressors, interactions they may have with other stressors, and a range
    of solutions currently available to mitigate the problems they cause. The last two chapters
    pull together key concepts to focus on the restoration of freshwater ecosystems and the im-
    portance of long-term monitoring"— Provided by publisher.
Identifiers: LCCN 2023058968 (print) | LCCN 2023058969 (ebook) | ISBN 9781604271690
    (hardcover) | ISBN 9781604278576 (epub)
Subjects: LCSH: Freshwater biology—Textbooks. | BISAC: SCIENCE / Environmental Science
    (see also Chemistry / Environmental) | SCIENCE / Earth Sciences / Hydrology
Classification: LCC QH96 .R53 2024 (print) | LCC QH96 (ebook) | DDC 578.76—dc23/
    eng/20240206
LC record available at https://lccn.loc.gov/2023058968
LC ebook record available at https://lccn.loc.gov/2023058969

Direct all inquiries to J. Ross Publishing, Inc., 151 N. Nob Hill Rd., Suite 476, Plantation,
FL 33324.

Phone: (954) 727-9333
Fax: (561) 892-0700
Web: www.jrosspub.com

CONTENTS

INTRODUCTION

Water is essential for all life and sustains freshwater ecosystems around the world. In addition to the myriad life-forms that freshwater ecosystems support, these ecosystems provide us with water and food, as well as opportunities for recreation and personal reflection. Unfortunately, our freshwaters have become seriously impacted and degraded by various stressors. Throughout my career I have witnessed some of this degradation firsthand and have been able to incorporate case studies into aquatic ecosystem courses I have taught.

This book began as an outcome of teaching aquatic ecosystems courses for over 25 years, along with the need to provide students with a comprehensive volume containing practical applications of the lessons learned in freshwater biology. Being unable to find a book that fit the way I want to approach these subjects, I was inspired to write one that follows many of the key themes from my courses. Each chapter addresses relevant environmental issues affecting freshwater ecosystems and provides a solutions-based approach to help mitigate these stressors.

It would be difficult, however, for any one textbook to address all the topics that are important to the protection and conservation of freshwater ecosystems. I have tried to include most current issues, but admittedly there are more. Where appropriate, I have included examples from most types of freshwaters, but some threats are specific to particular ecosystems. For example, I have not addressed groundwater ecosystems in any substantive way. Some freshwaters are less frequently considered in the literature, such as temporary waters, tree holes, springs of all sorts, and others, and this book reflects that dearth of studies in having few mentions of such ecosystems.

The first two chapters of this book are intended as a review of freshwater biology, hydrology, and freshwater ecosystems but will not constitute a full course on these subjects. I assume the student has some previous background in ecology and freshwater biology. If not, an excellent text for freshwater ecology is Dodds and Whiles (2019). There are many other good books on freshwater ecosystems, and those should be consulted for a thorough grounding in the basics of the field—for instance, Hynes (1970), Hutchinson (1983), Allan et al. (2021), Jones and Smol (2023), Hildrew and Giller (2023), as well as others on specific topics and taxa.

Chapters 3 to 14 are each set up to explore different stressors affecting freshwater ecosystems. The first half of each chapter provides background and a summary of the impacts of a particular class of stressors and how they can also interact with other stressors to compound problems. The second half of each chapter is designed to consider the range of solutions currently available to remedy the impacts of these stressors. Chapters 15 and 16 pull together key concepts that were previously discussed to focus on the restoration of freshwater ecosystems and the importance of monitoring these restoration projects. Importantly, these chapters all contain relevant case studies as well as student activities. While each chapter could very easily be expanded to be a book on its own, the materials are intended to fit a one-term course. Examples are all referenced so that a reader who is interested in further exploration of a topic has entry points to the current literature.

While much of this book deals with the practical application of science to mitigate impacts that degrade water and watersheds, it is important to understand that use of this science is governed by policy and law enacted by governments that have the legal authority to set rules, objectives, targets, and limits. Watershed management and policy differs globally, and that is a broad topic covered in political science, law, engineering, and other fields and not primarily the domain of scientists. In an ideal world, scientists working in freshwater ecosystems are actively sought after for their advice on how to align policy with practice and outcomes, from the local government level on up.

Today we need trained practitioners who have the understanding to protect and restore our freshwater ecosystems for both human and nonhuman use around the world. It is my sincere hope that this book will be a valuable resource to meet this end. The intended audience includes students in upper-level classes and graduate classes in biology, environmental sciences, and environmental engineering (as well as practitioners in these respective fields). Each chapter is intended to provide material for approximately two consecutive class periods. Although I have used case studies throughout, instructors may wish to augment the examples in the text with local cases to amplify certain topics. Instructors may also want to elaborate on specific examples included and use figures from the original articles for classroom instruction.

ABOUT THE AUTHOR

John S. Richardson, Ph.D., was born and raised in Toronto, Canada, and earned his first degree from the University of Toronto (B.Sc. 1979). From there, his academic career took him west as he earned degrees from the University of Alberta (M.Sc. 1983) and the University of British Columbia (Ph.D. 1989). Dr. Richardson spent three years at Simon Fraser University as a post-doctoral fellow before landing a faculty position at the University of British Columbia. He counts himself as very fortunate to have been at UBC all these years. Dr. Richardson has had many roles, including faculty member, Head of Department, member of the Peter Wall Institute for Advanced Studies, and a member of the editorial boards of several high-ranking journals. He has also been on many panels dealing with endangered species, riparian regulations, and other applications of his research.

Photo credit: UBC Brand and Marketing/Martin Dee

Richardson's research has focused on freshwater and riparian area ecology, primarily as a population, community, and ecosystem ecologist. Science can ask basic questions to provide for generalization, and at the same time use applied problems and systems as the specific context to give relevance to our studies. The application of science requires the best science available, as that is where we put our understanding to the test. He has successfully mentored many excellent graduate students and post-doctoral fellows who have been his great pleasure to work with. He adds that working in a global community of scholars with many outstanding colleagues has been a treat, and there is not space to name them all, but it will be obvious from his publication record that he has been fortunate to work with many outstanding people.

Beyond the academy, John enjoys travel. Many of his excursions have been to kayak or hike in other parts of the world. When not traveling, he enjoys cycling and running. Music has also been one of his hobbies throughout his life.

ACKNOWLEDGMENTS

I am grateful to the University of British Columbia for the support during the writing of this book and the research that helped shaped the content. I appreciate the research funding from the Natural Sciences and Engineering Research Council of Canada and other research agencies. I have benefited over the years from the many members of my research team, including post-doctoral fellows, graduate students, undergraduates, and colleagues who have stimulated my thinking.

I am thankful for the thoughtful reviews of the chapters in this text by Ana Chará-Serna (British Columbia Institute of Technology), Will Clements (Colorado State University), Kay Colletti (University of British Columbia–UBC), Alan Dextrase (Trent University), Catherine Febria (University of Windsor), Jennifer Fisher (UBC), Jennifer Harding (Department of Fisheries and Oceans Canada–DFO), Marwan Hassan (UBC), John Janmaat (UBC), Karen Kidd (McMaster University), Peter Kiffney (National Oceanographic and Atmospheric Administration, USA), John Kominoski (Florida International University), Lenka Kuglerová (Swedish Agricultural University), Neil Mochnacz (DFO), Fielding Montgomery (Nova Scotia Salmon Association), R. Dan Moore (UBC), Timo Muotka (Oulu University, Finland), Sean Naman (DFO), Liz Perkin, Dave Reid (DFO), Scott Reid (Ontario Ministry of Natural Resources and Forests), Ralf Schäfer (Institute for Environmental Sciences, RPTU Kaiserslautern-Landau), Doug Watkinson (DFO), and Alex Yeung (BC Ministry of Water, Land and Resource Stewardship). Their reviews greatly improved the clarity and provided references I hadn't come across. My thanks to Bill McMillan for his interesting information about Columbia River fish populations and the impacts of dams. I am grateful to all those who provided data, photos, and other information that aided in preparation of this text.

I am extremely grateful to Gwen Eyeington (J. Ross Publishing) for seeding the idea of making this book, her very thoughtful and thorough editorial work in making it become a reality, and for her guidance and patience throughout this process. Thank you, Gwen!

My parents, Kathleen and Ronald Richardson, were great supporters, and I am grateful for all their guidance. I have been fortunate to have excellent academic mentors who have taught me and supported me throughout my career, especially in my early years—most notably, Rosemary Mackay and Bill (W.E.) Neill.

This work is dedicated to my wife, Daphne Richardson.

At J. Ross Publishing we are committed to providing today's professional with practical, hands-on tools that enhance the learning experience and give readers an opportunity to apply what they have learned. That is why we offer free ancillary materials available for download on this book and all participating Web Added Value™ publications. These online resources may include interactive versions of the material that appears in the book or supplemental templates, worksheets, models, plans, case studies, proposals, spreadsheets and assessment tools, among other things. Whenever you see the WAV™ symbol in any of our publications, it means bonus materials accompany the book and are available from the Web Added Value Download Resource Center at www.jrosspub.com.

Downloads for *Applied Freshwater Biology* include selected graphs, tables, and images from the book.

HYDROLOGY AND FRESHWATER BIOLOGY

There has always been a dance between water and living organisms, with life being ever dependent on water. Most of the earth's water is contained in our oceans, seas, and coastal marshes as saltwater. Of the freshwater on the planet, most is locked up in glaciers (about 68.7%) and groundwater (about 30.1%) and therefore not available for species living on the earth's surface. That leaves only about 1.2% of freshwater for lakes, streams, wetlands, and other freshwater ecosystems (source: United States Geological Survey). Lakes and streams cover an estimated 2.3% of the nonglaciated land surface of the world, and very much less of the planet's total surface, and yet support about 9% of the world's animal species (Balian et al. 2008; Allen and Pavelsky 2018; Reid et al. 2019). Freshwater species are more than twice as likely to be in decline as species in marine or terrestrial ecosystems (WWF 2022). These are important statistics that demonstrate that the relation between freshwater and life is delicately balanced. The way water moves through ecosystems falls under the science of hydrology, and the way plants, animals, and microorganisms use freshwater is addressed in the science of freshwater biology. It is important to briefly review key concepts of both disciplines before moving forward.

HYDROLOGY

Hydrology is the study of the movement of water—from evaporation through to precipitation, pathways of water flows, and runoff generation (see Figure 1.1). Hydrology is a science of its own, and there are full courses and degree programs devoted to this topic (one suggested text is Dingman 2015). Hydrology of any particular region is influenced by climate, geology, topography, and biology; and hydrology provides one of the primary controls on the kinds of freshwater ecosystems you could encounter.

One basic spatial unit for studying freshwater ecosystems is the watershed or catchment, which is the area that contributes water to a stream, lake, or wetland (also known as contributing area). Watershed is commonly used in North America, whereas the term *catchment* is more typical in Europe, where the word *watershed* usually means the point at which water is shed one direction or the other. Water moves from higher to lower elevations, due to the force of gravity. Water flows on the surface in the form of streams and rivers or below the surface in the form of groundwater. The difference in elevation that water could move through is also known as the gravity head or hydraulic head, and indicates a potential amount of energy per unit weight of water if it falls over that distance, which we will return to in Chapter 7 where we consider

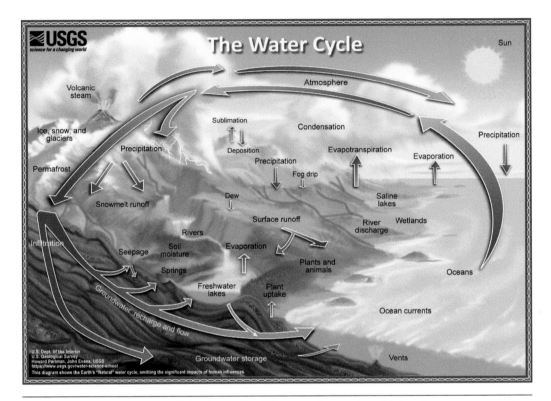

Figure 1.1   The water cycle. This shows the different pathways by which water moves around the planet and where it comes from, thereby creating the freshwater ecosystems that we study. *Source*: United States Geological Service; Howard Perlman and John Evans.

hydroelectric power production. The topography may be a good indicator of the direction that water movement might take, although in relatively flat regions with porous soils, groundwater may not follow the slope at the earth's surface. In some places, especially with little topography, the surface may not indicate the direction of water flow beneath the ground since water flow is also affected by the conductivity of the ground and how deep the soils are.

Water moves in other ways as well. It can be deposited from the atmosphere as precipitation in the form of rain, snow, or hail, and it can also condense as water droplets on various surfaces. In forests (especially cloud forests) and on other structures, condensation may be substantial (Berland et al. 2017), although this input may not be measurable directly as precipitation. These inputs from precipitation and condensation flow with gravity into the ground or overland. When water percolates below the earth's surface it is known as groundwater (deep or shallow) or the water table, and then it is referred to as the phreatic zone or the saturated zone. As water moves through the groundwater it re-emerges at lower points on the surface as run-off generation to springs, streams, lakes, and wetlands. Groundwater may percolate slowly, depending on the pore size of the materials it is working through, with it being slower through finer materials. Water stored in the ground may reside there (in *storage*) for time periods ranging from minutes to hours or days, and even sometimes up to many centuries. Hydrologists can age water to determine its residence time. In some cases, macropores develop in the soils as water erodes away particles or moves through earthworm burrows or spaces left by decaying roots to create preferential flow paths with lower resistance, which allows for faster movement of water through the ground (Sidle et al. 2000). In some cases, rock can be dissolved by slightly acidic water and particularly in limestone areas, underground streams can develop, and develop karst systems.

Water generally enters a stream from groundwater where the water table elevation is higher than the stream, and there may be preferential flow paths, as noted in the previous discussion about macropores. These can create focused areas of groundwater inflow, such as springs. As water reaches low points in the local environment, flow accumulates at the surface creating streams or filling areas such as lakes or wetlands. As streams flow downhill, they accumulate water from adjacent areas. Along a channel, there can be exchanges of surface flow with water underneath the channel in the gravel or sand bed, which is called *hyporheic* flow. The area beneath the stream bed is the hyporheic area where specific organisms are primarily found since the area can be used as a refuge from flooding. There is also parafluvial flow, a type of hyporheic exchange which is flow that is moving through gravel bars and sand bars at the edges or bends of a stream.

Water accumulates at the surface in streams, lakes, and wetlands and moves downhill toward the sea. Along the way, a large component of the net loss of water is through evaporation and transpiration (water that is taken up by plants and released through their leaves). Some estimates suggest that 55 to 67% of water is lost back into the atmosphere through transpiration by plants in temperate and boreal forests (Schlesinger and Jasechko 2014). Typically, as streams flow, they gain water. However, there can be *losing* stretches of a stream where the water table may be lower in elevation than the stream (due to extraction or change in geology) and where water leaves a stream and enters the groundwater, thereby reducing the stream's volume. A common example of this is where streams leave an area that is rocky and enter an area where soils are more permeable, such as desert areas where the water table may be much deeper and more permeable than the source areas.

There are many tools and methods specific to the study of hydrology. Two important measures are the inputs (precipitation, condensation) and outputs of water to a watershed or catchment. Inputs are often measured as rainfall using a gauge. However, precipitation can accumulate as snow. There are ways to figure out the snowpack water equivalent (depth of water resulting from melting the snow) based on the mass of water in the snow, which can be measured from a snow core or the mass of snow on top of pressure sensors on the ground beneath the snow (*pillows*). We also mentioned that inputs through condensation (usually included within the precipitation term) require special measurement methods.

Outputs from a known area via flow in a stream channel is usually measured as discharge (often L/s or m³/s) past some point. These estimates need to be calibrated from the depth of water impounded upstream of a weir (the *pond*) (see Figure 1.2) or from the depth of the water at a known cross section to calculate discharge. Weirs are typically found in smaller streams, and often in a research context. Both measures are usually based on a pressure transducer, which records the height of the water surface above the sensor. In some instances, particularly larger rivers (throughout this book we will mostly use the term *stream* for all moving waters), the water level known as *stage height* is measured against a kind of ruler that stands at the stream's edge. Whether with a stage gauge or pressure transducer, the measurement is just of water depth, which needs to be converted into discharge.

There are a few methods that can be used to create a calibration that will convert depth measures into discharge, also called a rating curve. One of those methods is an estimate of

Figure 1.2 A V-notch weir used to measure the height of the water surface; in this case, from a very small watershed. Water slows in the pond that was dammed just upstream of the weir, and the height of the water can be recorded on a pressure transducer, which measures the elevation of the water surface above it. Higher flows result in deeper water in the pond. The actual discharge in L/s still needs to be calibrated through a rating curve that relates the height of water to the flow, and this can be done by using a couple of methods, based on depth and current velocity profiles, or salt-dilution gauging. The calibration depends on getting a regression of discharge against the pressure reading. In very simple systems, there may be a staff gauge, which is often a big ruler that someone must read directly for water depth, but still needs calibration.

velocities and depths across a stream. In this case, a cross section of a stream is broken into vertical segments of a given width, and for each segment, the width and depth (area) and velocity can be multiplied to get the rate of flow for that vertical segment. One can then integrate all the verticals across the channel to get an estimate of instantaneous discharge. The selection of a point in the stream to measure is often associated with a natural narrowing of the stream width. Another method to calibrate the discharge is timing how long it takes to fill a bucket of known volume, which only works on very small streams. Yet another method is salt dilution gauging, where a known quantity of salt (either dry or dissolved in a brine) is injected at a point along the stream, and the salt concentration (usually measured as change in electrical conductivity) is recorded at a specific point that is a sufficient distance downstream where the salt has been fully mixed across the channel (BC Government 2018). Discharge is computed by the degree of dilution of the injected salt (lower concentration indicates higher dilution and thus, higher discharge). Stage height or pressure can be measured periodically by a direct reading of gauges, but are more commonly measured continuously given the widespread availability of data loggers. Discharge is a measure of total rate of water passing a point based on the rating curve developed from stage gauges. However, it may be useful to estimate the volume per second per unit area, the latter known as unit discharge, which accounts for the watershed area. Continuous records of flows can provide a very detailed understanding of discharge.

One tool used in hydrology to understand land-use effects is known as the paired-catchment approach; we will explore the results from such studies in later chapters. In this case, two (or more) catchments (hence paired) are matched for similarities in area, climate, vegetation cover, elevation, etc., and are measured for discharge for some period of time before a landscape-scale manipulation, such as forest harvesting or another land disturbance. By establishing the relationship of one watershed against the other (regression analysis) before the disturbance, we end up with a prediction of what the discharge would be from the watershed in the absence of a disturbance (see Chapter 8, Box 8.1). These studies usually involve alteration of one of the study watersheds through forest harvesting, pesticide application, agricultural expansion, or some other kind of land use (Moore et al. 2005). Such studies have been instrumental in advancing our understanding of how land-use practices affect the quantity and timing of flows, and often additional measures are taken, such as water quality (including temperature).

In summation, the movement of water in the study of hydrology is illustrated by the water balance equation. This equation is based on conservation of mass and has inputs and outputs (see Figure 1.1). The equation is usually shown as $P = Q + ET + \Delta S$, where P = precipitation, Q = discharge (streamflow), ET = evapotranspiration, and ΔS is the change in storage (such as in groundwater and aquifers). Water input to watersheds comes from the atmosphere (precipitation) in the form of rain or snow, and it can also condense as water droplets on various surfaces. In forests (especially cloud forests) and on other structures, condensation may be substantial (Berland et al. 2017); however, this input may not be measurable as precipitation and typically requires specialized sampling equipment and not simply a rain gauge.

IMPORTANT PARAMETERS OF WATER QUALITY

Water Temperature—The "Master Variable"

Temperature is an important physical property of water; it determines chemical reactions, water movements, and biological responses. Water is an unusual substance in several ways, but one that is tremendously important is that water *as a liquid* reaches its most dense state at ~4°C, and then gets lighter (less dense) again toward freezing. This means water that is about to freeze is at the surface, and hence, ice forms primarily from the top of lakes, wetlands, and streams, so that liquid water remains below and allows life to avoid being frozen. If this were not true, ice would form from the bottom and eventually an entire lake or wetland would be solid, leaving little to no refuge for life from freezing. This is not an issue in many parts of the world, but in north and south temperate regions, and at certain elevations, it is critical to how freshwater ecosystems change seasonally. In Chapter 2, we will explore more about how these patterns of water temperatures affect habitats.

The physics of thermal regimes is similar for all surface waters. However, the relative importance of each process for gaining and losing heat energy varies. These processes include gains and losses of short- and long-wave radiation, groundwater inputs, latent and sensible heat exchanges, and advection of water from upstream (see Figure 1.3). The relative rates of these processes differ with stream (or lake) size, climate, and other factors, but the physical processes involved are universal. A good introduction to stream temperature processes is Leach et al. (2023).

Flowing water does not easily crystalize to ice, and so under such conditions water can supercool to below 0°C. This supercooled water can crystalize around objects in the water, such as rocks on the bottom as the friction at the surface of an object slows water down sufficiently to crystalize, creating what is sometimes referred to as anchor ice. As ice accumulates on objects below the surface, it can create layers of ice that can break free from below the surface, float to the top of the water as small patches of ice, potentially contributing to *pancake ice*. Water can also freeze rapidly at the water's surface as air temperatures fall far below freezing—usually known as *frazil ice*—which can block flows and cause local flooding or block intake pipes.

In some parts of the world, lakes can stratify. For example, the surface water may be warmer and hence, less dense than deeper water, at least in summer, which many people will have experienced from swimming in lakes. As water heats from the surface by inputs of solar radiation, the ability of that warm water to mix throughout a lake is limited by either conduction (a slow transfer mechanism) or convection if there is enough wind at the surface. However, as the difference in temperatures diverges between surface and deeper waters, the difference in

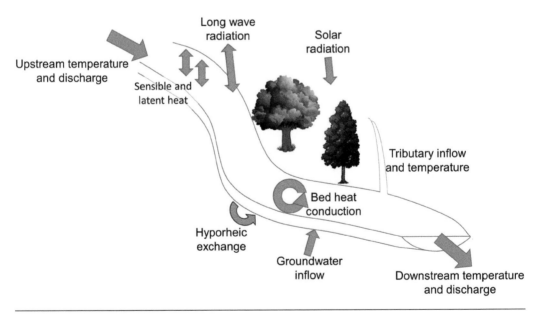

Figure 1.3 A schematic of the main thermal exchanges (inputs and outputs) with surface waters; redrawn after Moore et al. (2005). Blue arrows indicate direction of flow, and therefore inputs and outputs of water of a given temperature, which will be modified by the volume of water. In the case of a stream, the rate of heating is inversely related to discharge (more water takes more energy to heat). Orange arrows are the heat exchanges showing directions of gains or losses of thermal energy to the water.

density of the water along that gradient begins to require more and more wind energy to mix, which may not be present. Eventually, only the warmer surface water can be mixed (relatively similar density), leading to layering or stratification of a lake with a warm surface layer (epilimnion) and a cooler bottom layer (hypolimnion). Once stratified, a lake can resist deeper mixing unless there is an enormous storm that stirs the lake to depth, or more commonly, when the surface waters lose their heat energy as weather cools in the autumn. This process—called *turnover*—is when lake water mixes and thereby averages out temperatures, nutrients, and oxygen concentrations. In winter there can be a reversed stratification; that is, as the surface water loses its heat from the top and water cools to around 4°C, the 4°C water (most dense) will be on the bottom while the surface continues to lose its heat energy and form a colder surface layer. When turnover occurs twice a year, the lake is considered dimictic. However, some lakes never stratify and some stratify once per year (monomictic) if there is insufficient temperature difference between the surface and deeper water or if there is enough wind energy to mix the water to the bottom. Some lakes are permanently stratified (meromictic), which is often a consequence of a chemical gradient in density (such as salty water forming the bottom layer) causing stratification (Overmann et al. 1991).

Temperature is sometimes referred to as a controlling variable or even the *master variable* because of its large role in metabolic rates of organisms. Given that most species living in water are poikilothermic (or ectothermic), their body temperatures are determined by their environmental surroundings. Generally, biological activity is very low (or nearly zero) when the temperature is zero, although some organisms can tolerate and function with temperatures *at or below zero* by having high concentrations of solutes in their bodies, which reduces the point at which water freezes. Biological activity increases with increasing temperatures, although the slope of this relationship differs by organism and activity measured. However, biological responses do not continue increasing without bounds as temperatures increase, and at some point (optimal growth temperature), growth rates reach a maximum (see Figure 1.4). Beyond that maximum temperature, growth rates generally decrease as organisms' physiology (such as cardiac scope or energy intake) cannot cope with higher temperatures; eventually, temperatures may reach a point at which they are lethal.

There are many aspects of biology that modify the particular shape of the temperature-response curves. The specific shape and endpoints of these curves obviously differ by species, and by the particular measure (for instance, growth, activity, or metabolism) being plotted. Moreover, the details of these curves can differ between populations of a species because there is local genetic adaptation of populations, as seen in sockeye salmon (*Oncorhynchus nerka*) for example (Eliason et al. 2011). The shapes may also vary depending on the density of a population, the particular activities they are engaged in, and how much food they are getting. Often more stress (such as higher densities, less food, lots of predators) reduces the scope for activity and the threshold for peak activity.

The viscosity and density of water get lower as temperature rises—water becomes only 64% as viscous as temperature increases from 4 to 20°C. For organisms of a larger size, such as a fish, the water's viscosity relative to their mass is reduced only slightly. However, for a very small organism, such as an algal cell or a copepod (a small crustacean), the water is like molasses. The reason for this is that a small organism has very low mass and therefore little momentum, so that with each movement of something like a copepod, it stops after each stroke of its antennae, which act like paddles. Thus, temperature-dependent changes in viscosity can have a large effect in a small organism's ability to escape predators or chase their own prey. Reynold's number

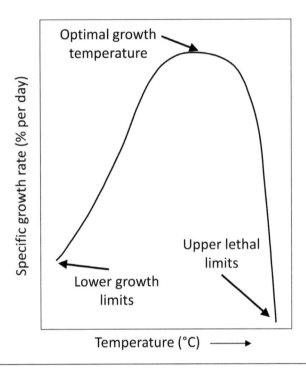

Figure 1.4   An idealized temperature-growth curve (or temperature-activity curve) for poikilothermic organisms.

and other means for scaling these mass-momentum relationships are helpful when trying to understand the way organisms use their aquatic environment and how much control they have over their movements relative to directional and turbulent flow of water. We will not address this topic further here, but for reference, an excellent source is Vogel (1996).

The Force of Flow

Flows of water exert forces, which can move materials in its way, reshaping the morphology of the environment, known as geomorphology (Church 2015). This erosive power of water is referred to as tractive force or shear stress for which there is a specific equation. In general, tractive force, *tau*, is denoted by its Greek letter, τ, for flowing waters. Without specifying the precise equation, there is an approximation that will simplify that relation:

$$\tau \propto average\ surface\ slope, water\ depth$$

Force from moving water can be found in stream flows, wave action, or even ice flows (as ice breaks up in spring). The majority of the force is exerted during high flows, such as floods, and the higher the discharge, the more work that can be done in rearranging morphology. The resulting geomorphology depends on the sizes of materials available to move and their specific gravity (Church 2002). For instance, small particles (sand, gravel) will move at lower tractive forces than large particles (cobbles, boulders). Lighter particles, such as wood and leaves, will move more easily than mineral particles. Movements of materials typically result in wave formations, such as dunes, which with bigger materials, look like riffle-pool sequences in streams. These will be described further in Chapter 2.

Oxygen Content

Oxygen concentrations in freshwater are on average about 0.0045% of that in the air. Solubility of oxygen is a function of temperature, and saturation concentrations decline as water warms. Consequently, while biological activity increases with higher temperature, available oxygen concentrations decline from about 14.6 mg/L (or ppm) at 0°C to ~8.2 mg/L at 25°C (at standard pressure). Oxygen enters through the air-water interface by diffusion or mixing by aeration by wave action or turbulent flows, and then moves through the water by diffusion and mixing. In most surface waters there is mixing of water that moves oxygen and other solutes through much of the water (however, see Chapter 2 on lake stratification). Oxygen is also produced through photosynthesis during daylight by algae or vascular plants, but then, during the dark, all organisms use up oxygen, so there can be pronounced day-night cycles in oxygen concentrations.

Oxygen in freshwater systems can be consumed through biological and chemical processes. Biological oxygen demand (BOD) is the amount of oxygen required by all organisms, including microorganisms that break down and decompose organic materials on lake or wetland bottoms. Chemical oxygen demand (COD) is the amount of oxygen used by nonbiological processes, including oxidation of breakdown products. As the amount of BOD and COD in a system increases, it can result in less oxygen available in the water if demand exceeds the renewal rate of oxygen. These low oxygen concentrations are a challenge to most organisms in water (in extreme cases, managers may resort to artificial aeration). One problem for life occurs when lakes or wetlands are covered in ice. With little or no diffusion of oxygen from above the ice, the remaining oxygen is used by active BOD organisms, including decomposers breaking down organic matter, and COD; thereby, oxygen can be depleted (anoxia). This depletion can result in *winter kill* where extremely low concentrations of oxygen are insufficient to support many species, and large numbers of individuals die, which is most often observed as dead fish. Similarly, anoxic conditions can occur when nighttime respiration rates (and no oxygen renewal from primary production) in productive waters can use up most of the oxygen. This can be common at outfalls of wastewater treatment plants (treated domestic or industrial water) or industrial sources of organic materials.

Oxygen can limit the activity rates of an individual, and concentrations might even be insufficient for an organism to survive in some environments. A large fraction of animals living in freshwater have gills, an expansion of surface area over which oxygen exchange (and other osmoregulatory exchanges) can take place. Most people are familiar with gills in fish, but gills are found in many animals. Some species rise to the water's surface to refresh air carried in specialized chambers, plastrons, and other mechanisms. Some species ventilate their bodies to create a flow of water to enhance oxygen availability, such as cased caddisfly larvae (Trichoptera).

Some groups of bacteria and archaea can function without oxygen using an alternate electron donor in respiration, and are responsible for anaerobic metabolism, which can generate methane and participate in denitrification. These anaerobic processes can occur in the organic sediments of wetlands and lakes, in large accumulations of organic matter in other freshwaters, and even in groundwater and hyporheic areas. Anaerobic processes often affect the forms of different chemical compounds, that is, going from oxidizing to reducing conditions.

Nutrients

All organisms require an external source of energy to survive. Primary producers primarily depend on solar energy. Other organisms require some form of fixed energy, either from the consumption of living primary producers, dead plant tissues such as leaves, or prey. Beyond energy, it is also critical to get nutrients from the environment or their food, primarily nitrogen (N), phosphorus (P), and potassium (K). Nitrogen is one of the building blocks for protein and enzymes. Phosphorus is part of DNA, adenosine triphosphate (ATP), bone, teeth, cell membranes, and serves other biological roles. Potassium is needed for nerve cell transmission, electrolyte balance, stomatal control in vascular plants, etc. These three nutrients are often considered to be the most limiting of elements for freshwater organisms, but are not the only substances required by life.

In addition to the previously mentioned nutrients, there are elements that are needed in small amounts. Calcium is needed for bones, nerve signaling, exoskeletons, etc. As humans, we are familiar with our need for iron (Fe) in hemoglobin, such that shortage results in anemia, but excessive amounts are toxic. Iron is similarly needed by all other vertebrates for hemoglobin, which is the mechanism for transporting oxygen to cells within the body. Iron is also needed in trace amounts for other functions in most other organisms. Copper is the basis of the molecule haemocyanin, which is used as an oxygen carrier in some freshwater invertebrates, just as the iron-based hemoglobin is used in vertebrates. Diatoms are a group of photosynthetic protists that require silica from the environment to construct frustules around themselves as protection (like shells) and are characteristic for taxonomic identification (see Behrenfeld et al. [2021] for an alternative hypothesis for the evolution of frustules). Organisms need other elements in addition to the aforementioned nutrients (for example, iodine and boron are needed by most organisms), and although other elements may be needed in only trace amounts, they are still essential.

Consideration of how ratios of different elements may be required for an individual organism to function, can be determined through a pillar of chemistry known as *stoichiometry* (Sterner and Elser 2002). One such stoichiometric ratio is known as the Redfield ratio, which states that on average, the molar (atom for atom) ratio of N:P in organisms is about 16 (there is a lot of variation between species). This can also be used as a quick check as to which nutrient is most in short supply in water; such as, if the ratio is >16, then P is probably limiting, and if it is <16, then N is probably limiting. This is not so exact that one can use this without additional information, but it is a good starting point. In general, organisms do not need *extra* nutrients, and so are most often limited by whatever nutrient or trace element is in the shortest supply—sometimes called Liebig's law of the minimum. However, organisms may be able to make trade-offs to maintain function. Phosphorus can often be in short supply in freshwaters, and if an organism cannot get all the P it requires, it may be able to compensate slightly by investing lower concentrations in some organs and even some proteins, although this ability is usually quite limited (Sterner and Elser 2002). Some species can also store nutrients, for example diatoms have a large vacuole to store materials, in what is sometimes called *luxury* uptake.

Nutrient status of a body of water can also affect the size spectrum of the species there. For instance, in very oligotrophic (nutrient poor) waters, it gives an advantage to small-bodied algae with a larger surface-to-volume ratio, and may favor picoplankton and nanoplankton (*pico* and *nano* being an indication of body sizes), whereas more nutrient-rich waters would likely not provide such a size advantage and one might expect larger-bodied primary producers (Stockner and Shortreed 1989).

In one study, Stockner and Shortreed (1989) showed that nutrient status could affect the numbers of trophic levels, and they compared some oligotrophic lakes to some lakes with higher nutrient concentrations. In their oligotrophic lakes, there were more steps required to go from the tiny primary producers to slightly larger species like protozoa, then crustaceans, then bigger species large enough to be consumed by young sockeye salmon. Remember that picoplankton are one million times smaller than the microplankton at the base of the food web in more nutrient-rich lakes. At each trophic level in a food web, energy is lost because the consumer uses a lot of energy for respiration, and thus the transfer of energy through each step may be only 10 to 50% of the energy entering each trophic level. In the oligotrophic lakes there was less primary production to begin with because of the limits to nutrients, and more trophic steps to reach the size of prey eaten by young sockeye salmon, and therefore there was less energy overall for the sockeye, meaning fewer and smaller salmon. Later on in this chapter, we will explore the roles of nutrient status and the numbers of trophic levels on ecosystem productivity.

Acidity

One measure of water acidity is pH. Rain is naturally at a pH slightly below neutral (pH of 7) due to carbonic acid in precipitation. There are a range of ion exchanges that occur in soils and in water that affect pH. The underlying geology of a catchment is one factor that can strongly affect pH. A lot of base cations, such as Ca^{2+}, can result in a basic or alkaline pH (greater than 7), whereas a lack of such ions and a high concentration of H^+ results in more acidic water pH (less than 7). The level of pH matters as organisms have to maintain ion exchanges to maintain their internal pH, often using calcium to buffer acids, which can deplete calcium from bones and exoskeletons. Other vital functions can also be affected if pH is strongly acidic. Experimental lake acidification demonstrated the range of impacts that acidity in water can have (Schindler et al. 1985) (see Chapter 4, Box 4.1). The pH of water also affects the ways ions behave, which in turn affects nutrient and other solute availability. In the 1970s, the northern hemisphere experienced decades of acid rain, caused by pollutants such as sulfur (as sulfuric acid) and nitrogen (nitric acid) transported into the atmosphere from industrial and agricultural sources, which we will learn more about in Chapter 4.

There are a number of other measures related to acidity, such as alkalinity, buffering capacity, and conductivity. Alkalinity is a measure of the capacity of water to buffer against changes in pH, and thus is equivalent to buffering capacity. Conductivity is a measure of the capacity of water to conduct an electrical charge (measured as μS/cm or mS/cm), and very solute-poor water may have conductivities of 10 μS/cm, while freshwaters in catchments with a lot of sedimentary rock may have conductivities in the 1000s of μS/cm.

Turbidity

Turbidity is one way to measure the concentrations of fine, suspended sediments, and is an important measure used to describe and monitor water quality. Turbidity is an estimate of the ability of particles to block light and is generally proportional to the concentrations of suspended particles. It is measured in a standardized way as NTUs (nephelometric turbidity units). However, turbidity measures are affected by the types of particles—whether they are organic or inorganic—their sizes, and even the *color* of the water created by the colored dissolved organic matter (similar in color to iced tea or coffee). Comparable measures include total suspended solids (TSS), which is generally reported as the dry mass of suspended particles per L. We will discuss turbidity further in Chapter 3.

Light

Light is essential for photosynthetic organisms, provides for visually oriented organisms, and also brings long-wave and short-wave radiation that heats water. The amount and quality of light reaching a water surface depends on latitude, solar angle (varies across days and seasons), the elevation of the water surface, any kind of cover such as forest canopies, cloud cover and vapor in the atmosphere, as well as any other particulates in the air, like smoke from forest fires. There are models available that can provide the predicted amount of incoming radiation for a given latitude and day of the year. Different wavelengths of light are attenuated with water depth, with red and infrared being absorbed faster than other wavelengths. The rate of attenuation and how it affects each wavelength is affected by whatever is in the water, such as color and suspended particles. Turbidity, as discussed before, affects transmission of light and absorbance of particular wavelengths.

Because light is critical to photosynthesis, we often speak of photosynthetically active radiation (PAR), which is the range of wavelengths corresponding to the main photosystems—generally 300 to 700 nm and depending on photopigments involved such as chlorophyll a, chlorophyll b, carotenoids, xanthophylls, and so forth. In limnology, there is a concept of *compensation depth*, which is the depth at which only 1% of incoming light remains, and it is sometimes considered the maximum depth from which rooted plants can grow. As discussed previously, the attenuation of light depends on a number of properties of water, so in reality, this depth is not quite so absolute, but rather, is a useful generalization.

Altitude and latitude can also affect the quality of light since some wavelengths are attenuated depending on the thickness of the atmosphere that light passes through or the solar angle. As a component of sunlight, ultraviolet radiation—especially ultraviolet B radiation (UVB)—intensity varies, with higher fluxes closer to the equator and at higher elevations. UVB can be harmful to life as it can cause genetic damage. Some organisms are more tolerant of UVB than others, but for some it can be fatal; for instance, it can cause deformities in amphibian and fish larvae (Blaustein et al. 1997; Alves and Agustí 2020). Dissolved organic matter (DOM) in water can also act as a bit of a sunblock from UVB, especially colored DOM (tea or coffee colored that is produced from leaching of soluble molecules from organic matter) as it absorbs energy from UVB, which lyses the molecules. However, as in most things the quantity matters, as UVB is also essential in small amounts to many vertebrates to produce vitamin D, which is essential in calcium uptake.

BIOLOGY OF FRESHWATER ORGANISMS

Freshwater is scarce and only makes up about 3% of all water on the planet (most is seawater), and only about 1.2% of that freshwater is actually available to support humans and freshwater ecosystems. However, estimates of the numbers of species globally suggests that even though freshwaters cover less than 1% of the earth's surface (or about 2.3% of the nonglaciated land), about 10% of species are exclusive to freshwaters—making freshwaters a key ecosystem in terms of biodiversity (Dudgeon 2020). It is not possible to cover all of freshwater biology here, so I provide suggestions for sources of information in the following paragraphs, but I also advise taking a course in the biology of freshwaters.

In freshwaters there are hundreds of thousands of species, and that count does not include most micro-organisms. Organisms come from all kingdoms of life, although some groups are better characterized than others. There are viruses, archaea, bacteria, cyanobacteria, algae of many kinds, vascular plants and mosses, protists, invertebrates (at least 12 phyla), and fish, along with other vertebrates. It has been estimated that there are at least 126,000 species of freshwater animals and 2,600 species of macrophytes (Balian et al. 2008). One estimate is that there are at least 13,000 species of freshwater fish (Lévêque et al. 2007). Groups such as protists, bacteria, and archaea lack good estimates of numbers of species. However, the tree of life is not as simple as once characterized in kingdoms such as animals, plants, fungi, bacteria, etc. For an example of relations between organisms see Wikipedia's entry for *tree of life* (*biology*). Evidence from many kinds of genetic analyses have shuffled what was known and indicates that organisms we once may have called protists are not monophyletic, and may be in completely separate clades of eukaryotes (organisms with a nuclear membrane). Keep in mind that there are many versions of the tree of life, with different names used for some major groupings. One conception of a modern tree of life for eukaryotes (see Figure 1.5) shows that even groups that we typically call algae are more distant from each other than animals are from fungi (after Keeling 2004). An examination of the bigger tree of life will reveal that eukaryotes are a relatively small component of total biological diversity.

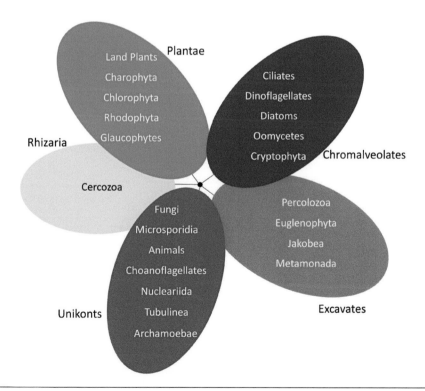

Figure 1.5   One modern conception of the relationships of major taxonomic groupings of eukaryotes based on genetic sequencing (following Keeling 2004).

There are many books on the biology of freshwater organisms (Dodds and Whiles 2019), and some of those include keys and general biology. Some of the best taxonomic keys, which also have great background information include Merritt et al. (2019) for North American freshwater insects, Tachet et al. (2010) for European freshwater invertebrates, and for freshwater invertebrates (other than insects) in North America there is Thorp and Rogers (2014) and Thorp et al. (2016). For algae, Wehr et al. (2015) is a good guide. There are an enormous number of guides for fish identification and biology in different parts of the world. In the following table, I provide a quick review of the kinds of organisms found in freshwaters of the world. For the purposes of this book, we will use an old system that may not reflect the true evolutionary relations among groups as previously noted. We will consider these as viruses, bacteria, protists, fungi, plants, and animals (see Table 1.1).

Table 1.1 List of the six taxonomic groupings (kingdoms) used in this book.

Kingdom	Characteristics	Nuclear Membrane
Virus	Only able to reproduce within a living host using host's gene replication machinery; generally considered parasites, but may be more influential in evolution of life than previously thought	None; a segment of RNA surrounded by a protein coat
Bacteria (including Cyanobacteria) and Archaea	Single celled; includes a range of decomposers, pathogens, and even photosynthetic form	Procaryote (no membrane around DNA genetic material)
Protists	Large polyphyletic group; some predatory and some photosynthetic	Eucaryote (genetic material [DNA] contained within a membrane, separated from the rest of the cell)
Fungi	Mostly decomposers and pathogens; a few taxa are considered mutualists	Eucaryote
Plants and algae	Typically photosynthetic, but some are non-photosynthetic parasites of other plants; single-cells, colonial, or multicellular. Now thought of as a polyphyletic group	Eucaryote
Animals	Multicellular; heterotrophic	Eucaryote

Micro-Organisms

We often lump viruses, archaea, bacteria, and fungi together as small organisms or *microbes*—mostly due to their size. Viruses are ubiquitous and yet there are relatively few studies of them in freshwaters. Bacteria are important decomposers of organic matter, and can also include pathogenic species. One group of bacteria, the cyanobacteria, are often called blue-green algae even though they are actually photosynthetic prokaryotes.

Freshwater fungi are generally referred to as aquatic hyphomycetes, but in fact, this grouping includes representatives of many taxonomic groups of fungi (Bärlocher 2012). Fungi, along with bacteria, are major decomposers of organic matter, while some species are pathogens of particular host species. One group of aquatic fungi that have become well-known is the chytrid fungi (*Batrachochytrium dendrobatidis*, *B. salamandrivorans*) that have afflicted many amphibian species around the world. They grow in the skin of amphibians, causing chytridio-mycosis, which sometimes leads to death.

Algae and Plants

What we call algae includes several diverse groups of photosynthetic eukaryote organisms that may be distinct at the kingdom or phylum level (see Figure 1.5). For instance, as mentioned before, the so-called blue-green algae are, in fact, photosynthetic bacteria. Most other *algae* are predominantly single-celled, photosynthetic organisms, although some can be colonial or chained, thereby creating something beyond single cells. In one modern classification, many algae are single-celled species of the kingdom Plantae, including green algae (Chlorophyta) and red algae (Rhodophyta). However, diatoms (Bacillariophyceae) are in a separate kingdom in this classification. While most algae are single-celled, some are in chains (such as *Cladophora*) and some even form erect structures as in the stoneworts (Charophyta: *Chara* spp.). For more on algal biology see Stevenson et al. (1996) or Wehr et al. (2015).

Multicellular plants or vascular plants (sometimes called macrophytes or hydrophytes) are grouped in various ways, and make up about 1 to 2% of all angiosperms (Chambers et al. 2008). One way of grouping them is in terms of whether they are rooted or not, and also whether they are floating, emergent, or submerged. Duckweeds (Lemnaceae) are floating plants. Most fresh-water plants are rooted in lake, wetland, or stream bottoms and have flowers that reach the water surface, as their flowers need to be pollinated (for instance, pondweeds, milfoil, lily pads). The depth to which plants can root is determined by the amount of light reaching the bottom, and the *photic zone*, defined on average as where light is only 1% of incident light at the water surface. However, there are some plants that can tolerate less light and some need more. The actual depth of the 1% limit varies depending on the various factors that limit light penetration such as turbidity and colored dissolved organic carbon (DOC). Many plants are found around the edges of freshwaters, including cattails (*Typha* spp.) and sedges (*Carex* spp.), which are not fully aquatic, since they cannot survive being permanently submerged.

Animals

Invertebrates are represented by at least 12 phyla in freshwaters, although a few phyla have only a small number of species (for example Cnidaria—*Hydra*; Porifera—sponges) whereas phyla such as the Arthropoda (insects, crustaceans) are represented by tens of thousands of species. Other phyla include the Annelida (worms), Mollusca (snails, clams, limpets), Bryo-zoa (moss animals), Nematoda (roundworms), Gastrotricha (hairy bellies), Rotifera (wheel animals), Platyhelminthes (flatworms), Nematomorpha (horsehair worms), and Tardigrada (water bears).

Freshwater fishes may number more than 13,000 species. Most of these species are in the tropics, and the carp order (Cypriniformes) is considered to have the most species at ~4,250 species, followed by the catfishes (Siluriformes) with an estimated 3,000 species globally. Fishes can be found in almost all freshwaters. They are limited primarily by access, such as waterfalls and other impassible barriers and water permanence. Fish are rare in temporary waters, but even then, there are some that can make do, such as lungfish, snakeheads, and others that can breathe air and are able to move themselves across terrestrial landscapes.

Among the vertebrates associated with freshwaters, some groups breathe air for oxygen (birds, mammals, reptiles), while fish and many amphibians depend on gas exchange across their gills or skin for oxygen. Of course, there are some fish that are capable of breathing air and certain amphibians may use different methods depending on their developmental stage (larvae versus adults). Some of these are almost exclusively aquatic, such as river dolphins and crocodilians (alligators, crocodiles, caiman, gharial). Mammals such as beavers, muskrats, coypu, platypus, and otters spend most of their time in water. Some other species such as Desmans of Eurasia (*Galemys pyrenaicus* and *Desmana moshata*) or water shrews (*Sorex* spp.) forage exclusively in the water, but mostly live on shorelines. The same association exists for many birds that feed on fish or other aquatic organisms, nest near water, or use water as escape habitats. There are many organisms that are associated with water and interact with freshwater food webs, whether we would say they are an aquatic species or not. We would say many of these species are riparian dependent; that is, they must be near water for parts of their life cycle. This includes organisms of all sorts (Richardson et al. 2005; Ramey and Richardson 2017).

Beyond taxonomy, we can consider organisms in other ways that are more easily comparable across geography and ecosystems, including total biomass, biomass size structure, feeding groups (trophic levels), and other groupings. This is referred to as trait analysis (Statzner and Bêche 2010; Verberk et al. 2013). For instance, we would expect that two streams in different parts of the world that look the same (same gravel-cobble bottom and same size) would probably have similar types of organisms (feeding groups, trait groups), even if the actual species are different.

Food Resources

We often refer to the food resources at the base of any food web as basal resources. Two primary categories are primary producers and detrital materials (dead organic matter, such as leaf litter or wood). Primary producers, including a diverse set of photosynthetic organisms such as algae, protists, cyanobacteria, mosses, and vascular plants, use the energy of sunlight to fix CO_2 into complex carbohydrates and take up nutrients to form proteins and other internal chemistry. Primary producers are also known as *autotrophs*. There are also *chemoautotrophs*, which are widespread globally, but not common. For instance, iron bacteria oxidize ferrous iron (Hildrew and Townsend 1976) and purple-sulfur bacteria use hydrogen sulfide under reducing conditions (Overmann et al. 1991). Because primary producers are within the aquatic realm, we refer to that as *autochthonous* production, that is, from within. Detrital materials are the remnants of primary production outside of the system (terrestrial) and therefore called *allochthonous* materials. Sometimes these two categories are referred to as the green part of the food web (primary production) and the brown part (detritus).

Most of the organic materials that enter freshwater from the terrestrial realm are referred to as detritus, as they are often the dead, senescent tissues of plants. Detrital materials can include leaf litter, seeds, flowers, branches, and whole trees (wood), etc. These are generally referred to by particle size: coarse particulate organic matter (CPOM, >1 mm diameter), fine particulate organic matter (FPOM, <1 mm, >0.63 μm), and dissolved organic matter (DOM, <0.63 μm). Most DOM is not technically dissolved, but is operationally defined as particles that pass through a glass fiber filter paper, nominally with pore size ~0.63 μm. DOM is also essentially the same as DOC, except one is based on the entire molecule (DOM) and the other is solely the mass of carbon (DOC).

Detrital processing is one term given to the decomposition of organic matter. Much of the organic matter from plants is cellulose based, and animals do not independently have the ability to digest cellulose. Therefore, it is up to fungi and bacteria to digest organic matter—the actual process of decomposition—making it more available for animals by converting it into consumable microbial biomass and by partially digesting plant matter. The surfaces of decomposing leaves, wood, and animal matter are covered by a biofilm of these fungal and bacterial decomposers, and it is often this biofilm that constitutes the food of many detritus consumers. Some animals that eat plants or plant detritus have symbiotic microbes that help with the digestion of cellulose; such microbes are found in other kinds of herbivorous animals, for example, termites and ruminants. Detritus-consuming animals include many kinds of invertebrates, as well as amphibian larvae and some fish species.

Fungi as a decomposer needs detrital particles large enough for them to develop their mycelia (*roots*) on, and hence, the division between particles greater than one mm in area and those particles that are even less. CPOM can support fungi and bacteria, whereas smaller particles are digested by bacteria.

This dichotomy between primary producers and detritus is too simple, as there are complex interactions between resource types. For instance, algae can produce excess molecules of organic carbon and lead to the *priming* of decomposition of organic matter by enhancing microbial growth through the extra energy provided by *leaky* algae (Halvorson et al. 2020).

Allochthonous inputs include detritus as just discussed, but it also includes living organisms, such as invertebrates falling from the surrounding land, and it might also include seeds and fruits that fall from vegetation nearby (Richardson and Sato 2015). Another kind of allochthonous input can include organisms like salmon, eels, and other species that return to freshwater and die, thereby importing large amounts of energy and nutrients gained in the marine environment (Wipfli and Baxter 2010). These flows of energy and nutrients across ecosystem boundaries are known as cross-ecosystem resource subsidies. Such flows are the basis of the concept of meta-ecosystems, that is, separate ecosystems linked by these flows of energy and nutrients.

FOOD WEBS

Food webs are one way of describing the feeding relations and flow of energy between organisms. Organisms are typically divided into trophic levels or feeding levels. After the *autotrophs* or primary producers, such as autotrophic plants and algae, most other organisms get their energy and nutrients by eating other organisms or parts of other organisms; such consumers are referred to as *heterotrophs*. Furthermore, primary consumers are those that eat primary producers, secondary consumers are those that consume primary consumers, etc. We sometimes refer to top predators as those consumers at the apex of the food web, which typically are large-bodied species. Each of these groupings constitute a trophic level.

Trophic levels are not always tidy and discrete, since there are predators that feed on a range of prey sizes and types; this is called omnivory. Omnivory includes organisms that feed on different combinations of food. Some organisms consume detritus and animals, others detritus and plants, plants and animals, or other combinations of a range of tropic levels. Also,

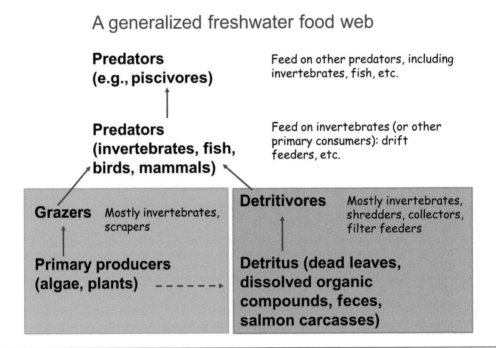

Figure 1.6 A generalized food web from freshwater ecosystems. Arrows indicate the flow of energy and nutrients upward from basal resources. There is a greater emphasis on detrital-based pathways of energy flows and less structural biomass from woody tissue than in most terrestrial ecosystems. We often refer to the green side of the food web as primary production and grazers, as opposed to the brown side of the food web which is based on detrital materials and detritivores—note the two distinct colored boxes at the base of the food web.

organisms can exhibit *life history omnivory* where different life stages of a single species feed at different trophic levels.

Food webs are structured with basal resources at the bottom, primary consumers that feed on those resources, and several levels of predatory animals (see Figure 1.6). Thus, there can be primary consumers feeding on autochthonous production (the *green* side of the food web), grazing on algae, or feeding on plants. Primary consumers feeding on detrital materials (the *brown* side of the food web) are known as detritivores (Zou et al. 2016). This detrital-based part of the food web has many similarities to food webs in organic soils. Organic materials that constitute detritus can include wood, leaf litter, and even dissolved organic matter. The quantitative contribution of detritus varies depending on the size of the waterbody, how much it is covered by vegetation, and other characteristics of the watershed. Heavily shaded streams or forested wetlands can be mostly dependent on detritus, and large, deep depositional areas such as deltas can also be based on detritus. Lakes and wetlands also receive a large amount of dissolved carbon from groundwaters and inflow streams, and while there are debates about the relative amounts, it is still significant.

Freshwater food webs can be further complicated by organisms that do not fit nicely into a particular trophic category, such as predatory plants and photosynthetic animals. Some non-plants in freshwater are photosynthetic due to symbiotic algae, mostly via the green algae *Chlorella* that can exist as cellular inclusions (Venn et al. 2008). These organisms include some

Figure 1.7   A freshwater sponge growing on rocks in a stream; notice the green photosynthetic tissue that comes from symbiotic algae within the sponge.

single-cell species, such as *Euglena* spp., and sponges such as *Spongilla lacustris* (see Figure 1.7). Thus, these species can be both heterotrophic and autotrophic. Predatory plants, such as bladder wort (*Utricularia* spp.), are typically associated with oligotrophic (low nutrient) environments. *Utricularia* have small traps, or *bladders*, that capture bacteria, protists, or small animals in a manner much like the terrestrial Venus fly trap (*Dionaea muscipula*). The edges of low nutrient lakes and bogs often have other partly carnivorous plants, including pitcher plants (*Sarracenia* spp.) and sundews (*Drosera* spp., in the same family as the Venus fly trap).

COMMUNITY ECOLOGY

The interaction of species within an ecosystem can include more than consumption and predation found within food webs. These interactions include competition, facilitation, and mutualism. Moreover, food webs rarely include the study of feedbacks, such as how predators may deplete their prey and result in density-dependent feedbacks to the predator's reproductive rates, as in predator-prey dynamics. All of these dynamical relations are the domain of community ecology.

In terms of the dynamics of food webs in freshwaters, there are a number of general processes described that have largely developed from studies in freshwaters. One idea is the concept of top-down or bottom-up effects. Bottom-up effects are largely due to more energy entering the base of the food web, leading to overall greater productivity and often greater taxonomic diversity (more species as rarer species get included). Top-down effects are largely a consequence of top predators (or even intermediate-level predators) often being capable of suppressing the abundance and productivity of their prey populations. Top-down control can have an outcome that is especially obvious in freshwater food webs through trophic cascades, where a high-level predator can depress numbers of its prey, which might free up the prey of that first prey (Carpenter et al. 1985; Shurin et al. 2006). A trophic cascade is a community response where a top predator has effects on at least two trophic levels below, hence cascading from the top of the food web down through the web.

Life history omnivory was previously discussed, but organisms can also switch habitats and feeding behaviors dramatically throughout their life history. Some obvious examples include amphibians and aquatic insects that spend their larval stages in water and adult stages out of water. However, these life-cycle transitions between habitats can also include using different features of freshwaters at different stages of their life cycles. A good example would be the bluegill sunfish (*Lepomis macrochirus*), where in the presence of largemouth bass (*Micropterus salmoides*), which is predaceous on small bluegills, the young bluegills remain in among the macrophytes in order to reduce predation risk and forgo eating planktonic crustaceans. Then, when they are large enough to not be eaten by the largemouth bass, they switch habitats and food choices (Carpenter et al. 1985).

ECOSYSTEM PROCESSES

Ecosystem processes occur when biological and nonbiological components of the ecosystem interact. Nonbiological components include nutrients, inorganic sediments, etc. Some examples of ecosystem processes include production (primary and secondary), decomposition, ecosystem metabolism, and nutrient cycling and mineralization.

We often refer to the amount of living tissue in an ecosystem at any given time as standing crop or biomass, usually in units of g/m^2. Depending on the question, biomass might be wet mass, dry mass, or even ash-free dry mass (AFDM). However, biomass is not the same thing as productivity, which is a rate that is expressed as the change in biomass through time, such as $g\ C/m^2/y$ or in smaller time units like per day or minute. Production is different yet, and is the net carbon produced over an interval; but there may actually be very little biomass present if a lot of it is eaten, so production includes productivity and losses. Even a species with low current biomass, such as algae or bacteria, may have high production and also high turnover (essentially how quickly biomass is being replaced), so the biomass is getting replaced at a high rate such that productivity and consumption may be balanced (Carpenter et al. 1985).

2

FRESHWATER ECOSYSTEMS: HABITAT TYPES AND GEOGRAPHIC VARIATION

INTRODUCTION

In ecology we often specify the scale of study in several ways, such as ecosystems, communities, populations, and individuals. Ecosystems include the biological, chemical, and physical processes that provide for the full range of functions, including nutrient cycling, production, decomposition, and others. Thus, ecosystems can also be described as the suite of interactions between the biological and abiotic aspects in a place, often with some kinds of boundaries. In freshwaters, the boundaries between terrestrial and aquatic are relatively distinct, but not absolute, as there are many strong connections between the land and the water, and vice versa. Some of these connections have been consolidated into ideas about *metaecosystems* (Gounand et al. 2018) and cross-ecosystem fluxes (Richardson and Sato 2015). Sometimes we consider the biological entities alone, and we refer to that as an ecological community, and if we are only interested in the trophic relations (who eats whom), then we talk about food webs (see Chapter 1). Communities are defined by a group of organisms (any organisms) that occur in the same place at the same time, and includes species from the scale of viruses through bacteria, protists, and others, up to complex multicellular plants and animals.

Within an ecosystem, we refer to specific habitats because organisms typically only use a part of the whole ecosystem. A habitat needs to provide for the full range of requirements for that species, including sources of energy and nutrients, potential mates and breeding places, places to hide from predators, etc. Thus, within freshwater ecosystems, we can define an organism's habitat more narrowly, based on specific features or depending on the process of interest. Habitats can be defined at multiple scales for every species (Addicott et al. 1987). For instance, an individual may use different habitat features for feeding, for avoiding predators, or for reproduction. As a group of individuals of a species, that is, a population, it may include a number of habitats (Addicott et al. 1987). Over the life cycle of a species, individuals within the population may move between habitats seasonally. Thus, we can think of a hierarchy of habitat types at different spatial scales depending on the function, whether it be feeding, mating, or other functions that the organism needs (Frissel et al. 1986). In addition, most freshwater organisms change their habitat requirements as they progress through their life cycle, known as life history shifts (Werner and Gilliam 1984). Some species, such as adult aquatic insects and amphibians, also require terrestrial habitats at some point in their life cycle. The requirement for multiple habitat types, or even hosts for many parasites, is often referred to as a complex life cycle, that is, different life stages use distinctly different habitats (Wilbur 1980).

To expand on how we define habitat at a number of different scales, consider this example. A given fish species might occur in a particular geographic zone or ecoregion. This species might only occupy lakes or streams of a given size, but not all streams or lakes in their geographic range. Populations of a certain species in a particular lake or stream may have limited gene exchange with other populations if there is unsuitable habitat between; for instance, when a lake species is not able to navigate through a large stream to get to the next lake where there are others of its kind.

At an even finer scale, individual organisms can be discussed in terms of their microhabitats, which might include the specific position on or under rocks, being in or under wood, or swimming in the open water, etc. All of these different ways of describing habitats depend on the scale of the biological process you are interested in, such as energy flow, population dynamics, foraging, mating, or others (see Addicott et al. 1987).

TYPES OF FRESHWATER ECOSYSTEMS

Freshwater ecosystems are generally broken into two categories based on how the water moves—*lotic* for flowing waters (streams) and *lentic* for still waters (lakes and wetlands). As with any classification, there are variations such that flow varies through time and across habitats. For instance, wet season flooding of places like the Amazon River floodplain create large, seasonally flooded lakes and wetlands. In a general sense, we can define freshwater habitats as streams (the general term in this book for all flowing waters), lakes, and wetlands (and others). Another habitat classification is to distinguish benthic (bottom) habitats from those in the water above the sediment surface (limnetic or planktonic). Benthic habitats may be separated further into epibenthic zones which distinguish organisms that live on the bottom surface as opposed to those that burrow into the bottom sediments.

There are many kinds of freshwater ecosystems, and each will be introduced further in the upcoming text (for a review of types of freshwater ecosystems read Cantonati et al. 2020). In addition to streams, lakes, and wetlands, there are other specific habitats. Springs, including geothermal springs, occur in many parts of the world where groundwater reaches the surface in noticeable quantities, and sometimes under pressure. Groundwaters themselves constitute an often-overlooked freshwater ecosystem, and organisms can be highly specialized; these are referred to as *stygobionts* (Ercoli et al. 2019). Very small, natural water containers (tree holes, plant leaves) can also be functioning ecosystems, and are given the name *phytotelmata*. Although a great deal of emphasis is on permanent waters, temporary pools or intermittent streams can be important ecosystems that can be highly productive and biodiverse.

There are general similarities (water chemistry and temperatures) and differences (flow, structures) between freshwater ecosystem types. For instance, mechanisms to deal with low oxygen levels in water and abilities to move through water is a common feature of most aquatic organisms, as discussed in Chapter 1. Water chemistry varies geographically, but less so by ecosystem type within a geographic region. All freshwaters show differences depending on climate, geology, topography, latitude, elevation, and catchment vegetation. Climate affects precipitation regimes and thermal patterns. Geology affects the kinds of soils, their chemistry, porosity, and other properties, which of course, affect the quantity and quality of the water in surface waters as well as how fast water gets to a stream, lake, or wetland. Latitude and elevation influence the characteristics of freshwaters, for instance by affecting the range of temperatures, whether or not there is ice cover or snowmelt contributions, the seasonal distribution of day length, and the angle of sunlight hitting the water's surface. All of these aspects will affect the biome, that is, the living organisms in an area, and the vegetation.

STREAMS

For streams we can think of habitats at multiple scales, as outlined by Frissel et al. (1986), that is, at the scale of watershed, reach types, channel units within reaches, and microhabitats within channel units. At a reach scale, we often have a floodplain, channel (active, wetted, bankfull), thalweg (central line of the main channel or deepest point), and riparian area (terrestrial ecosystem influenced by being near the water or the terrestrial area that influences water, for instance, by shading).

Classification of Streams

Streams can be classified in a number of ways. We often have common names that confer an idea of stream size, such as brook, beck, creek, stream, or river. Another might be seen in the French language where *rivière* means a large stream not entering the sea, whereas a similarly sized stream that flows to the sea is called a *fleuve*. A common classification method is the stream order attributed to Strahler (1957). The first stream in a network that appears on a topographic map at a map scale of 1:50000 is defined as a first-order stream (see Figure 2.1); when two first-order streams meet, it becomes a second-order stream, and so on. However, this classification depends a lot on map scale and how maps were developed. Since the middle of the 20th century, these were all based on photointerpretation of aerial photos, and of course, misses about 80% of stream length in some landscapes, especially forested regions (Leopold et al. 1964). Meyer and Wallace (2000) estimated that 80% of the perennially flowing streams they worked on would not be considered streams under this definition. Thus, it is really important to pay attention to the map scales that are used to define stream order (Richardson 2020). A related classification scheme is to count stream *links*, which is the number of first-order streams found above a given point in the stream network.

Many classifications of streams are based on management purposes. A common division for management reasons would be fish-bearing versus non-fish-bearing streams, which is often used for forestry regulations. Another might be whether a stream is *navigable*—a vague term that is used in the U.S. Clean Water Act. A stream might also be classified based on its bankfull width, which is a measure of the bank-to-bank width, and is the point on the streambank that shows evidence of enough erosion that there is little perennial vegetation. Many management agencies divide streams into classes with somewhat arbitrary divisions based on some breakpoint, such as streams less than or greater than 5 meters in width, or those above or below some channel gradient, which have no scientific basis other than simplicity of measurement.

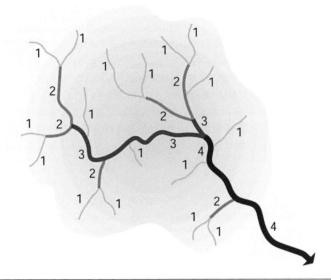

Figure 2.1 Defining stream sizes as stream orders or stream link numbers. First-order streams are the smallest streams and numbered 1, then the confluence of two first-order streams become a second-order stream, numbered 2, and so on. The definition as to which streams count as first-order depends on selection of map scale (see text). Stream link number is the summation of all the first-order streams upstream of the point of interest. *Source*: U.S. Department of Agriculture, NRCS. "Stream Corridor Restoration: Principles, Processes, and Practices, 10/98, by the Federal Interagency Stream Restoration Working Group (FISRWG)."

Given that streams really form a continuum and that there are no universal break points in sizes, we can use continuous measures such as size (bankfull width or hydraulic radius [similar to average depth]) or even discharge (a rate of flow, for instance m³/s of water passing a particular point). The latter is difficult to determine for any given location unless there is a gauging station there since discharge varies considerably throughout the year and across years. Often when discharge is used to categorize a stream, it is based on annual average discharge, and might even reference the number of years of data that were used to calculate that annual average discharge.

At the next scale below the stream itself, we can describe different habitats within streams as reach types, which are a repeating sequence of units, and are named based on those units such as riffle-pool, step-pool, and glides, and typically carry on as long as the channel gradient (average slope of the stream channel), size, and bed materials are similar (see Church 2002, 2015). *Riffles* are the units within streams where rocks may break the surface and where the water moves quickly, whereas *pools* are slower-moving and deeper, and these two units alternate—riffle-pool-riffle-pool, etc.—along a riffle-pool reach. When the gradient changes sufficiently to alter the channel characteristics, we often refer to it as a reach break, and we will have a different reach type (see Montgomery 1999). For instance, riffle-pool reaches typically have a gradient of somewhere between 0.1 to 2%, and step-pool reaches typically have about a 4 to 8% gradient. Within reach types, the channel subunits are pools, riffles, steps, cascades, and others. The alternating of channel subunits is generally a consequence of the force of water at high flow, thereby creating wave forms with the stream bed materials. The peak forces typically occur during floods, whether that is an annual snow-melt-associated freshet or during intense storms. These flows create wave forms that lead to undulating bed morphology, like riffle-pool sequences, where pools are scoured out and that material gets deposited in the slightly shallower riffles (recall tractive force from Chapter 1). That also means that flood peaks are a time when extreme forces are acting on bed materials, and most organisms will die if they cannot find places of refuge. To understand the physics of these morphologies, one should consult a text on stream (fluvial) geomorphology.

Streams vary in size from headwaters (Richardson 2020) to the scale of the Amazon River—the world's largest stream. There are differences in streams that are related to size. The River Continuum Concept was introduced (Vannote et al. 1980) in order to consolidate a generalized framework for how stream ecosystems varied from the smallest to the largest of streams (see Figure 2.2). This concept has provided a useful heuristic for how stream networks function on average, although most streams do not fit exactly. Stream size is a useful predictor for position along the continuum, although not sufficient by itself. Many aspects of streams are related to stream size, on average, such as grain size (negative), slope (negative), sediment storage (positive), dependence on allochthonous resources (U-shaped), light (hump-shaped), and production-to-respiration ratio (hump-shaped). Small streams are often heavily shaded and get much of their energy from organic inputs such as leaves; intermediate-sized streams are more open (more light and photosynthesis) with a smaller per-area contribution of organic inputs; and large streams are often turbid (less light and photosynthesis) and more dependent on the transport of organic particles from upstream. This produces the U-shaped relation with stream size, where small streams and very large streams are more dependent on organic matter inputs than medium-sized streams. These are very broad generalizations, but they are a helpful starting point to consider how stream size and other characteristics influence the biology.

As mentioned before, reach types are made up of channel units, such as riffle-pool, bedrock, plane bed, and step-pool. These units each foster distinct habitat types within streams. At an even finer spatial scale within channel units, we can define microhabitats where an individual organism might live based on aspects of the hydraulic forces, depths, substrate type, exposure to sun, etc. (see Figure 2.3) (Frissel et al. 1986; Hawkins et al. 1993). Species such as algae or moss might live attached to the tops of rocks where they can receive light or filter-feeding invertebrates that capture particles from the water flow passing by. Other species might live in or on wood or in accumulations of organic matter, perhaps as a place to hide or as a food source. Yet other species might live in the stream bed where gravels or cobbles accumulate small particles of food. There are many kinds of microhabitats, and some organisms may live in the groundwater or the hyporheic zone, variously defined as the mixing zone of groundwater and stream water (see Figure 2.4). One can also define microhabitats by the hydraulic forces there (see Rempel et al. 2000). Particular species may be highly selective of specific forces—such as hydraulic jumps where prey or other particles may get trapped in nets or anatomical structures—in order to capture food. For example, Wetmore et al. (1990) found that the predaceous larvae of a caddisfly were more likely to attach themselves at a specific part of a rock where the flow was near critical flow. Flow velocity might also provide a refuge from some kinds of predators; for instance, black fly larvae may select sites where the water velocity is too high for a predaceous flatworm to reach them (Hart and Merz 1998). There are many ecological reasons that an organism might select a specific microhabitat.

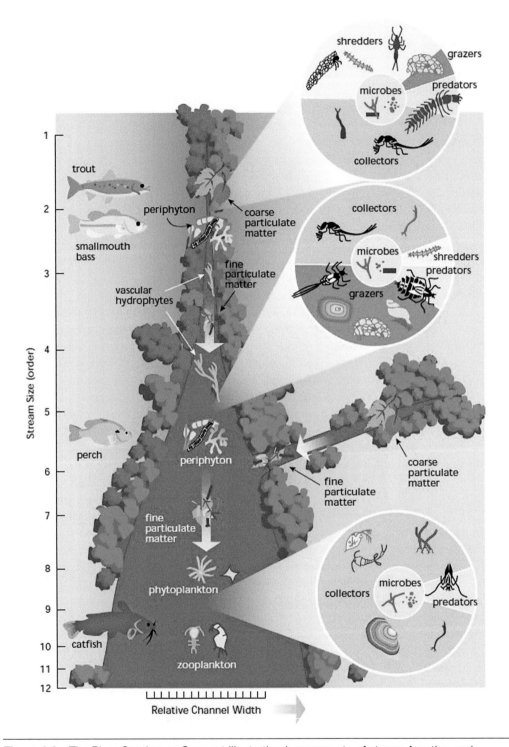

Figure 2.2   The River Continuum Concept illustrating how aspects of stream function and structure may change with stream size (after Vannote et al. 1980). Changes in physical structure are conceptualized to scale with changes in shading, organic matter inputs, biological composition, and ecosystem functions (production). Graphic courtesy of Stroud Water Research Center. *Original source*: U.S. Department of Agriculture, NRCS.

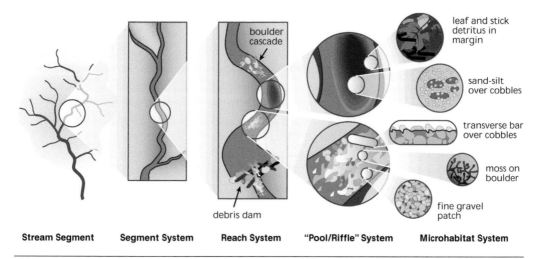

Figure 2.3 Scaling and names of habitats in streams at decreasing spatial scales, from segments down to microhabitats. *Source*: U.S. Department of Agriculture, NRCS.

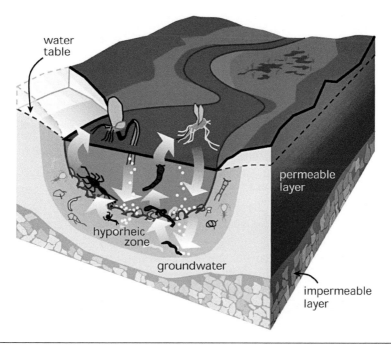

Figure 2.4 Organisms may live on the stream bottom, among the materials forming the bottom, or even in the hyporheic zone separate from the surface materials. These would all be considered different microhabitats. *Source*: U.S. Department of Agriculture, NRCS.

LAKES

Lakes are distinguished from wetlands (see wetland definitions in the next section) as having a zone of water that is deeper than the photic zone (or compensation depth), which is generally defined as the depth to which 1% of light can penetrate. Lakes are often classified by the process that forms them (a geologic boundary, landslide, glaciation, etc.) or by features such as their volume-to-surface area ratio. There are six major classifications (or types) of lakes: (1) geological sills or mass movement dams; (2) tectonic lakes, such as rift lakes, for example, the African rift lakes and lakes found in Iceland; (3) volcanic lakes such as Crater Lake in Oregon; (4) glacial lakes; (5) fluvial lakes, such as are formed by oxbows; and (6) reservoirs, like Williston Lake in British Columbia (Hutchinson 1983). Other lake types can include solution lakes (in karst topography, where sediments are dissolved to form basins) and meteorite lakes, which, as the name suggests, are formed by meteorite strikes and resulting depressions in the surface (for instance, Lac Wiyâshâkimî, formerly known as Lac à l'Eau Claire, in Québec, Canada).

As a still-water (lentic) environment, lakes are generally heated by short-wave and long-wave radiation at the surface. The ability to transmit this heat deeper into a lake depends on wind energy available to mix the water. In temperate lakes, warm water at the surface in summer can create a separate layer or stratification where warmer, less dense water remains at the surface, with cooler water below. The density gradient created can be resistant to water mixing any deeper leading to a warmer surface layer, the epilimnion, and a cooler deep layer, the hypolimnion, separated across a thermocline. As a lake's surface cools in autumn, the density gradient gets less, and wind is able to mix the entire lake volume; this is known as *turnover*. A dimictic lake will experience turnover twice a year—once in spring and once in autumn when the temperature differences between the surface and bottom layers become negligible. In cold climates, further cooling can lead to surface waters being cooler (water density is highest at 4°C), and can stratify in reverse with water less than 4°C at the surface and water of 4°C or warmer in the bottom layer. Such lakes often have some ice cover for periods of winter if the air is cold enough. In lakes that stratify, the cooler bottom waters (hypolimnion) in summer can provide habitat for species that need cool water year round. Where the hypolimnion is not sufficiently cool to create a temperature gradient, lake waters may mix thoroughly, more or less, all year round, which is to be expected in tropical and subtropical regions.

Within a lake, habitat zones include the benthic zone where benthos (organisms living *on* or *in* the bottom) dwell, and those species that live above the bottom's surface are referred to as pelagic or limnetic (see Figure 2.5). A term often used for pelagic organisms is *plankton*, which includes bacterioplankton, phytoplankton, zooplankton, and ichthyoplankton. Shallow and deep environments are a relative term, and most often shallow bottom areas are called *littoral*, while deep areas are called *profundal*, with the boundary commonly based on the light compensation depth, which is the depth at which gross photosynthesis is equal to respiration rate, or where <1% of incoming solar radiation reaches the bottom (photic zone). Above this depth, plants can live on the bottom; however, in practice, this is not quite so simple since some species can adapt to very low light levels and others cannot, and not all wavelengths of light penetrate to the same depth. The specific depth that forms the boundary between littoral and profundal zones will, of course, vary depending on the color or clarity of the water—perhaps affected by dissolved organic carbon, turbidity, and planktonic algal density.

Classification of surface waters can use many criteria. As previously discussed, lakes might be classified through the process by which they are formed. They can also be classified by how productive they are. Productivity is determined by their trophic status, which relates to their nutrient supply. These categories include oligotrophic (very low nutrients), mesotrophic, and eutrophic (excess nutrients) or hypereutrophic. Lakes may also be dystrophic (nutrient poor,

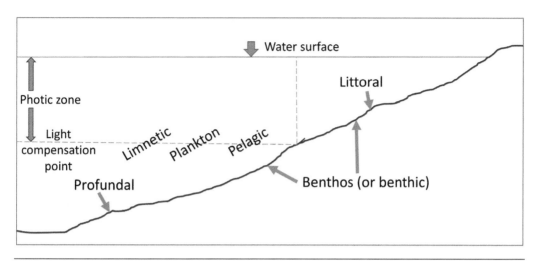

Figure 2.5   Habitat zones within lakes and wetlands. The photic zone is generally defined as that depth to which 1% of incident light (light compensation point) can reach (note that not all frequencies penetrate to the same depth). The bottom is called the benthic zone, and is often divided into the littoral zone (above the light compensation point) and the profundal zone. The open water habitat is variously called the pelagic, limnetic, or planktonic zone. In the majority of temperate lakes, there can also be stratification into an epilimnion (surface layer) and a hypolimnion (bottom layer) with a transition zone referred to as the thermocline, or in some situations, a chemocline (water density differs by chemical composition).

mildly acidic, and/or cold), similar to bogs (as defined in the following section on wetlands), which means that dissolved organic matter is not broken down as fast as it enters and can result in water becoming brown colored.

Lakes can also be defined by their size, such as surface area or volume. One estimate is that there are 304 million natural lakes around the world, and most of these are <1 km² (Downing et al. 2006). The global area of freshwater that is made up of small lakes and ponds (<1 km²) has been largely underestimated and better estimates show that they make up half the area of global surface waters and are probably more productive than larger lakes per unit area (Downing et al. 2006; Downing 2010). This means we need to pay more attention to small lakes and ponds.

Lakes can be classified based on their stratification patterns (see Chapter 1). A lake that stratifies twice a year (winter and summer) is dimictic, a lake that only stratifies once a year is monomictic, and a lake that never stratifies is polymictic. In rare circumstances, lakes can become chemically stratified; for instance, when there is a concentration of salt water (high ionic strength) at the bottom that creates a density gradient that cannot be overcome by mixing via wind energy. The boundary in this case is known as a chemocline, and such lakes are referred to as meromictic. At the chemocline, there can be unique organisms that can benefit by the nutrients available and can use hydrogen-sulfide as an electron donor. Hydrogen-sulfide is produced as a breakdown product of decomposition in anoxic environments, and typically, water below a chemocline is anoxic since it does not mix with the surface layer. One kind of microbe that thrives in these situations includes purple sulfur bacteria, which are photosynthetic, but use hydrogen sulfide instead of oxygen (Overmann et al. 1991).

WETLANDS

Wetlands are distinguished from lakes in that wetlands have most of their area shallow enough that at least 1% of the incoming light reaches the bottom, but there is no absolute distinction between some types of wetlands and lakes. Five categories of wetlands have been defined (Keddy 2010): (1) bogs; (2) fens; (3) swamp; (4) marshes; and (5) shallow, open water (such as ponds). Keddy (2010) further defines wetlands based on their flooding frequency and duration, depth, fertility, disturbance, competition, herbivory, and burial. Bogs and fens are peatlands and can also be referred to as mires. However, bogs and fens differ as two categories of wetlands based on water chemistry, with bogs being nutrient poor (dystrophic) and acidic, and fens being solute rich and usually slightly alkaline. Bogs usually have brown (or black) water resulting from low rates of decomposition of resistant organic compounds (dissolved organic matter [DOM], of which coffee or tea would be examples). Bogs occur in most parts of the world, and depend on at least one of three things to result in the accumulation of DOM—cold temperatures, low nutrient concentrations, or low pH, all of which reduce the ability of microbes to grow and use DOM. Swamps have trees and shrubs distributed throughout, whereas a marsh has no trees, but is characterized by rooted, aquatic plants, such as lily pads (Nymphaceae) and pond weeds (*Potamogeton* spp.) over much of the area, and are rarely deeper than ~90 cm (Keddy 2010). Marshes are the most common type of wetland in much of the world. Shallow, open-water wetlands are deep enough that rooted aquatic plants do not cover the whole bottom, and are generally intermediate between marshes and lakes, with no discrete separation.

ESTUARIES

Estuaries are often left out of books on freshwater ecology, but they are transition zones from freshwater to saltwater, and have interesting properties of their own (Hodgson et al. 2020).

Estuaries are often stratified, most often with freshwater on top of denser saltwater. There is usually a mixing zone that can create flocculation of organic materials. Since they are adjacent to marine systems, there are tides that enter twice a day (or in some places just once—diurnal tides), with a salt *wedge* penetrating upstream under the freshwater on top as tides go in. This is a complex environment for organisms on the bottom that experience large changes in salinity and other conditions twice a day. Estuaries can also create tidal conditions in upstream freshwaters, as the rising of the tide backs up water that was flowing downstream. This can even result in stream water flowing back upstream, even without a salt wedge below. A variety of estuaries exist, and they cover a gradient of salt content, tidal height fluctuations, and other differences. Some seminal work on ecosystem ecology was from freshwater tidal marshes, much of which was done by Odum (1988) on the Atlantic coast of the USA.

OTHER FRESHWATERS

There are other freshwater ecosystem types that we will only briefly mention in this book, such as springs (hot and cold), cave streams, temporary pools, rock pools, containers (tree holes, bromeliads, etc.), and groundwater ecosystems (see Cantonati et al. 2020). Also, there are a number of specialized organisms that live within the groundwater, but we will not address those further here (see Gilbert et al. 1994). Some of the previously listed ecosystems are at risk for a range of reasons.

Springs come in many forms (Cantonati et al. 2020), depending on where and when the groundwater emerges at the surface (sometimes at the bottom of pools or streams)—sometimes under pressure and sometimes not. As springs are the expression of groundwater, they usually have a relatively constant temperature, and in temperate regions are often cooler than nearby surface waters in summer and warmer in winter. Springs may also be the source of many streams in non-ice-dominated regions (Cantonati et al. 2020). The biological diversity of these sites is specialized to live in these habitats. Springs are defined and classified based on geomorphology, chemistry, biology, and other characteristics. Spring types include hotsprings and geysers, obviously originating where geothermal warming heats the water. In these hot waters there are specialized bacteria, brine flies (Ephydridae), and other species that are adapted to such extreme environments. Springs, in general, are imperiled by human exploitation for water supplies, isolation in the landscape (specialized spring organisms cannot easily move to another suitable spring), and high degrees of species endemism.

Temporary pools (Wiggins et al. 1980; Williams 2005) are seasonally flooded wetlands, and a specialized community can exist, including fairy shrimp, tadpole shrimp, etc., with a complete food web—so these are not just rain puddles. The dry phase provides habitat for many plant species and associated species, which are unique. In California, temporary vernal (filled in spring) pools are considered an ecosystem at risk. Temporary pools can be forested (Stoler and Relyea 2011) and nonforested. These pools provide accumulations of organic matter and can serve as a refuge from many kinds of freshwater predators that cannot survive the dry period (for instance, most predatory fish and amphibians). As such, many species are only found in these habitats, including some amphibians, unique crustaceans and insects, and many other taxa. Some species can live there and fly away as pools dry up in late spring, such as giant water bugs in the family Belostomatidae (Lytle and Smith 2004). There are many other interesting adaptations for life in temporary waters (Wiggins et al. 1980; Williams 2005).

As with temporary pools, many streams are also intermittent and can lose surface flows for long distances and for various durations. This has become a major topic in terms of why this occurs, how organisms and their communities cope with it, and the ways in which human use of water supplies influences stream intermittency (Datry et al. 2017; Bruno et al. 2022). In the USA, about 58% of stream length across the country is made up of streams that can go dry at the surface, and these include streams that are important water sources to users downstream (Marshall et al. 2018). While they are flowing, intermittent streams provide high rates of nutrient cycling, productivity, and connections between other stream sections and floodplain wetlands.

There are a surprising number of freshwater species that can complete their life cycles in very small habitats, such as tree holes (Kitching 2001; Juliano et al. 2019), bromeliad wells (Srivastava 2006), pitcher plants (Heard 1994), and others. These plant-based habitats are sometimes called Phytotelmata. Each of these small habitats have been studied for their relatively complex and replicable communities and ease of manipulation. The same species that survive in such small habitats are also capable of thriving in small, human-furnished containers, such as old tires, rain gutters, catch-basins, buckets, and others. These can also harbor mosquitos that transmit diseases. Public health agencies work on educating the public about draining these containers (for example, see the website of the U.S. Center for Disease Control and Prevention).

DISCONTINUITIES

Although we can describe major habitat types as streams, lakes, etc., there are often discontinuities in these features. For instance, where small streams flow into larger streams, as seen in tributary junctions (Kiffney et al. 2006; Rice et al. 2008), there may be local differences if small streams add cooler water or more particulate organic matter (see Wipfli and Gregovich 2002). Tributary junctions are one example of variation within freshwater networks, and can occur where any bodies of freshwater meet, whether it is two streams, a stream flowing into a lake delta or a wetland, or even lakes and wetlands flowing into streams. In this section, we will learn about a few examples of these discontinuities.

Within a stream network there may be lakes or wetlands that change the character of a particular stream. The majority of the suspended load (inorganic and organic) or bedload sediments carried by streams are deposited in lake bottoms. This results in streams that are downstream of lakes often being sediment starved, but also with very clear water, possible moderation of flood peaks, an altered temperature regime, reduced amounts of particulate organic matter, and other differences. Observation of these differences resulted in the development of the serial discontinuity concept (see Stanford and Ward 2001). The idea in this concept also applies to reservoirs, a common feature on many of the world's rivers (Nilsson et al. 2005).

Lake outflow streams, for some distance downstream of a lake outlet, will have features that are distinct from streams that are not affected by lakes (Richardson and Mackay 1991). As mentioned before, organic matter and inorganic sediments are trapped in lakes, so water that is flowing from lakes (and reservoirs) contains mostly planktonic organisms (bacterioplankton, phytoplankton, zooplankton), which constitute a high-quality food resource. As a result, population densities of filter-feeding organisms are extremely high in outflow streams, but also rapidly decrease as this food source is depleted (Richardson and Mackay 1991; Wotton et al. 1998). Other changes that are related to hydrology, temperature, or other features may also be involved, and very few studies have attempted to separate out these processes.

Lake deltas (inflows) are features that have not been studied extensively, but occur wherever streams flow into lakes (or wetlands) (Tanentzap et al. 2017). The inflows create a plume of organic matter that is transported from stream catchments, but also a delta of mixed sediments differing from most of the shores of a lake. There may also be different thermal properties and nutrient inputs from inflow streams that create special conditions, either as a persistent or a seasonal feature (Richardson et al. 2021).

Even surface films of water and the water-air interface may be a separate habitat or resource type. Hydrophobic organic molecules can accumulate at the surface of any water body (Wotton and Preston 2005). These molecules can include lipids and other calorie-rich organic materials that some organisms can take advantage of as a food source (Wotton and Preston 2005). When these organic surface films are mixed with air (in stream rapids or generated by wind on lakes) it can produce spume, a frequently observed foam, but usually misinterpreted as pollution, on the surface of waters (see Figure 2.6). The surface of water also provides resources for *trophic interceptors*, those organisms that live on or under the water surface to feed on particles or prey trapped on the surface (Marczak et al. 2007). Some examples of these species are backswimmers (Notonectidae), water striders (Gerridae), some fishes (for example cutthroat trout [*Oncorhynchus clarkii*]), whirligig beetles (Gyrinidae) (see Figure 2.7), and others.

Figure 2.6   Spume on a natural stream, a product of fatty acids from bacterial production and other sources that combine with air to create a *foam*. Spume usually contains an abundance of fungal spores, other particulates, and mucopolysaccharides (from microbial metabolism).

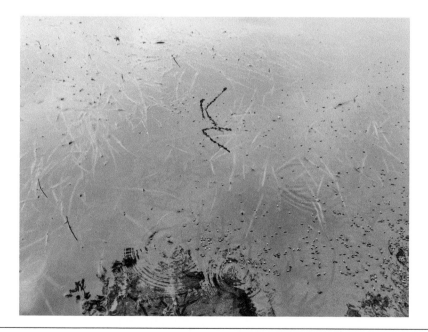

Figure 2.7   A mating swarm of Whirligig beetle adults (Gyrinidae) on the St. John's River in New Brunswick, Canada. Adult whirligigs feed on prey trapped in the surface film and are unique in having their eyes paired with an above-water lens and a below-water lens to watch for prey from either side of the water-air interface. They are one example of a species that captures food at water surfaces.

SUBSTRATE

Within each of our freshwater environments, we often describe the materials that organisms live on or in as *substrates*. Earlier, we described these substrates as types of microhabitats. For instance, some worms live in the mud at the bottom of a pond, and we would call that mud a substrate. Or an amphibian egg mass attached to a log, where the log would be a substrate. Many organisms live *within* the sediment, either in between gravel particles or burrowed into finer materials. Worms and larval chironomids in lakes and wetlands burrow into the fine bottom sediments and can undulate their bodies within the burrows to create a current for oxygenation, and then poke their heads out of the tube to feed on fine particles deposited around the entrances to their burrows.

Mudfish and other species may seal themselves in mud. Some organisms live on top of the sediment, and we call that epibenthic, or in more specific cases epipelic (on mud) or epipsammic (on sand). In the next chapter we will learn about movements and storage of sediment, and how that creates and influences habitats for species. Many stream organisms move about within the matrix of cobbles, gravel, and coarse sand at the bottom. Some species may dig into the coarse substrates for nests as salmon do, or a place to hide for creatures such as crayfish.

Wood can form an important substrate for some species, whether that wood is big or small. We typically define large wood as pieces both >10 cm diameter and >1 m in length, with small wood being anything smaller than that. Some species attach themselves to wood, such as larval Hydropsychidae, a family of filter-feeding caddisflies. As an organic surface, fungi and bacteria use the carbon in wood as an energy source, which was discussed in Chapter 1. In turn, this biofilm is consumed by some relatively specialized consumers (Cranston and McKie 2006; Eggert and Wallace 2007), such as the beetle *Lara* spp. and the fly *Lipsothrix* spp. (Dudley and Anderson 1987), and also by generalist grazers of biofilm. Some fish, amphibians, or other animals can attach their eggs to wood so they will not wash away, often on the underside to hide them. In some places wood provides the only large attachment sites, often called *snags*, when the remainder of the habitat might be unstable, soft sediment. In low-gradient streams most of the instream production occurs on wood by organisms attached to the wood (Benke and Wallace 2003). Some species burrow into wood as a food source, but also a safe place to hide. Other species, such as some Amazonian fish, hide in crevices within large wood. Finally, some species of caddisfly larvae (including *Amphicosmoecus* and *Heteroplectron*) will burrow into small twigs and other pieces of wood to provide their larval case.

There are many species that live attached to plants for some or part of their lives. Many animals attach their eggs on or insert them into vascular plants under the water, such as some salamanders, dragonflies, and fish. Other species, such as algae (periphyton), attach themselves to larger plants, and might even be referred to as epiphytes. Sit and wait predators—for example, larvae of some dragonflies and damselflies, *Hydra*, or filter feeders (such as bryozoans)—often hold on to plants to access food. Algae and bacteria may also get an advantage living on plants through priming. *Priming* is the secretion of carbohydrates produced by the plant in excess of the plant's needs that is used as an extra food source by algae and bacteria (Halvorson et al. 2020). There are also many species of herbivorous animals feeding on vascular plants and their fruits and seeds, including many fishes, Chrysomelidae beetles (Wallace and O'Hop 1985), and others.

This chapter cannot possibly give a full background to all the kinds of freshwater habitats. As mentioned in Chapter 1, there are many good books on freshwater ecosystems that address the basic biology of these ecosystems, such as Dodds and Whiles (2019) and others. There are also many online resources available, with one example being https://www.nature.com/scitable /knowledge/library/rivers-and-streams-life-in-flowing-water-23587918/.

This book has free material available for download from the
Web Added Value™ resource center at *www.jrosspub.com*

3

SEDIMENT

INTRODUCTION

In aquatic ecology we often refer to *sediment*, which includes an array of particles, inorganic and organic, and of many sizes and densities, that can be deposited on the bottoms and sides of water bodies. These particles share a common feature in being subject to movement by the erosive force of water, and subsequent deposition to form different kinds of benthic habitats. Sometimes there can be confusion as the beds of lakes, streams, and wetlands are also called sediment or substrate, but that is just while those particles are stored on the bottom, as opposed to being transported by the force of water. As such, sediment and sediment transport are natural features and important elements of habitat for the many organisms living in freshwaters (see Chapter 2). The movement and rearrangement of sediments creates new habitats for some organisms, and is a natural part of the dynamics of stream environments. However, as we will see in this chapter, human activities greatly increase the rates of sediment movement and deposition, and change the size spectrum of particles making up sediments (Walling and Fang 2003).

Sediment is made up of inorganic (mineral) and organic particles that range in size from microscopic up to the size of large trees. Particles can be defined by their size (often according to the Wentworth Scale) or how they are transported. Characteristics of mineral particles depend on the types of rocks they came from and can range in size from fine silt and clay (less than 62.5 μm) to boulders (greater than 256 mm) and bedrock. Particles such as *rock flour*, which gives glacial streams and lakes their bluish and turbid appearance, are silt and clay; they remain suspended in lakes because the particles are so small and even the slightest turbulence keeps them suspended (see Figure 3.1). Often, we see the term *fine sediment*, which can be defined as inorganic and organic particles that are less than 2 mm in diameter and as small as fine silt and clay particles (Jones et al. 2012b). Excess amounts of these small particles are most often what people are concerned about when they speak of sediment as a problem since these tiny and mobile particles can cloud the water (turbidity) and potentially bury organisms under a layer of material.

In order for sediment to move, the two most obvious ingredients are a supply of mobile sediment and a force to move those sediments. Exposed soils, hillslopes, and shorelines provide the sediment supply and the force of moving water, particularly during floods, or storms (waves), provide the force. The amount, size, and distribution of the particles available will determine the range and amounts of habitat types (pools, riffles, braided, backwater pools, etc.). Sediment also gets deposited along the stream length and into lakes and wetlands creating a variety of habitat types. This dynamic reworking of stream channels creates and modifies habitats.

Figure 3.1   Lower Joffre Lake, a glacial lake in British Columbia, Canada.

Particles enter water through erosion by the force of moving water, slope failures (such as landslides) that fall by gravity into water (see Figure 3.2), and large forces such as volcanoes and glaciers. Organic materials (leaves, seeds, flowers, woody stems, etc.) fall into water from the land, and trees fall over into (or over) freshwaters due to wind, decay, or land movements. However, all of that sediment does not enter freshwaters at once, as some gets stored temporarily in reservoirs, at the base of hillslopes, or in floodplains, which can be mobilized in pulses when floodwaters exceed the streambanks (Hassan et al. 2017). As discussed in Chapter 2, moving water exerts a lot of force, which is proportional to the volume of water and the slope of the stream. During floods, a lot of power can be exerted to move sediments, a big reason to maintain natural landscapes that can help keep peak flows lower than in managed landscapes (Moore and Wondzell 2005).

Fluvial geomorphologists define three movement pathways as (1) suspended particles or wash load—carried in the water column (see Figure 3.3); (2) saltational load—particles bouncing along the bottom; or (3) bedload—particles moving along the stream bed. Suspended particles are sufficiently small (such as silt and clay) that they can be suspended in the turbulent flow of water and this is what contributes to turbidity. Particles in saltation, typically medium sand particles, are too big to be suspended, but small enough to bounce along and be above the stream bed for short periods. Bedload are larger particles that get rolled along a stream bottom, again mostly during high-flow periods. The rates of movement of particles depends on their particle size and specific gravity, hence it takes more force to move larger particles, or particles that are heavier for their size. For instance, organic particles will have lower specific gravity than mineral particles and hence, will be more easily moved by the force of water. When the force of stream flows (or lake shorelines) is not sufficient to move particles (as during low-flow periods), they are stored in various configurations, such as gravel bars or cobble bars, riffles, and pools. Typically, we refer to materials that are reworked by high flows as alluvium, or to the

Figure 3.2 A natural debris flow (slope failure) in Banff National Park, Canada. Natural slope failures occur delivering coarse sediments to freshwaters in steep landscapes. Such slope failures tend to increase in frequency where there is land use, such as road construction or forest harvesting.

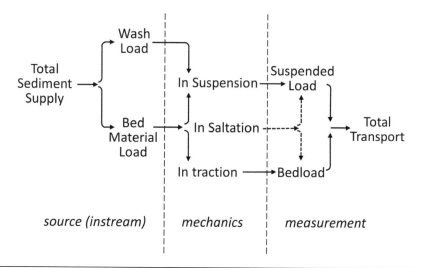

Figure 3.3 Components of sediment transport (figure modified after one by Michael Church, University of British Columbia).

structures as alluvial; for instance, alluvial floodplains and streams flowing through their own sediments as alluvial streams. Smaller streams that rarely, if ever, have discharge strong enough to move bed materials are called colluvial (Montgomery 1999).

Magnitudes of sediment movements through the largest rivers in the world show large rates of transport, but some of their rates are increasing and some decreasing (see Walling and Fang 2003). Generally, land use, especially agriculture, leads to increased supplies, but some sediment is deposited in floodplains. Many of the world's rivers have been dammed (see Chapter 8), causing the sediment export to the sea to be lower from some rivers. Estimates suggest that ~15% to ~25% of sediment yield from land globally is trapped in dams (Vörösmarty et al. 1997; Syvitski and Kettner 2011) (see Box 3.1). Soil conservation measures around the world have also reduced sediment yield to rivers in many regions relative to rates of several decades ago.

Sediments as Habitat

Sediments are the basis of stream structures and form the bottom of lakes and wetlands. Stream structure is highly dynamic as it is moved around and reworked by the force of water, especially during higher flows, and the organisms living there also need to respond in real time to changes. The variations in physical structures created by sediment movement (inorganic and organic particles) provide a variety of habitats for different organisms.

Many organisms live between or on sediment particles, and as such, this provides an important component of habitat. While larger organisms, such as some fish and mammals, swim around in the water, most do not. Most organisms attach to, burrow into, or live between particles of sediment, including wood. In flowing waters this is especially true, otherwise they would be washed away to potentially uninhabitable areas. Many organisms attach themselves to sediment particles (rock, wood) as a place to grow (Wallace and Benke 1984; Fonseca and Hart 2001), which can aid them in avoiding being washed away by flowing water, including wave action. Organisms may be very selective of particle sizes, or even positions on particles. For instance, black fly larvae or filter-feeding caddisflies select very specific flow velocities on top of cobbles and boulders (Wetmore et al. 1990; Fonseca and Hart 2001). Bacteria can grow on just about any surface, and this includes some of the finest particles of mineral or organic materials. Many kinds of algae attach themselves to mineral particles, and to some extent, organic particles. Invertebrates, such as filter-feeding caddisfly larvae or blackfly larvae attach to larger particles where they can gain access to food particles that are suspended in the water.

In lake bottoms, worms, chironomid larvae, bivalves, and other species select fine sediments in which to burrow and some of the species may stabilize their burrows using silk or mucus to bind sediment particles. The particular choice of particle sizes depends on the species and age class. These species then feed off the top of the sediment, where fine particles of organic matter, algae, and biofilm develop, all the while the invertebrates retain a hold in their burrow. In some lake bottoms, such as Iceland's Lake Myvatn, chironomid midge densities can exceed 500,000 larvae m^{-2} (Thorbergsdóttir et al. 2004), and thus sediment supply may restrict distributions. The organic content of the sediments may also be important for some organisms that feed on organic particles in bottom materials—for instance, worms, chironomid larvae, or gathering mayflies. A great number of small and large organisms burrow into fine sediments (such as chironomids, oligochaetes, mussels, crayfish, dragonfly larvae, and fish).

Living organisms can also contribute to sediment movements and stability. Stream-spawning salmonids, like sockeye salmon, dig redds (nests) into which they deposit their eggs, and in so doing, can displace large amounts of sediment downstream (Hassan et al. 2008). Chinook salmon (*Oncorhynchus tshawytscha*), the largest of the Pacific salmon, can reshape a stream bed by creating dunes from the coarse sediments displaced by their nest building activities (MacIsaac 2010). In streams that are downstream of lakes, salmonids can be the most important agent of sediment movement since lakes dampen flood flows and restrict sediment inputs into those streams (Gottesfeld et al. 2008). Other species contribute to movements of smaller particles, such as burrowing invertebrates—for instance, crayfish that create tunnels and other openings in stream beds, dislodging sediment (Statzner 2012). In fine sediments of lakes and wetlands, invertebrates can move sediments around in ways that influence the biogeochemistry of benthic environments (Mermillod-Blondin and Rosenberg 2006). Net-spinning caddisflies, like *Stenopsyche* spp. can *cement* together rocks with their silk filtering nets, increasing the stability of sediments, thereby creating increased resistance to movement (Takao et al. 2006).

Larger aquatic plants or macrophytes can also interact with sediment movements in interesting ways. Macrophytes can slow water velocity and enhance deposition of fine sediments (Jones et al. 2012a). Thus, macrophytes can promote their own growing environments if those fine sediments increase rooting zones and nutrient accumulation. Fine sediments also come with organic particles and often higher nutrient concentrations, which also enhance productivity (Jones et al. 2012a). At the same time the increased flow resistance from dense macrophyte beds may displace the energy of moving water and cause more erosion beyond the macrophytes (Jones et al. 2012a). However, when stream macrophytes die back in the autumn in temperate regions, it can lead to large pulses of sediment movements (Dawson 1981).

Sediment movements are exacerbated by land use, such as agriculture that exposes bare soils to the erosive force of rain, forestry that removes streamside structure that otherwise reduces erosion, and other activities that elevate sediment supply or increase exposure to erosion. Estimates of erosion rates from plowed crop agriculture indicate 10 to 100 times higher rates of soil loss compared to pre-agriculture or nonagricultural situations (Montgomery 2007) (see Box 3.1). Other sources of sediments come from urbanization and development, mining, road construction, and other land uses. Unpaved roads provide sources of sediments, and also help to channel sediment transport along a simple route, often through ditches, with little impedance (see Reid et al. 2016).

Removal of large wood and other changes to stream structures can reduce storage, and by definition increase transport. One demonstration of this was an experimental removal of large wood from a small stream in New Hampshire, which increased export of inorganic materials by 6.8-fold and organic materials by 5.5-fold in the year after removal (Bilby 1981). Similarly, another experimental removal of large wood from a small stream resulted in a 4-fold increase in sediment transport of bed materials (Smith et al. 1993). These are examples of one of the important roles of large wood in freshwater ecosystems.

IMPACTS

There are several mechanisms by which sediments of various kinds affect life in water. These include increased turbidity, smothering (burying) of organisms on the bottom, abrasion, suffocation, and reduction of light availability (see Table 3.1 and Figure 3.4). Good summaries of the various impacts of sediments have been published and provide further details and references (Ritchie 1972; Waters 1995; Kemp et al. 2011; Jones et al. 2012b).

Table 3.1   Possible impacts of elevated rates of sediment movements in freshwaters.

Sediment Form	Impact
Turbidity	Reduces visual capacity; may impair foraging, makes it difficult to detect predators, and may make it easier for predators to catch prey
Smothering	Coats gills and reduces their efficiency
Burial	May cover eggs or nonmotile species (e.g., mussels) in bottom sediments and kills them by suffocation
Abrasion	Scratches tissues, such as gills, or may dislodge biofilm organisms or other attached species
Infilling of spaces	Fine sediments (organic and inorganic) infill space between rocks reducing habitat availability; e.g., salmonid spawning gravels
Light attenuation	Reduces primary productivity
Ingestion	Organisms that collect fine particles as filterers or gatherers may have more inorganic particles, which would dilute the food value they collect

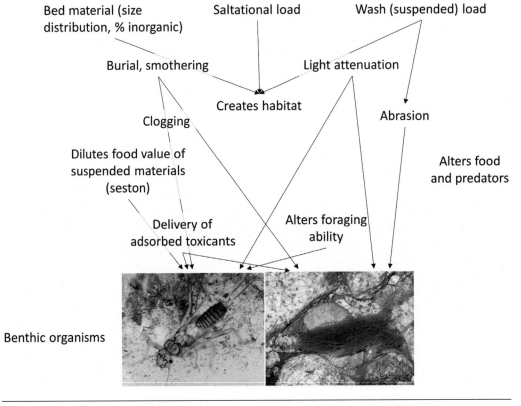

Figure 3.4   Pathways of sediment influences on an aquatic invertebrate (left) and filamentous, red alga (right) (redrawn after Jones et al. 2012a).

Turbidity

Turbidity or total suspended solids (TSS) (see Chapter 1) can impact organisms and ecosystems in several ways (see Figures 3.4 and 3.5). The impairment of vision can affect organisms, such as many fish that use sight to locate their prey (see Nieman and Gray 2019). At the same time, this can reduce predation rate on smaller organisms that might be able to use turbid areas to hide from their predators, assuming they are not similarly affected by the turbidity. Some species use elaborate mating displays, often with bright nuptial colors, and turbidity can impair the ability of a mate to see these displays and colors. For example, in the fish called threespine sticklebacks (*Gasterosteus aculeatus*), males typically have bright red on their ventral and lateral edges during breeding, and use those colors along with a mating display to attract females to their nest. In one case where a sympatric pair of sticklebacks had evolved, the increase of turbidity in a lake apparently interfered with the visual component of mating selection and resulted in the two forms hybridizing and losing the two distinct species (Taylor et al. 2006). Increased turbidity is also thought to be one of the explanations for the collapse of some of the famed species' diversity of fishes and other freshwater organisms in Lake Victoria, Africa (Seehausen et al. 2008). There are many other examples of turbidity altering mating interactions based on attenuation of visual discrimination. Turbidity is also a major problem for water quality (see Chapter 4).

Many filter-feeding invertebrates are nonselective of particle types within a given size range, and consume inorganic and organic particles indiscriminately, so that higher concentrations of inorganic particles will dilute food acquisition rates (Kurtak 1978). Freshwater mussels filter particles from the water passing above them. In experimental manipulations of the

Figure 3.5   Tributary to the Tarn River near Bruniquel, France, showing heavy suspended sediment load contributing to turbidity. Note the cloudy water (high turbidity) from the tributary to the left flowing from an agricultural area where precipitation has eroded soils into the stream.

concentration of suspended, inorganic sediment across a range of values, reproductive rates in mussels were significantly reduced, especially at higher concentrations (Gascho Landis and Stoeckel 2016). The mechanisms for these effects may have been the disruption of fertilization or lower energy assimilation (Gascho Landis and Stoeckel 2016). This indiscriminate ingestion of particle types can also predispose these organisms to feed on microplastics and nanoparticles, in addition to food particles, thereby reducing the growth, survival, and reproduction rates of freshwater organisms (see Chapter 4) (Haegerbaeumer et al. 2019). Experimental additions of fine sediment to ponds reduced survival rates of toad tadpoles by 20% and growth rates by 24%, relative to controls. Examination of gut contents of these toad tadpoles showed they were directly consuming the inorganic, fine sediments as part of their food intake (Wood and Richardson 2009).

Smothering, Abrasion, and Suffocation

Particles in suspension can affect gill function and oxygen uptake. Particles may abrade the gills of fish and other organisms, reducing their effectiveness (Sutherland and Meyer 2007). Furthermore, fine particles may adhere to the sensitive oxygen exchange and osmoregulatory surfaces of other organisms, lowering their efficiencies (see Donohue and Garcia Molinos 2009). Organisms that are not mobile may be smothered as they are buried by deposition of particles, and this can extend to the restriction of oxygen supplies and anoxia.

Suspended sediments during flooding can also lead to the abrasion of biofilms, namely, the thin, but complex, film of bacteria and algae on the surfaces of mineral and organic particles. When organic particles are large enough for fungal mycelia to develop, then fungi can be part of biofilm, adding to the bacteria and algal mix. Biofilms can be highly productive and are grazed on by a variety of consumers. Abrasion can reduce the biomass of biofilms, and some species making up the biofilms are more sensitive than others to physical damage (Francoeur and Biggs 2006). In the study by Francoeur and Biggs (2006), the combination of higher velocity and high concentrations of suspended sediments were able to remove ~80% of biofilm from colonized substrates. Fine particles in moving water can also abrade leaves of macrophytes, reducing productivity (Jones et al. 2012a).

Fine or small sediment particles can fill in spaces between rocks, reducing space where animals can hide. Welsh and Ollivier (1998) found densities of tadpoles of tailed frogs were reduced by about half in streams that had received a pulse of sediments from a highway project, which they interpreted as having filled in habitat space. The addition of fine sediments is typically associated with a reduction in numbers and species richness of stream invertebrates (Larsen et al. 2011). In one experiment, additions of fine sands resulted in large decreases in stream invertebrate numbers (mostly through drifting of invertebrates) and reduced juvenile rainbow trout (*Oncorhynchus mykiss*) growth by over 60% (Shaw and Richardson 2001). Additions of fine sand in another stream system likewise lowered the density of invertebrates by more than 50%, mostly by drift; reduced algae biomass by 50%; and lowered decomposition rates by 30% (Louhi et al. 2017).

Fine sediments deposited on top of rocks, lake beds, and even wetlands can smother organisms, particularly those unable to move effectively. For example, fish eggs might be covered and unable to get sufficient oxygen resulting in death. Eggs of other species would suffer a similar fate. Organisms with limited or no mobility—such as rooted plants, attached algae, clams, and sponges—could be buried and unable to recover. In some cases, being covered in fine sediment (silt) can result in short-term reductions in algae. However, rapid recovery over the course of days can occur for those algae that are capable of growing up through the silt layer (Kiffney et al. 2003; Izagirre et al. 2009). Nevertheless, recovery of productivity from photosynthetic activity in those examples came with changes in algal species composition.

The amounts, sizes, and distribution of sediment sizes can potentially limit the distribution and abundance of some organisms. There can be too little sediment (bedrock), or too little of the right size of sediment (sorting by flow regimes). One way in which sediment sizes can be altered is by sediment-supply limitation, as is found downstream of reservoirs and lakes. Downstream there is little to no renewal of sediment, and high flows can wash out finer sediments leaving a layer of larger, and hard to move sediment in a process called armoring. Even though these larger sediments may overlay finer sediments, there is often no possibility of fish moving large stones to breed in such places. In other places, a higher supply of small or fine sediments can fill in pools, fill in spaces between rocks, or bury larger substrate. This will cause changes in the habitats of organisms, and for those living within the sediment layer, there can be reduced flow and lower oxygen and nutrient renewal, thereby reducing the value of these areas.

Food-Web Effects

Fine sediments affect some organisms more than others. For example, fine sediment that accumulates on stone surfaces following forest harvesting seems to benefit algae growing up through the fine sediment where it may have had a refuge from grazing by invertebrates (Kiffney et al. 2003). Light interception as a result of turbidity or covering by fine sediments could lead to reduced light and changes in light quality (wavelengths) reaching algae or plants, and thereby reducing primary production (Jones et al. 2012a) (see Figure 3.4). A reduction in the amount of light reaching the bottom of streams, lakes, and wetlands can also occur if concentrations of dissolved organic matter (DOM) increase; and this might be an issue for the brownification of lakes noted in some northern latitudes (such as, Kritzberg et al. 2020). This has been demonstrated experimentally in ponds where a gradient of DOM concentrations was applied, and while bacterial productivity increased (more DOM), the attenuation of light reduced algal production much more than the increase in bacteria (Jones and Lennon 2015).

Elevated turbidity can also alter predator/prey relations since the ability to see prey (or a predator) is reduced. Higher turbidity may be due to greater inputs of fine, inorganic sediments in suspension, or increased biological contributions through algal blooms. Turbidity interferes with the ability to see and reduces foraging efficiency, for instance, as shown experimentally for the predatory walleye and its potential prey, the emerald shiner (Nieman and Gray 2019). If one species in a predator/prey relationship is disproportionately affected, this can have food web consequences. The range of prey types consumed may also be affected by decreased visibility created by suspended materials of different kinds and concentrations.

INTERACTIONS WITH OTHER STRESSORS

Elevated sediment loads can interact with other stressors to exacerbate impacts. Elevated sediment concentrations affect water quality, alter habitats, interfere with species interactions within communities, and can damage infrastructure. The following paragraphs include a few examples of how sediment supply might interact with other stressors—as we will see throughout the rest of this textbook for even more stressors.

Changes in sediment supplies can have curious interactions within biological communities, some of which have been previously mentioned. Many pollutants can adsorb to the surfaces of organic or inorganic particles and may be more easily incorporated into food webs by consumption by organisms (see Wang and Wang 2018). Sources of sediments frequently carry contaminants as well, so it may be difficult to discern whether sediments, contaminants, or both are affecting aquatic life. Infilling of spaces between rocks by fine sand can reduce escape space for small invertebrates from their predators, increasing predation rates (Louhi et al. 2017). Organisms already affected by other stressors, such as warm water, etc., may be more susceptible to additional stresses.

More frequent and more intense storm flows that are associated with climate change will increase both erosive force and duration but will also transport more materials. However, it is difficult to disentangle climate-related effects from those of land-use changes, given the poor resolution of most flow and sediment flux data (Syvitski 2003). This is a common challenge with environmental problems since many of these stressors are increasing simultaneously and can interact with each other to compound the impacts—a theme we will explore more in most of the subsequent chapters.

In addition to providing functional freshwater ecosystems, humans also depend on ecosystems that are in good condition to provide water for human consumption. Elevated concentrations of suspended sediments in these systems can cause various problems. Measured as turbidity or TSS, it impacts water quality by increasing concentrations of particles, some potentially carrying adsorbed contaminants. Elevated suspended particle concentrations create problems for water treatment for several reasons: (1) it increases the amount of filtration needed to process potable water (for health and aesthetics), (2) it supplies more bacteria living on particles, and (3) more sediment provides additional surfaces that adsorb disinfection products and thereby require large amounts of chemicals (De Roos et al. 2017). The existence of high concentrations of suspended particles is also associated with higher concentrations of disease-causing protists, such as *Giardia* and *Cryptosporidium* (LeChevallier and Norton 1992). This may lead to requirements for high concentrations of disinfection chemicals (such as chlorine), which can lead to a variety of disinfection byproducts, some of which can be carcinogenic (Gopal et al. 2007).

SOLUTIONS

Turbidity is a commonly measured aspect of water quality that essentially equates to the inverse of water clarity, with the clearest water having almost no turbidity and with turbidity increasing as particles in the water increase in concentration. There are commonly available field instruments that measure turbidity as the ability of light to pass through a water sample. This is an easy measure to obtain. The turbidity measurements are used for monitoring water quality by agencies and water supply companies, and are often measured by biologists. Detailed studies of movements of larger particles are complicated and are mostly done by fluvial geomorphologists.

Limiting the Erosion of Sediments

Once sediments enter freshwater, they cannot be removed, except perhaps by dredging and capture within reservoirs and lakes. These sediments may settle in slow-moving parts of streams, lakes, and wetlands, causing problems there. The most obvious solution is to reduce inputs caused by erosion which will minimize sediment supplies. These solutions can take many forms (see Table 3.2). Reducing the availability of sediments caused by erosion through soil retention and good management of vegetation as land cover along streamsides and other shorelines will help (Zaimes et al. 2019). Streambanks need protection since these are the most erosive places, and vegetation along streambanks reduces erosion of sediments into streams (Millar 2000). Shores of larger lakes can also be highly erosive if there is sufficient wave action. Planting shore-lines is an important restoration tool (see Chapter 15). Nevertheless, reducing sediment supply and mobility in the first place is the key to reducing sedimentation.

A good solution for agricultural operations is reducing the exposure of soils to being washed away, which has benefits to landowners who grow crops. Farming practices in some parts of the world have changed dramatically in past decades (Montgomery 2007). This is partly through enlightened self-interest because farmers depend on fertile soils and do not want to lose them. The use of cover crops (as opposed to commercial crops) during times of the year that the fields are not under cultivation can slow rates of erosion of soils and deter the delivery of soil into the streams (Daryanto et al. 2018; Zaimes et al. 2019). There are also best management practices for forest harvesting and other land disturbances to minimize exposure of erodible soils (see Chapter 8). This also can involve good maintenance of drainage ditches or streams to reduce the amount of bare sediment and erosion potential.

Table 3.2 Methods in use for managing sediment supplies to freshwaters.

Source	Actions
Reducing Sediment	
Agriculture patterns	Direction of furrows parallel to hillslope and water; buffers of vegetation alongside water
Forest road maintenance (and other roads)	Better road construction; culvert checking to avoid blockage and erosion
Silt fencing for construction	Fencing and hay bales to reduce fine sediments washing into local waterbodies
Livestock exclusion	Fencing to keep livestock off banks and out of water (providing an alternative source of water)
Reducing boat speeds	Speed limits for boats and ships to reduce shoreline erosion from waves
Armoring banks	When other solutions to land erosion do not exist, hardening banks may be a viable solution
Enhancing Sediment	
Additions downstream of dams	Replacing spawning-sized gravels that have been washed away and not resupplied because they are trapped in reservoirs
Experimental flows to move sediments	Use of *flushing* flows to rearrange geomorphology and to permit development of cobble bars for species dependent on such habitats
Additions of channel altering pieces, such as boulders	Increasing complexity of stream channels with the addition of difficult-to-move boulders

It is frequently recommended to reduce direct *pipes* of sediments from landscapes (ditching, etc.). In urbanized areas (see Chapter 9), stormwater runs off roads, into stormwater pipes, and into downstream areas very quickly. The high peak flows downstream where all of that stormwater interacts with sediment sources and can have enormous erosive power that often requires engineering to reinforce streambanks. In many locations around the world, this requires turning natural environments into what the Japanese call *three-sided streams*, where streams are converted effectively into concrete flumes.

Gravel roads, used in forestry and agricultural areas, can be large sources of fine sediments. Good surfacing and maintenance of gravel roads can reduce the amount of sediment runoff. Keeping the drainage crossings, such as culverts, clear is important as well, in order to avoid clogging and failure. When culverts or small bridges are clogged by wood and other materials, they can cause the water to back up, which initiates erosion around and over the culverts and roads. Often culverts fail because the water running across the road has enough erosive force to create a new channel, frequently carrying portions of the road with it. Roadside ditches are designed to yield water quickly from the landscape and can carry a lot of fine sediment, but blockages in these ditches can also lead to water eroding its way around the blockage and sometimes capturing even more sediments to move downstream.

As we will see again in Chapter 8 about forestry and agriculture, one solution to reducing erosion is fencing out cattle and other livestock. This normally requires fence maintenance and the provision of alternative water supplies for cattle and other livestock. This has multiple benefits in terms of direct erosion, and also for the protection of riparian vegetation that contributes to bank stability. One study from Iowa indicated that suspended sediment loads could be reduced by 75% by fencing pasture away from streams to restrict cattle (Zaimes et al. 2019). However, fine sediments stored in floodplains from land-use practices of past decades continue to be available for mobilization during floods (Knox 2006).

Protecting Infrastructure and Reducing Bank Erosion

In many situations where infrastructure and properties are at risk of erosion, armoring of banks is used (see Figure 3.6). This engineering approach protects shorelines but reduces habitat since floodwaters get deeper and do not spread out across the floodplain, so organisms have nowhere to go to seek refuge from high forces and velocities. In addition, such armoring transfers the water's energy downstream where it affects the location of other organisms. This armoring can include large rocks, rip-rap (wire baskets filled with rock or pieces of concrete) (see Figure 3.7), large wood, or concrete. Building such armoring further away from the active channel with more emphasis on space for the river to expand can have many benefits for the river ecosystem and floodplain wetlands but makes the land less directly useful to humans. In some countries, such as the Netherlands (*Room for the River*), moving dikes further away from river channels provides wider floodplains, which were designed to help reduce the erosive force of floodwaters and to help store water to reduce flood risk downstream (see Chapter 7). Similar actions are underway in other parts of the world, including along the Sacramento River and the Mississippi River where they have one of the longest dikes in the world. Some types of farms can be designed to elevate their infrastructure, such as barns, above flood levels, while still leaving the floodplain subject to flooding in some years. Of course, another solution is to stop building and farming in floodplains, which is happening in a few places.

Figure 3.6   Manawatu River, New Zealand, with a heavily armored bank to reduce erosion and protect a park, but eliminating the river access to its floodplain.

Figure 3.7   Gabions to restrict streambank erosion along a small creek, thereby protecting nearby paths through a park.

In urban and suburban areas, various designs of stormwater detention ponds serve to delay water inputs to streams and help reduce the erosive force of floods. These ponds are generally constructed as designed wetlands, bioswales, or other depressions in order to store water and also trap sediments (see Figure 3.8). These features can vary in size from small to large, and can also serve as productive freshwater environments called stormwater retention ponds that can

Figure 3.8 Bioswales, also known as stormwater retention ponds, detention ponds, or constructed wetlands, for stormflow and sediment trapping. These are usually associated with housing developments to capture stormwater, reduce peak flows in nearby streams, and capture sediment and contaminants that may flow from surfaces like roadways and parking lots. (A) Retention pond in Denver, Colorado. (B) Central Park, Denver, Colorado. *Source*: Stacey Eriksen and Greg Davis, U.S. Environmental Protection Agency.

hold water through much of the year. Often these artificial ponds require maintenance, which might include cleaning of sediments and excess vegetation. Such designed wetlands can be considered a form of green infrastructure that is sometimes referred to as a nature-based solution.

Boat traffic in larger waterbodies can cause erosion of shorelines, including dikes (Bauer et al. 2002). Typically, governments have attempted to instigate speed limits for boats close to shorelines to reduce the energy from boat wakes that erode lake and river margins. These vary tremendously around the world in terms of regulations, enforcement, and outcomes. Vegetated buffers around lakes and wetlands can be effective at protecting shorelines, especially from the effects of waves.

During construction, sediment fencing or silt fencing is commonly seen at sites to attempt to reduce runoff of sediments into local waterways (see Figure 3.9). However, water can erode under the fence or around the fences in many instances if they are not well maintained, allowing for direct flow of sediment to streams. Designs to keep water from backing up and going around fences have been demonstrated to increase the efficiency of sediment capture, but may not be completely effective (Zech et al. 2008). Construction projects also use constructed swales and other means of capturing sediment that has washed from exposed areas during projects (McCaleb and McLaughlin 2008).

Construction works in or near streams often have a limited window of time each year during which operations can occur to avoid impacts on important fish stocks and their habitat. These timing windows are governed by local rules. In the case of pipe installation (utility lines, gas pipelines, domestic water supplies), it is possible to use horizontal, directional drilling beneath streams and rivers, as is often done with oil and gas pipelines (Skonberg 2014).

Figure 3.9   Silt fencing used to reduce sediment flows from construction projects into nearby surface waters.

Too Little Sediment

Man-made dams across a river are constructed to make artificial lakes called reservoirs. Reservoirs are built to generate hydroelectricity, to store water for human use, or to reduce flood risks. However, reservoirs also change sediment flows by capturing most of the sediments in the storage *pond* (the water backed up behind the dam), which starves downstream reaches of new sediment supplies and both *hardens* and *simplifies* river structure (Staentzel et al. 2020). This can lead to extreme requirements being necessary to remedy the situation. As we learned earlier in this chapter, removal of small and large dams is one solution, but that is not going to be done everywhere because we need reservoirs, thus, additional solutions need to be devised. Experimental releases of water from the Glen Canyon Dam in Arizona (Korman et al. 2023) and the addition of gravels in the Sacramento-San Joaquin River in California are two examples of progress being made.

In Arizona's Grand Canyon, several large dams alter natural flow patterns in order to store water for hydroelectricity production and other human uses. One of the largest is the Glen Canyon Dam, which causes alterations to natural flows (see Chapter 7) and also traps sediments in the reservoir, such that the streams that are downstream of the dam are sediment starved. The channel has been deepening, and gravel bar habitats have been disappearing. Trials with experimental high flows have been attempted to try to recreate gravel bar and sandbar habitats, and return channel complexity (see examples in Yao et al. 2015; Mueller et al. 2018). The channel complexity observed has mostly been linked to natural flows of large sediments from tributaries of the river downstream of the dam that provide coarse sediments to the otherwise sediment-starved river; thus, the peak flows have materials to rearrange.

In the delta of San Francisco Bay, large efforts to restore aquatic habitats have attempted to repair damage from development, gravel extraction, and damming of the major river systems (Kondolf et al. 2008). Among the many actions taken has been the addition of materials to replace spawning gravels that are required for the endangered Chinook salmon—sometimes as gravel tips (self-seeding piles of gravel) and as re-engineered channels (Harrison et al. 2011). Enormous movements of materials were needed to complete these projects, but have successfully created a more mobile channel that may still be adjusting. This has also resulted in changes to habitats for invertebrates and fishes (Romanov et al. 2012).

PERSPECTIVES

The impacts of alterations in sediment supply have long been known, and methods for adjusting these are still developing. Once sediment reaches freshwaters the main processes for its removal are deposition on floodplains (now often alienated) or in lakes and reservoirs, or it is exported to the oceans. Reservoirs have received enormous amounts of sediment, which also reduces their capacity to store water or produce power. Best practices at land management are key to reducing supplies of sediments to freshwaters.

The next steps for research and management include more specific testing of protection and restoration measures, which are often implemented with little evidence of their effectiveness. This would include practices such as vegetated buffers, bank stabilization, agricultural practices, and others. Measurement of sediment movements is difficult, and proxy measures, such as turbidity, are not perfect substitutes but do provide some guidance. While this chapter has focused largely on flowing waters, it is important that there be more study of lakes and wetlands for effects of sediment fluxes since most of this material ends up suspended and deposited there.

Box 3.1   The Mississippi River, USA

The Mississippi River is the largest watershed in North America at about 3.2 million km^2, and drains around 40% of the United States (Benke and Cushing 2005). The river's basin is subject to intense land use, and the river itself is a critical transportation corridor that has been channelized and diked, and its flows controlled. A large proportion of the watershed's wetlands have been drained, and around 30% of the area converted to agricultural lands (crops and livestock) (Hassan et al. 2017). Losses of soil in the early part of the 20th century affected productivity and impacted streams. Other large streams of the world, such as the Amazon, Ganges, and Yangtze rivers, similarly move impressive amounts of sediment from the land to the ocean each year. Soil conservation measures in the Mississippi River watershed began in the 1930s and have reduced sources of sediments entering streams.

In addition to being a huge river, the mass of sediment movements into and through the Mississippi River is also enormous. The yield of sediments from the entire Mississippi River watershed is about 210 Mt/y for the recent years (post-dam) and a value of 400 Mt/y in the early years of the 20th century (Walling and Fang 2003; Milliman and Farnsworth 2011). Dams have a huge effect in reducing flows of sediments, with about 46% of sediment flux stored in reservoirs averaged across the Mississippi watershed (Hassan et al. 2017). There are also sediments stored between locks. Globally, it is estimated that during the 20th century, dams have reduced the annual sediment movements to the ocean coasts by 15% or more (Syvitski and Kettner 2011), which represents an enormous amount of storage in reservoirs. That decrease in sediments reaching the oceans is despite estimates that sediment losses from land to freshwaters have increased over the same time frame.

The amount of sediment yield per unit area is about 100 times higher for cultivated lands than areas remaining as grasslands or forested (Hassan et al. 2017) (see Table 3.3), which gives a good indication of the impacts of agriculture. The yields of sediments amount to over 100 t/km^2/y from agricultural lands, in particular in the lower parts of the Missouri, Mississippi, and Ohio river valleys. The Mississippi Delta downstream of New Orleans, Louisiana, receives sediments, nutrients, pesticides, etc., from the entire basin, and in recent decades, part of the Gulf of Mexico by the delta has been considered a *dead zone*.

Table 3.3   Estimates of sediment yield in tons/km^2/y from land uses of different sorts in the Mississippi River watershed. Note that cropped farmland, arable, has over 100 times more sediment yield than grassland or rangeland. Data from Hassan et al. (2017).

Land Use	Median	25–75% quartiles	Maximum	Minimum
Grass dominant	1.83	0.57–4.59	141.42	0.11
Mixed grass	1.41	0.86–8.48	1241.86	0.04
Mixed arable	210.4	22.13–723.76	30,000	0.45
Arable dominant	227.02	18.34–500	35,909.38	0.55

Box 3.2   Case Study: Removal of the Elwha River Dams

Over the lifetime of a dam, sediments can accumulate in the basin of its reservoir, reducing the amount of water stored and reducing its effectiveness for hydroelectric power generation. As noted in Box 3.1, reservoirs store enough sediments that delivery to the ocean coasts has been reduced in the past century. Dams also restrict access of anadromous fish and other species from areas upstream of dams where they historically occurred. Alienation from upstream areas can be through lack of passage (for instance, absence of fish ladders) or conversion of streams to unsuitable reservoir habitats. In some parts of the world, dams are increasing in numbers and storage capacity (Dudgeon 2011). However, in the United States, work to remove some dams has begun with more than 1,400 removed since the 1970s, including some very large dams, and this is a learning experience for biologists and engineers (Hart et al. 2002; Bellmore et al. 2019).

The biggest dam removal project to date was in Washington State's Olympic National Park, where two large dams have been removed (see Figure 3.10A). There were decades of planning for the dismantling of the Elwha (built in 1913) and Glines Canyon Dam (built in 1927). Nearly 21 million m^3 of sediments (inorganic and organic) had accumulated in the two reservoirs over the period of just around a century (Duda et al. 2019) prior to dismantling (see Figure 3.10A). The project team had to anticipate the huge plume of sediment from the finest to coarsest materials, as well as determining what to do with fish populations that were unlikely to persist while the large amount of sediment was being moved. One estimate was that sediment movement rates were over 1,000 times higher than background. In order to protect the fish, in consideration of the amount of expected sediment movement over time, they used hatcheries (especially for Chinook and coho salmon [*Oncorhynchus kisutch*]) and adult relocation, along with natural recolonization to enable fish to persist and expand into newly available habitat (Duda et al. 2019). Over 20 million tons were released in the first five years after dam removal (see Figure 3.10B). There was also fear for infrastructure, such as bridges and roads, since sediment was allowed to escape. Thus, dam removal was staged over several years to avoid one extreme event.

continued

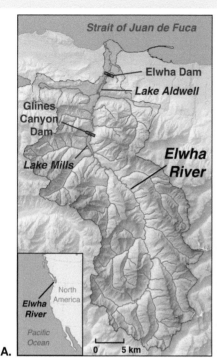

Figure 3.10 (A) Map showing the location of the Elwha River in the state of Washington, including the sites of two dams and their reservoirs, since removed. (B) Estimates of sediment flux from Elwha River reservoirs five years after dam removal began. Mt = millions of metric tons (1,000 kilograms) (*Source*: Ritchie et al. 2018 and USGS Public domain, and Creative Commons Attribution 4.0 International License).

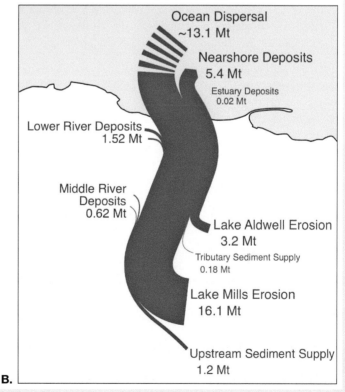

continued

One of the big learning outcomes from the dam removals was the speed at which fish, including several anadromous salmonid species, colonized the areas upstream of the dams' position (Duda et al. 2019). Chinook salmon were observed spawning upstream of the former Elwha Dam within three years of removal. Moreover, marine-derived nitrogen from anadromous salmon was detected in food webs upstream of the dam sites within a year of removal (Bellmore et al. 2019). Other elements of the food web of the ancient stream beds upstream of the dams may take longer to recolonize, especially species lacking a flying stage or strong swimming ability. The very large amounts of stored sediment released from the two dams were reshaped by hydraulic processes in the streams below the dams. Ultimately, a large plume of this sediment reached the Strait of Juan de Fuca, which is being reshaped by coastal water flows. The experience with this dam removal shows the importance of a detailed response to sediment accumulations and release, and also the success that dam removals can have.

ACTIVITIES

1. Examine local water bodies for signs of turbidity or unusual amounts of fine sediments along the shores and try to determine where that sediment comes from. Is your local stream, lake, or pond turbid?
2. What do you think are the possible impacts of turbidity in your local ecosystem?
3. Use measures of suspended sediment (mass per liter) from your region (for instance, data on U.S.-EPA websites), then multiply by annual average flows (L per year) in a river in your region to estimate how much material is washed away each year. Can you estimate the amount per area of the drainage basin?

4

NONPOINT SOURCE POLLUTION

INTRODUCTION

Pollution of water by elements and materials that are natural or unnatural occurs globally. Sediment was discussed in the last chapter, which is one of the most pervasive kinds of pollution through turbidity and sediment deposition. The most obvious kind of chemical pollution is from excessive nutrients including nitrogen (N), phosphorus (P), and potassium that spill off the land and enter the water. Other kinds of pollution include contaminants such as pharmaceuticals, heavy metals, plastics, and nanoparticles. We can even include heat pollution from excessive warming of water, which often occurs at industrial sites, from a runoff of hot road surfaces in cities, and from some power plants. In most cases, pollution is about *excess* concentrations because nutrients and other elements such as iron (Fe) and copper are required at some level for biological functions and are naturally occurring at *background* concentrations. Over 100 million distinct chemicals are known, but fewer than one percent are regulated (Gessner and Tlili 2016). Many chemicals can be spread through the atmosphere or from the land and end up distributed in freshwaters, even in remote locations. Once upon a time, there was a perspective that *the solution to pollution is dilution*, but humans very rapidly overwhelmed the abilities of most natural earth systems to deal with pollution that way. We will not deal with all polluting chemicals in this chapter, but you will learn about the major classes, why they cause problems, and what can be done about them.

Pollution sources are often referred to as point sources or nonpoint sources (also called diffuse sources). A point source is a discrete place where discharge goes into a surface water. Point sources include discharge points from industry, outflows from wastewater treatment plants, and pipes from combined sewage and storm drains. Such locations are easier for measuring outputs to surface waters, and we will deal with point source pollution in Chapter 5. These two categories (point source or nonpoint source) are not absolutely distinct, and there are some contaminants that enter freshwater ecosystems from a range of spatial scales, some discrete and others from many small sources. There are many chemicals that can pollute our waters, for instance, manure from livestock operations (see Chapter 8) or road salts applied in many temperate regions to melt snow and ice during winter (see Chapter 9). In this chapter, our focus will be on nonpoint sources of pollution and how they affect freshwater systems.

Nonpoint sources of pollution include contributions of contaminants flowing through groundwater, or from dispersed sources that vary in size and distribution (see Box 4.1). For instance, pesticides (insecticides, fungicides, herbicides) and antibiotics from agricultural or forestry sources can all be found in surface waters. Other examples might be chemicals from agricultural runoff, from septic tanks in small communities or from lakefront cabins, outhouses in parks, among others. Septic systems may be sources of pharmaceuticals and personal care products (PPCPs) in surface waters without point sources. These PPCPs can include anti-depressant drugs, birth control hormones, and even artificial sweeteners. Note that PPCPs are also found in point source pollution coming from wastewater treatment plants and will be discussed in Chapter 5. Nonpoint sources also include runoff of nutrients and pesticides from private gardens and lawns, or even people washing their cars. Another example would be runoff from roads that transport heavy metals, hydrocarbon residues, cigarette butts, road salt (more on salt in Chapter 9), and trash that has been improperly disposed of.

A less often considered nonpoint source is from the atmosphere—including acid rain (see Box 4.1), a major environmental issue of the 1970s (Menz and Seip 2004)—as a consequence of elevated rates of inorganic N (as nitric acid) and sulfur (as sulfuric acid) deposition in rain (Fowler et al. 2013; Stevens 2016). Mercury is a serious health hazard because it can be transported long distances in the atmosphere. Studies of atmospheric deposition in high Arctic lakes (the only source of mercury there) show that it has been accumulating at rates higher than natural background rates since the industrial revolution (Kirk et al. 2011).

IMPACTS

Eutrophication

Elevated nutrients levels (*external loading*) often come from agricultural applications, homes, wastewater treatment facilities, or even atmospheric deposition. The amount of nutrient use and consequent deposition into the environment has dramatically increased over time and correlates with an increase in population (see Figure 4.1). This is one of the leading causes of surface water degradation. In the United States, about 46% of stream length, greater than 21% of lakes, and over 32% of wetlands are classed as being in poor biological condition, with the nutrients as a major stressor (U.S. EPA 2017). A similar state of degraded freshwater ecosystems exists in Europe and elsewhere. Excess nutrients lead to higher rates of production of algae (eutrophication) and faster rates of decomposition, which causes changes to food webs and water quality. Phosphorus is often the most limiting nutrient in freshwaters (but not always) and its excess increases productivity of undesirable algae—in particular, N_2-fixing Cyanobacteria that can produce their own N (although this is not the only explanation for blue-green algae blooms; see Pick 2016).

Increased productivity can result in problems with algal blooms, changes to food webs, oxygen depletion, and even toxicity from breakdown products of Cyanobacteria, such as microcystins (Conley et al. 2009; Wood 2016). Nitrogen deposition from the atmosphere doubled in the 20th century and has contributed to increased productivity in places where it is limiting. Experimental, whole-lake additions of P, N, or carbon (C) demonstrated conclusively that P was typically the most limiting of nutrients (Schindler 1974) (see Box 4.1). That was a first step in having P (as phosphate), a known surfactant, removed from detergents, a move that was heavily resisted by the industry in its day. In the United States, phosphates were first banned in several states in the 1970s, and by 1994 there was a nationwide ban of phosphates in laundry detergents. However, it was only in 2010, that some states in the U.S. finally banned phosphates from dishwasher detergents.

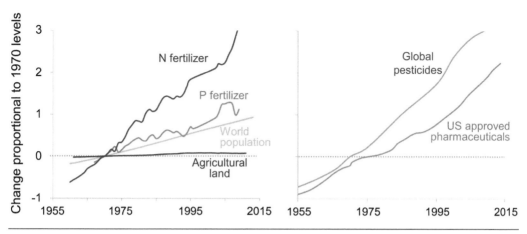

Figure 4.1 Proportional rates of increased use of nutrients and synthetic chemicals relative to 1970. Also indicated is the relative increase in the global human population. Redrawn after Bernhardt et al. (2017).

Increased production from high nutrient concentrations can lead to high amounts of algae, which causes aesthetic problems, but it also exceeds the capacity of aquatic consumers to eat so much biomass. Light can be attenuated (as in turbidity, discussed in Chapter 3), reducing the growth of plants on the bottom. At nighttime, the respiration of high biomass of algae and bacteria can deplete oxygen concentrations, sometimes to levels that are lethal for fish and other species. Oxygen depletion due to decomposition of biomass in high productivity systems can also lead to P release back from sediments (*internal P loading*) where it otherwise would be sequestered (Conley et al. 2009). The western basin of Lake Erie has had some of the worst algal blooms recorded during the decade of the 2010s (Michalak et al. 2013). While eutrophication of the Laurentian Great Lakes was seriously reduced in the 1960s and 1970s through international agreement, nonpoint source runoff has increased in recent decades. These blooms of harmful algae were attributed to the enormous amount of agricultural fertilizers, particularly P, that run off to the lake, as well as water circulation and temperature patterns that retained the blooms locally (Michalak et al. 2013). In winter, the death of a large biomass of algae can lead to oxygen depletion, particularly under ice for temperate lakes, and the re-release of stored P.

Toxic Metals and Synthetic Chemicals

Heavy metals such as lead, nickel, zinc, and cadmium often get into water through mining effluents (see Chapter 5), but can also result from deposition from vehicles, old pipes, and other sources that can be washed into surface waters. There are many types of heavy metals, and these elements are toxic to organisms at elevated concentrations. Some of these metals can enter water through road runoff carrying the particles coming from the wearing down of car brakes and other sources, through groundwater, and through wastewater. Heavy metals can be directly toxic through neurological means or other processes, or by displacing certain ions. For instance, lead is toxic and can replace Ca^{+2} throughout the body, including in blood and bones—in humans and in aquatic life. One heavy metal of particular note is mercury, which can be easily volatilized by forest fires, the burning of fossil fuels and garbage, metal smelting, and other means, thereby causing aerial transportation. When mercury enters water or even wet soils, it can be converted to its biologically active form (methyl mercury) through methylation by bacteria under anaerobic conditions.

These pollutants can accumulate within individuals and move through food webs, often referred to as bioaccumulation and biomagnification, respectively. Depending on the way that pollutants interact with an organism's physiology, a chemical can have many different effects. In some cases, metals such as lead can displace calcium (Ca) in the body, reducing physiological capacities. One particularly notable contaminant that can bioaccumulate through food webs is mercury, which is highly neurotoxic, and can lead to advisories to limit fish consumption in many places (Blanchfield et al. 2022). Fat-soluble organic compounds may accumulate in tissues of the body leading to a potentially toxic load of chemicals that can impair body functions. Pollutants can affect freshwaters through toxicity or physical impairment (such as plastics affecting food supply and digestion) (see Box 4.2), or through food web alterations.

Many kinds of synthetic chemicals are released into the environment, including pesticides, pharmaceuticals, personal care products, and others (see Figure 4.1) (see more in Chapter 5). A widespread group of pollutants include pesticides from many agricultural, forestry, and even urban applications. Pesticides for control of insects, bacteria, fungi (Zubrod et al. 2019), and plants can have adverse effects on freshwater ecosystems through toxicity to nontarget organisms or through alterations to food webs.

Synthetic particles from human use also pollute waters. Microplastics (<5 mm dimension), including microbeads and breakdowns of larger plastic items, are well known from publicity about their impacts in the oceans, but are also found in freshwaters (see Box 4.2). Nanomaterials (1 to 100 nanometer size) of many sorts are finding their ways into surface waters from industrial sources, wastewater treatment, washing of consumer goods, and even storm runoff systems (see more in Chapter 5). Once in the freshwater system, these particles can be ingested by freshwater organisms causing internal tissue damage and interfering with respiration if they make their way into the gills of fish.

Genetic Alterations

Some crop plants have been genetically modified to produce their own pesticides, such as the case of corn varieties that have had the gene for the insecticidal protein from the bacterium *Bacillus thuringiensis* (*Bt*) added to the corn genome (Tank et al. 2010; Venter and Bøhn 2016). While this constitutive pesticide reduces the need for application of other insecticides, this *Bt* protein can reach freshwater ecosystems in corn leaves and potentially harm aquatic insects.

Hormone Disruption and Other Effects

PPCPs are a group of compounds becoming more broadly detected in significant concentrations in our waters. These can include very obvious chemicals like caffeine from coffee, drugs for birth control, antidepressants, antibiotics, illicit drugs, and a host of other substances that are used by humans and passed through the waste stream (Kolpin et al. 2002; Rosi-Marshall et al. 2015), through septic systems, or via wastewater treatment plants (more on the latter in Chapter 5). Dr. Karen Kidd demonstrated how birth control hormones such as estrogen and derivatives are detectable in surface waters and have hormone-disrupting effects on some species, including the feminization of male fishes, which can severely disrupt fish populations (see Box 4.1).

Broadly used antibiotic compounds that are embedded in consumer goods to reduce odors, such as triclosan and silver nanoparticles, are prevalent in most freshwaters that are affected by humans. Triclosan is toxic to bacteria, algae, zooplankton, and fish, and has endocrine-disrupting properties. The enormous amounts of triclosan released into the environment (between 110 to 420 tons per year in the U.S.) result in antibiotic resistance and shifts to biological communities (Carey and McNamara 2015) (see more in Chapter 5).

Radiation Exposure

Radiation is another source of pollution in freshwaters globally, albeit usually at extremely low levels that are below levels of concern, and can be transported via the atmosphere. Most studies of radiation effects on freshwater organisms have been at sites of catastrophic leaks at nuclear power plants, but there are also potential impacts from leaks and waste sites (Hevrøy et al. 2019). Pathways of exposure include direct ionizing radiation, radiation absorbed into food materials (such as leaf litter), or through food web interactions (including indirect effects). In a microcosm experiment using multiple species, Hevrøy et al. (2019) found impacts along a gradient of radiation intensities on individual species' traits, but not on integrative measures of ecosystem function, such as respiration and primary production, indicating compensation of ecosystem rates (where some species made up for reduced function of others). Some species, especially primary producers, responded more sensitively to radiation exposure than others, such as consumers (see Figure 4.2A) (Hevrøy et al. 2019). In a Japanese stream downstream of the Fukushima nuclear plant accident (a point source), radioactive Cesium concentration was over a million times higher on fine organic particles or clay particles than in the water, as Cesium ions adsorb to surfaces of particles. Consumers of these particles, such as the net-spinning filtering caddisfly *Stenopsyche marmorata*, had significantly elevated body burden of radioactivity (Fujino et al. 2018). Such pathways into food webs through consumption have not been extensively studied.

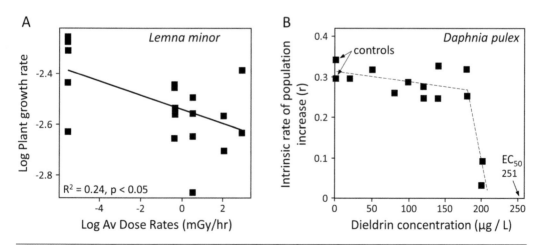

Figure 4.2 Two examples of ecotoxicology studies using criteria other than lethal concentrations. (A) Growth rates of the plant *Lemna minor* exposed to a gradient of radiation (redrawn after Hervøy et al. 2019). (B) Response of intrinsic rate of increase (at r = 0 a population is just replacing itself) for *Daphnia pulex* exposed to a concentration gradient of the pesticide dieldrin (redrawn after Daniels and Allan 1981).

INTERACTIONS WITH OTHER STRESSORS

Many stressors beyond pollutants cause problems for biological processes at species and eco-system levels, and these can act *in combination* to create larger impacts. For instance, in some cases the combined effects of contaminants, such as pesticides, may result in effects greater than the supposedly protective levels specified in regulations based on standard ecotoxicology. Many chemicals can bioaccumulate or bioconcentrate, thus organisms can end up with con-centrations of chemicals even higher than the environmental concentrations. Also, the process of biomagnification results in higher trophic levels having higher concentrations than their prey. The combined effects of multiple pesticides in European waters had deleterious effects on species and biodiversity at concentrations that were deemed safe based on individual contami-nant tests (Beketov et al. 2013). This points to the need to consider the full range of stressors and how they interact to affect individuals and populations of aquatic organisms. We will ad-dress this issue of cumulative effects and multiple stressors in detail later (see Chapter 13).

One interesting effect of contaminants is what is called hormesis. This occurs when very low levels of some stressor actually enhance performance in some organisms. Note that this is at low levels, but sometimes these stimulate biological processes to enhanced levels of performance. As an example, one snail and one algal species exhibited hormesis to low concentrations of heavy metals (higher rates of growth and reproduction), and another snail found those same concentrations toxic, but all three species showed toxic effects at higher concentrations (Lefcort et al. 2008). This has been observed for a number of pollutants, but note that some organisms exhibit hormesis and for other species toxicity occurs even at the lowest of concentrations.

SOLUTIONS

One of the main ways to deal with pollution is through legislation. Most countries have laws that protect freshwaters from pollution. In the United States there is the Clean Water Act (1972), the European Union has the Water Framework Directive (2000), in Canada it is the Water Act (1985), and other countries have similar laws. These rules protect the quality and quantity of water. Typically, there are rules around water quality, but regulated limits may dif-fer between water for human use and for protection of freshwater ecosystems. Laws also make provisions for withdrawals of water, degradation of habitat, and other aspects of water, such as temperatures.

It is important to try to control inputs of pollutants at their source. For instance, the impacts of acid rain in the 1980s was solved with strong controls on air pollution from industrial and municipal emitters (see Box 4.1). Similar success needs to be attained for N. Nitrogen and sulfur deposition from the atmosphere doubled in the 20th century and is a result of industrial emis-sions and applications of inorganic fertilizers (Stevens 2016). Controls on industrial sources have led to large decreases in emissions of sulfur, but only slight declines in N deposition in recent years. Sustained high levels of N deposition is linked primarily to agricultural fertilizer applications, therefore, more controls are needed. Another current problem is the measurable concentrations of pesticides and other chemicals that are being transported long distances, in-cluding from other countries—requiring better control on pesticide applications. Regulations require most companies to reduce potentially toxic compounds in their waste streams, and many processes have removed certain chemicals to replace them with less toxic alternatives.

Some controls include what we do individually, including more responsible disposal of phar-maceuticals and avoiding overuse of antibiotics. Other actions include less use of disposable plastics, washing synthetic fibers in cold water, adding a filter to washing machines, and not dumping potentially hazardous materials down the sink. In some parts of the world, including many recreational properties, sewage is dealt with by septic tank systems, and thus proper management of these is needed to ensure limited release of nutrients, PPCPs, and microbial contamination.

One improvement to water quality in our freshwaters is an engineering solution, which would be to fix drainage systems. Leaks in drainage pipes that carry domestic and industrial sewage to wastewater treatment plants yield large amounts of pollution. Another engineering solution would be to separate stormwater systems from sewage systems since sewage can overflow into surface waters through such pathways. Better wastewater treatment and improvements in septic drain field management would be very positive as well. However, these are only reasonable options in fully developed countries, and even there, huge local problems in remote or small municipalities result in poor control when it comes to pollution.

The Field of Ecotoxicology

In the modern world we have regulations for allowable levels of many pollutants in drinking water and in aquatic ecosystems. To screen new chemicals, the field of ecotoxicology tests for the concentrations at which effects can be observed on a selection of freshwater species. Heavy metals, organic compounds, and others are toxic at elevated concentrations, although some heavy metals are essential for life at lower concentrations (for instance Fe and copper). The well-developed field of toxicology and its more environmentally focused arm called ecotoxicology (many schools have whole courses on ecotoxicology), help to set standards for regulatory limits regarding the concentrations of chemicals for the protection of aquatic life. There are many texts and articles on this field, such as Walker et al. (2014) and Newman (2019). These tests are most commonly based on standardized, single-species laboratory procedures. Other ways to set standards can include results of field surveys and larger-scale experiments. Toxicology is the study of the fates and toxic effects of contaminants, and ecotoxicology adds in an ecological component of effects on genetic, population, community, and ecosystem processes (Gessner and Tlili 2016).

The advantage of standardized toxicology tests is that they are simple, repeatable, and understandable by nonbiologists. One of the most common tests for toxicity is an LC_{50} (lethal concentration) or LD_{50} (lethal dose), which is based around the concentration that kills 50% of test organisms in a specific time period, generally 48 or 96 hours. While there are a number of sophisticated methods for determining safe levels, a probit analysis is common. In a simplified description, the y-axis and x-axis are log-scaled, and the slope of the line through the 50% mortality point is extrapolated to the point of zero mortality (see Figure 4.3A). Some toxicants are not entirely water soluble in laboratory tests, therefore, a solvent is often used, which requires its own testing to separate effects of the solvent from the toxicant of interest. However, this is just the starting point in many regulatory processes. One possibility is to consider the lowest observable effect concentration (LOEC) or no observable effect concentration (NOEC)—based around the lowest levels at which any impact has been noted on any biological process (see Figure 4.3A, B). There are many safety factors applied in regulations, often one or more orders of magnitude below the NOEC concentration. See the example for the insecticide imidacloprid in Figure 4.3B. These safety factors also depend on the kind of chemical and its potential effects on living organisms. Keep in mind that many of the test organisms used in toxicology are relatively tolerant species, as they need to be able to be cultured in the laboratory, and so they may

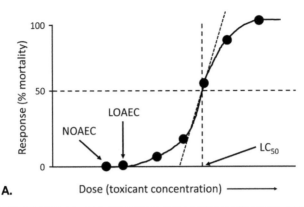

A.

Species	Toxicity endpoint

Acute

Leuciscus idus	96-h LC$_{50}$
Oncorhynchus mykiss	96-h LC$_{50}$
Lepomis macrochirus	96-h LOEC
Oncorhynchus mykiss	96-h LOEC
Daphnia magna	48-h LC$_{50}$
Daphnia magna	48-h LC$_{50}$
Daphnia magna	48-h EC$_{50}$
Chydorus sphaericus	48-h EC$_{50}$
Chironomus tentans	48-h LC$_{50}$
Hyallela azteca	96-h LC$_{50}$
Hyallela azteca	96-h EC$_{50}$
Aedes aegypti	48-h LC$_{50}$
Cypretta seurati	48-h EC$_{50}$
Simulium vittatum	48-h LC$_{50}$
Chironomus tentans	96-h LC$_{50}$
Cypridopsis vidua	48-h EC$_{50}$
Ilyocypris dentifera	48-h EC$_{50}$

Chronic

Oncorhynchus mykiss	60-d LOEC
Daphnia magna	21-d LOEC
Gammarus pulex	28-d LOEC
Chironomus tentans	10-d LOEC
Chironomus riparius	28-d LOEC
Chironomus tentans	10-d LOEC

Critical value

Concentration (µg / L)

Canadian water quality guideline
for imidacloprid   0.23 µg/L

B.

Figure 4.3   (A) A generalized graph of different endpoints used in ecotoxicology. LC$_{50}$ (or LD$_{50}$) is the concentration at which 50% of test organisms were dead within 48 or 96 hours. Other levels used are the no observable adverse effect concentration (NOAEC) or NOEC or the lowest observable adverse effect concentration (LOAEC) or LOEC. (B) An example of the range of toxicity endpoints using imidacloprid, showing the wide range of tolerance levels and acute-versus-chronic endpoints. Shaded sections are for fish and nonshaded are for invertebrates. The LOEC for larvae of the midge *Chironomus riparius* was set as the critical value and the water quality standard was set as one order of magnitude below (x 0.1) that concentration (the lowest value recorded was not used due to some uncertainties in the study with *C. tentans*; see further explanation in Canadian Council of Ministers of the Environment. 2007. Canadian Water Quality Guidelines for the Protection of Aquatic Life—Imidacloprid. Redrawn from the above article.

not be entirely representative of all freshwater species. Standards for water quality also depend on the chemistry of the receiving environment as water hardness and pH can have large effects on actual toxicity. Standards for the protection of drinking water are derived in a somewhat similar way, but there are many regulatory processes for determining allowable concentrations and is a whole field unto itself.

Ecotoxicology tests are usually done with one species at a time exposed to a concentration gradient (usually geometric) to measure survival rates at each concentration—an acute measure of toxicity. While the organisms used are generally robust enough to live in laboratory chambers, and thus may not be representative of all organisms, people recognize that fact and take it into account. Some of the common freshwater test species are juveniles of rainbow trout, fathead minnow (*Pimephales promelas*), zebrafish (*Danio rerio*), and other fishes, along with invertebrates such as *Daphnia*, *Ceriodaphnia*, *Chironomus riparius*, and others. These procedures are well established and are accepted by legal authorities and regulatory agencies.

Given that laboratory tests do not include the reality of sublethal effects of mild toxicity that may lead to higher rates of predation or lower birth rates, other measures are considered (see Figure 4.2B). Therefore, there are methods looking at chronic (or subchronic), as opposed to acute effects on study organisms, and can contribute to determining LOEC or NOEC. Studies in microcosms with suites of organisms in a functional test ecosystem can be used, and field studies can even be included. One endpoint for NOEC can include the frequency of developmental abnormalities (for example, the mouthparts of chironomid larvae) because early developmental stages are often the most sensitive in any organism (Di Veroli et al. 2014). There is always a trade-off between the realism of a study versus the costs and ability to replicate, so there are relatively fewer ecosystem-scale studies (see Box 4.1) and many test-tube or beaker/tank scale lab studies (see Figure 4.4). Whole-ecosystem experiments are relatively rare and costly (see Box 4.1), but mesocosms (smaller than whole ecosystems) can include a high degree of realism, while allowing for replication and experimentation (see Figure 4.5 and Chapter 5, Figure 5.1).

Toxic chemicals are rarely tested in combination, mostly because of the reality of the enormous numbers of chemicals—and even pairwise tests would be impossible, let alone

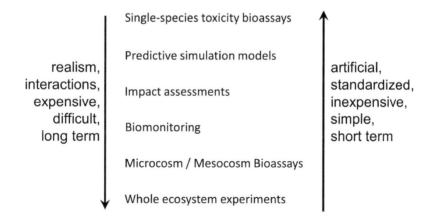

Figure 4.4 Trade-offs in realism versus cost for toxicology studies, ranging from test-tube scale, to mesocosms, to whole ecosystems.

Figure 4.5   Photograph of a mesocosm used for large-scale, replicated experiments within a lake. Photo courtesy of International Institute for Sustainable Development's Experimental Lakes Area, Ontario, Canada.

higher-order interactions. However, more and more, it is being appreciated that there are complex interactions among chemicals. Multiple stressors are addressed in greater detail in Chapter 13 of this book.

Controlling or Removing Nutrients

Managing loading rates of N and P is a major goal of wastewater treatment (see more in Chapter 5), as well as controls on other nutrient sources. Removal of P from detergents in the 1970s was considered a major success toward controls on eutrophication; however, total loadings to freshwaters are increasing as humans increase in numbers, with higher global food needs, and increasingly higher average standards of living. The work to control loading rates of nutrients into freshwaters takes many forms, and there are regulations around allowable rates of fertilizer applications in many situations. Avoiding excess fertilizer is both good for our freshwaters and saves money for farmers. Rules to reduce nutrient *loadings* from developments around waterways, including recreational properties with septic systems, are based on some simple models. An early attempt to set limits was based on nutrient budget models of runoff per cottage (cabin) around lakes (Dillon and Rigler 1974), and that has been refined over decades. Leaving riparian vegetation along streams and wetlands in agricultural landscapes can also be helpful at processing and storing nutrients before they reach freshwaters (Peterjohn and Correll 1984). Setting regulations and monitoring of nutrients and other chemicals in surface waters are done by various agencies, although determining the specific sources can be difficult.

A challenge occurs when, even if the external loading of P is reduced, it can be rapidly recycled within lakes especially under the hypoxic conditions created by high productivity of algal blooms. Therefore, to control nutrient-related productivity problems, it is relevant to reduce both N and P (Conley et al. 2009). Focus has been on P removal, as removal of N may lead to dangerous blooms of Cyanobacteria, as they can fix their own N if there is surplus P. One solution might be dredging, but this is typically very destructive, expensive, and may just redistribute excess nutrients from sediments. Artificial aeration of lakes to avoid the redox conditions that lead to P mobilization have been tried, but this is difficult to sustain, especially under ice in temperate regions (Prepas and Burke 1997). Decades of research has gone toward evaluating a variety of chemicals, most often based on aluminum (Al), Fe, or Ca, which could be added to surface waters of eutrophic lakes to sequester P and reduce its bioavailability (Wang and Jiang 2016). Such chemical approaches require testing to assess their efficacy, to determine dose quantity and frequency, and to avoid any adverse effects on ecosystems—and there is a large array of literature evaluating these types of chemicals. For instance, polyaluminum chloride, a frequently used chemical to remove P from internal circulation, is not expected to cause any issues with Al toxicity in hardwater lakes with circumneutral pH, but still may cause other ecological issues from changes in stoichiometry or altering rates of some ecosystem functions (Pacioglu et al. 2016).

Controls on Pesticides

Pesticides of all sorts (insecticides, fungicides, herbicides, etc.) are a problem for freshwater ecosystems since the pesticides or their breakdown products can easily run off into groundwater or surface waters. There are many impacts on nontarget species, along with changes to food webs. In many countries, the use of pesticides is regulated and often requires a license. Regulations can apply to the timing and rate of application and what kinds of pesticides are authorized for use. In many circumstances there are rules forbidding the use of pesticides within a certain distance of water, and aerial application is subject to such rules as well. Pesticides also can have high costs and run the risk of pest populations adapting to tolerate those chemicals, so users are motivated to limit applications. Nevertheless, widespread application of pesticides and dispersion of those chemicals across the landscape remains a challenge for protecting freshwater ecosystems.

Regulation of Effluents

A great many of the contaminants in freshwater are a result of urban wastewater, industrial effluents, mine waste (see Chapter 5), and agricultural runoff reaching the surface water through a variety of pathways. These routes have been previously described. One of the challenges of regulating water quality is the enormous number of contaminants to be measured and the costs of doing so on a regular basis. Few contaminants are measured continuously in a systematic way in order to evaluate the many kinds of chemicals across many sites. When toxicity is assessed, one is usually interested in attaining the lowest level of observable effects, and even then, it is typical to apply a safety factor of at least 10, that is, standards for water quality that are 10 times lower than the LOEC (see Figure 4.3B)—and often an even greater safety factor for drinking water standards.

Reduce Use of Plastics

Microplastics can be reduced by less use of plastics, especially those that degrade slowly. A major focus of control on plastics is a result of attention to the situation in marine environments, but most of those same controls assist freshwaters (see Box 4.2). Less plastic use for grocery bags and straws along with filters on washing machines to catch microfibers are some good individual actions that will help to reduce microplastics. Nanoparticles are pervasive in industrial effluents and in consumer products, therefore, increasing attention to these will likely result in stronger controls. Recent studies suggest that some of the so-called biodegradable polymers that are used in consumer goods do not break down rapidly, and so should not be considered as a complete solution to plastics pollution.

PERSPECTIVES

There are many emerging contaminants and most (hopefully all) are tested for toxicity before they are approved for use in products. In general, the chemistry is sufficiently well known and there are predictions about how contaminants might interact, but never with complete certainty. A range of methods for testing is needed, but it is expensive to test a large variety of chemicals with enough frequency and in enough places to obtain good data. (Also see: https://oceanservice.noaa.gov/education/tutorial_pollution/welcome.html.)

Box 4.1   Case Study: Ecosystem-Scale, Whole Lake Experiments at the Experimental Lakes Area

The Experimental Lakes Area (ELA)—run by the International Institute for Sustainable Development—in Ontario, Canada (see Figure 4.6), has been an important facility for testing whole-ecosystem effects of pollutants, largely under the leadership of the late Dr. David Schindler. One of the most celebrated studies was to show that P was the biggest contributor to pollution in our lakes and other surface waters, resulting in the removal of phosphates from laundry detergents, along

Figure 4.6   One of the lakes at the ELA in Ontario, Canada. Photo courtesy of International Institute for Sustainable Development's Experimental Lakes Area, Ontario, Canada.

continued

with other controls on P. In the 1960s, eutrophication of lakes was emerging as a big environmental issue. At the time, phosphates were a component of laundry detergents that contributed to cleaner clothes. Large corporations pushed back on regulators' claims that phosphates were part of the problem, saying that N (or even C!) was the cause. Laboratory experiments had not convinced corporate lawyers that P had a role in eutrophication. The solution was a whole-ecosystem test in a natural lake, with other lakes serving as reference sites. Lake 223 in the ELA was divided in two by an impermeable curtain at a narrowing of the lake. One side had N and C added and the other side had N, C, and P added. The result is a now iconic picture of a *slimy* green half of the lake having received N, C, and P, while the other half of the lake showed no increase in productivity (Schindler 1974). As a consequence, phosphates were removed from detergents.

Along similar lines, another set of lakes at the ELA was used to examine the impacts of acid rain during the 1970s and 1980s (Schindler et al. 1985). Over an eight-year period, a poorly buffered lake in the ELA had its pH reduced from 6.8 to 5, with the addition of sulfuric acid (one of the main contributors to acid rain due to sulfur emissions from factories and coal-fired power plants). The lake was studied for two years prior to additions and was also compared with untreated, reference lakes in the ELA. This study showed a range of responses, including direct toxicity of the acid, food web effects, alterations in biogeochemistry of the lake, and behavioral changes (see Figure 4.7). Some species declined due to toxicity and changes in food supplies, and others increased as a result of relaxation of competition or predation. Crayfish and other arthropods were unable to adequately harden their exoskeletons due to the loss of Ca (Ca is part of the natural buffering against acid), with impacts including higher parasite loads and other problems. The lake trout (*Salvelinus*

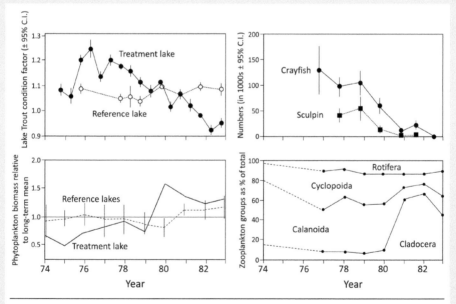

Figure 4.7 Surface water acidification, as a result of acid rain, was a big environmental issue of the 1970s. To isolate the effects of acidification from other changes in the environment, a lake acidification experiment was completed where sulfuric acid was added, which reduced lake pH to about 5.0 over the course of several years. Other lakes in the vicinity were used as reference sites—that is, there was no acid added. There were many direct and indirect (food web) effects of acidification, and only selected results are shown here. Redrawn from Schindler et al. (1985).

continued

namaycus), as the top predator, was seriously affected with loss of body Ca^{2+}, including extreme loss of bone mass and disruption of the food web (Schindler et al. 1985). Their paper summarizes the many aspects of the changes to this lake ecosystem.

In another whole-lake ecosystem study, Dr. Karen Kidd studied the effects of a synthetic estradiol, a birth control hormone, on lakes (Kidd et al. 2007, 2014). As a PPCP, birth control hormones such as estrogen and derivatives are detectable in surface waters and have hormone-disrupting effects on some species, including the feminization of male fishes. Ecosystem-scale experiments with contaminants are relatively rare, but a *before-after, control-impact* study at Canada's ELA tested the lake-wide impacts of synthetic estrogen (17α-ethinylestradiol) additions by adding the chemical for three summers and assessing the effects during those years and for seven years after the additions; then contrasted with data from several nearby reference lakes (Kidd et al. 2007). The estrogen additions led to an almost complete collapse of the fathead minnow population, mostly from reproductive failure and the rapid loss of young fish (Kidd et al. 2007) (see Figure 4.8). There were large declines in the quantity of other fish species, including pearl dace (*Margariscus margarita*), slimy sculpin (*Cottus cognatus*), and lake trout in the treated lake since the males of some species also lost reproductive function due to these exposures (Palace et al. 2009). In that study, male lake trout, and male and female white sucker (*Catostomus commersonii*) had weakened body conditions in the treated lake, which occurred after prey populations collapsed, indicating an indirect effect of the synthetic estrogen on these species (Kidd et al. 2014). These decreases in populations of fish were likely responsible for the increases in invertebrate populations, including

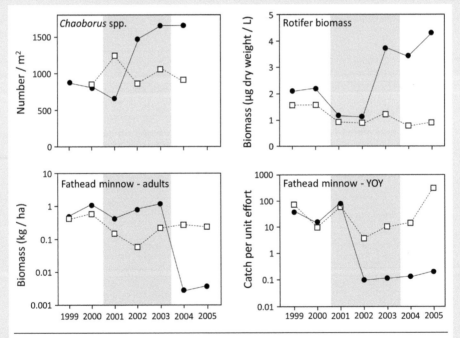

Figure 4.8 Selected results from whole lake additions of estradiol. Open squares represent the control lake measures, and filled circles are results in the estradiol-addition lake. The grey area in each graph indicates the period of estradiol additions. Note that each of the x-axes is different, and that the panels for fathead minnow are $\log_{10}$-scale. Redrawn from Kidd et al. (2014).

continued

Chaoborus, other aquatic insects, and zooplankton (see Figure 4.8), because of their reduced consumption by fish as synthetic estrogens have low direct toxicity to nonvertebrate species (Kidd et al. 2014). However, they noted no significant changes in density or composition of phytoplankton or microbial communities in the treated lake (Kidd et al. 2014). Subsequent work showed that within several years after ending the application of the hormone, the fathead minnow population had returned to pretreatment abundances (Blanchfield et al. 2015).

These kinds of whole-ecosystem experiments are costly and difficult to sustain, but they have been exceedingly influential in leading to protection of our freshwaters from contaminants. A recently concluded study of mercury additions showed that the amount of mercury in fish and other organisms can rapidly decline once loadings decrease (Blanchfield et al. 2022). This whole-ecosystem experiment would probably not be approved to be done elsewhere, a benefit to having a dedicated landscape devoted to such studies at the International Institute for Sustainable Development's Experimental Lakes Area in Ontario, Canada. Coupled with other kinds of smaller-scale studies, these ecosystem experiments—at natural scales with the full suite of ecological processes—demonstrate just what can happen to our environment.

Box 4.2   Microplastics in Freshwaters

Microplastics (defined as particles <5 mm—and here including nanoparticle-size plastics) are receiving enormous attention from the marine realm, but they are also causing an emerging issue for freshwaters. Such particles can be from the breakdown of larger pieces of plastic, synthetic microparticles or nanoparticles, microfibers from synthetic cloths, etc. A survey of invertebrates from streams in Wales found that half of all individuals contained some microplastics, and that streams generally have very high concentrations of such particles—streams which further serve as a conduit to marine systems (Windsor et al. 2019). Microplastic particles can take up space in the guts of animals, thereby reducing feeding capacity; they may interfere with digestion and may also have contaminants adsorbed onto their surfaces that cause additional toxicity (Foley et al. 2018; Wang and Wang 2018). However, it is important to remember that this grouping of particles includes a vast range of compositions, sizes, and shapes, and that all particles are not alike. Moreover, the mechanisms of impacts of different kinds of particles will vary, so there is a need for more nuanced considerations (Rochman et al. 2019).

Experiments on toxicity of microplastics have had a range of outcomes. Laboratory experiments on a range of invertebrates and fishes have found diverse organismal responses, from direct toxicity to impairment of metabolic functions to no detectable effects (Foley et al. 2018). However, many of these laboratory studies use sterile, uniformly sized, and commercially supplied spherical particles, which may not have the same impact as the irregular particles (especially fibers) with adsorbed contaminants that the organisms would encounter in the field. A meta-analysis of impacts of microplastics on a range of aquatic organisms (freshwater and marine) from 43 published studies showed that microplastics had significantly negative effects on average on consumption, growth, reproduction, and survival (Foley et al. 2018). The shape of the plastic particles also affected the responses, with round sometimes being more negative than fiber (see Figure 4.9). However, there was a lot of variation between studies and organisms, and some studies showed no effects, so there is much more to learn about this as an emerging pollution problem in freshwaters.

continued

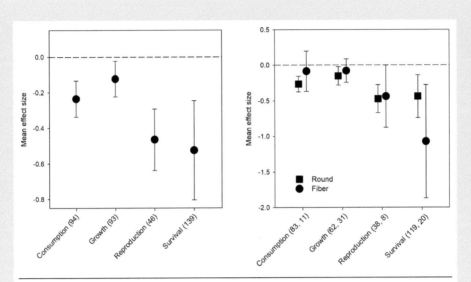

Figure 4.9 Overall effects from a meta-analysis of microplastic effects on freshwater organisms. The first panel shows the overall effects of microplastics in general, and the second panel shows the same data partitioned by shape of the particles (numbers in parentheses refer to sample sizes for each measure). Hedge's g is a standardized effect size based on the proportional difference of the means and the magnitude of the standard deviations of the estimates (a common comparison used in meta-analyses). Negative values of the effect size indicate reductions relative to their respective reference points. Redrawn from data provided by Dr. Carolyn Foley and Figures 3 and 4 in Foley et al. (2018).

5

POINT SOURCE POLLUTION

INTRODUCTION

As a special case of pollution, there are many point sources—typically from industrial or municipal industries—that require different consideration than pollution from more dispersed sources. We addressed aspects of sediments in Chapter 3 and nonpoint source pollution in Chapter 4. Point sources are typically an identifiable, single source, such as a discharge pipe. This chapter will address these more concentrated and identifiable sources of pollution. Of course, some of the worst instances are catastrophic, unplanned spills of wastes and contaminants, but we will not deal in detail with those here. Nevertheless, there are some well documented examples of spills of pesticides or other industrial chemicals into surface waters that have had lasting effects, although some have mostly recovered within several years (see examples in Reiber et al. 2021). Point source pollution may be from municipal wastewater treatment plants (WWTPs), mines, thermoelectric power plants, industrial sites, livestock feedlots, oil refineries and extraction sites, etc. (see Table 5.1). All of these point sources have impacts on freshwaters, but most are of a scale that involves big industry and engineering to solve. Hopefully it also involves biologists, at least in the identification of impacts and monitoring of the effectiveness of solutions.

Domestic sewage and storm runoff (from roads, etc.) carry various contaminants, and many cities have combined sewage overflow (CSO) so that during large storms, domestic sewage may directly overflow into receiving waters because capacity to treat the sewage is overwhelmed. These CSO systems are still common, even in many wealthy cities, discharging raw wastewater into surface waters, although some cities are trying to isolate stormflows from sewage conduits. When this type of overflow is not occurring, stormwater runoff from streets, parking lots, fields, and lawns still carry heavy metals, hydrocarbons, nutrients, pesticides, and other contaminants. The distinction between point source and nonpoint source pollution is not absolute, so some may consider these stormwater outfalls as nonpoint sources. Many municipalities lack WWTPs with advanced processing, and those can be large point sources of pollutants to freshwater ecosystems. Domestic sewage contains organic waste and a host of domestic chemicals, pharmaceuticals and personal care products (PPCPs), detergents, silver nanoparticles, and other wastes (see Chapter 4). Wastewater treatment may be very basic (screening of large particles) or very complex, including the oxidation of much of the nutrients and organic contaminants (discussed in more detail later). However, many WWTPs are not designed to remove many of the PPCPs, and that is typically related to a country's economic status since

Table 5.1 Some sources and types of point source pollution, and situations that modify their impact.

Source	Types of pollution	Mitigating circumstances
WWTPs	PPCPs, nutrients, heavy metals	Incomplete processing of many organics, but work is underway to upgrade to full oxidation systems in wealthy areas
Mines	Acid mine drainage, heavy metals	Storage in tailings ponds and treatment in situ
Thermoelectric plants: nuclear or fossil fuel	Heated water	Cooling towers and ponds
Hospitals	Pharmaceuticals	Often into domestic water treatment plants, but may have their own treatment
Leaks in sewage systems	Same as WWTPs, but unprocessed	Municipalities fix urban infrastructure to stop leaks
Pulp mills	Organic wastes, bleaching chemicals	Regulation of chemicals used, sometimes biodegradation used
Manufacturing plants	Industrial chemicals, heavy metals	Regulations imposed on contamination of water
Landfills and waste dumps	Organic chemicals, petrochemicals	Generally constructed with impermeable liners and overflow ponds

that is an expensive process. Moreover, increasing human populations stress the capability of some facilities. Regardless of the degree of treatment, these WWTPs remain as important point sources of pollution.

Landfills are often found near where humans live and contain an enormous range of materials. In wealthier cities, the systems are designed to minimize water flows through the landfill and into groundwaters, using plastic or clay liners, as well as leachate collection systems. However, not all landfills are well designed or managed, and can therefore release concentrated, local sources of nutrients, heavy metals, pharmaceuticals, and other pollutants into the surface waters or into groundwaters that eventually reach surface waters.

Mines for metals (copper, zinc, lead, lithium, etc.) and coal are usually large sources of heavy metals, acid mine drainage (many rocks are rich in sulfur, hence sulfuric acid), and other contaminants (Clements et al. 2021) (see Box 5.1). Mines often have tailings ponds where waste rock or other materials are stored in ponds with slow rates of release into surrounding water sources. The allowable discharge rates from tailings ponds into surface waters are often agreed to with governments. One profound impact of most mine operations is the large amount of water used, which can locally deplete groundwater and reduce the recharge of surface waters. Water from tailings ponds may be reused within the mining efforts. Thus, tailings ponds that are associated with mines and refining operations can be sources of low-level pollution, but there are many examples of failures of these storage facilities that release large amounts of contaminants. Tailings ponds are designed to have an impermeable layer beneath to stop water from draining into groundwater, but these can fail or be overwhelmed, ultimately leading to groundwater contamination by heavy metals, cyanide, and acid mine drainage. Tailings ponds are not always well managed, and can fail catastrophically by releasing large volumes of contaminants into streams and lakes, and there have been many such events around the world—often with many human deaths. There is a Wikipedia page devoted to tailings pond dam failures.

Power production by nuclear, coal, gas, and other fuels generates hot water that requires cooling before re-entering the natural network. Even though water is allowed to cool somewhat in cooling towers and other facilities, it is often still warm (or hot) relative to the receiving environment. This creates a point source of thermal pollution that can impact a range of species and ecological processes. Some effluents from these power plants can also carry many contaminants. Another type of pollution that can be included here is radiation, either from leakage from nuclear facilities, meltdowns, fallout from bomb tests, or naturally occurring sources. There are few such studies in freshwaters, and most are confidential government reports. However, some recent examples include streams near the nuclear plant in Fukushima, Japan (Sakai et al. 2014, 2015).

IMPACTS

There are multiple pathways of exposure of organisms to contaminants, both through contact (water, sediments) and in food. Some contaminants can reside in sediments for a long time, so there can be a legacy from past contamination, especially associated with point sources from former practices. Both current and legacy sources of pollution have impacts on freshwater species and ecosystem function.

Domestic Water Supplies and Wastewater Treatment Plants

Domestic water distribution systems can also be a source of contamination of water that eventually reaches freshwater ecosystems. A sad example was in Flint, Michigan, in 2014 when the city switched to using the Flint River as a source of water (heavily contaminated by industry and other wastewaters) and a water distribution system with lead pipes. This switch resulted in elevated lead in drinking water and the blood of residents. Lead is known to be toxic, especially to children. Other water-related issues arose due to the switch in water sources and management, and as a result, Flint, Michigan, has become a case study in how *not* to manage water distribution systems. Domestic water is also disinfected with chlorine, which is toxic to aquatic life (it can also create potential carcinogens, but this book is not focusing on human water supplies). In many municipalities, aging infrastructure includes pipes that are leaking domestic water into waters that reach our streams and lakes. These sources of contaminants, such as lead, eventually reach surface waters through wastewater and may impact aquatic life.

Wastewater treatment plants discharge the highest volume of effluents to surface waters relative to other industries in many countries (see Box 5.2). The impacts of this discharge include eutrophication, depletion of oxygen (through biological [organic materials] and chemical oxygen demand), endocrine disruption, other physiological changes to aquatic organisms (from other pharmaceuticals, heavy metals, and other contaminants), contamination by fecal bacteria including pathogens, and alterations to food webs (diversity, composition, productivity)

(Schäfer et al. 2016; Hicks et al. 2017). As a specific example, birth control hormones are excreted and found in municipal wastewater streams, and this synthetic estradiol, as well as natural estrogens, can affect fish and other species in freshwaters (see Chapter 4, Box 4.1; also see Box 5.2). In addition to discharge from WWTPs, many of these contaminants can reach freshwater ecosystems through leaks in wastewater pipes, septic systems, CSOs, and other unregulated releases of wastewater.

Another example of PPCPs that are widely found in surface waters downstream of WWTPs are antidepressant drugs (Kolpin et al. 2002). Studies with individual species and antidepressants have shown a number of toxic effects at low levels (reviewed in Richmond et al. 2016). A stream mesocosm study testing the effects of two antidepressants on population and ecosystem processes found that at low concentrations, these chemicals strongly reduced rates of primary production and respiration in rock biofilms, advanced the timing of emergence of an aquatic insect, but did not affect the rates of ecosystem metabolism (Richmond et al. 2016). Illicit drugs (or misused prescription drugs) enter water systems, particularly near urban centers. Some of these chemicals are even found in source water for drinking water supplies. These biologically active compounds, such as opiates, cocaine, and others, can have direct and indirect effects on freshwater organisms (Rosi-Marshall et al. 2015; Wilkinson et al. 2022). The impacts of PPCPs on freshwater organisms are not well studied, but many are expected to have toxic effects (Rosi-Marshall et al. 2015). At environmentally relevant concentrations, these drugs potentially affect metabolic functions and the behavior of freshwater species, such as susceptibility to predation, but few studies have been conducted. As a metabolite from human consumption, caffeine is commonly found in freshwaters, and has detectable effects on some organisms. Caffeine and sucralose have been used as tracers for PPCP loads, usually from point sources of pollution, including WWTPs (see Box 5.2).

The microparticles and nanoparticles present in WWTP effluent can be ingested in place of food particles by consumers of fine particulate organic matter, whether from the water column or benthic surfaces. Larger particles can also block digestive systems when ingested. Ingestion of such small particles may lead to reduced growth and survival rates by reducing intake rates of actual food particles or by having toxic effects. However, some laboratory experiments have detected no significant effects on survival or growth of *Gammarus pulex* and other freshwater invertebrates that ingested microplastics (Redondo-Hasselerharm et al. 2018; Weber et al. 2018), while some have shown large negative impacts (see Chapter 4, Box 4.2). The magnitude of impact on the growth and survival of an organism will depend on the particular species ingesting the particles along with the composition, size, and concentration of those particles.

Many pesticide compounds (antibiotics, insecticides, fungicides, herbicides) end up in surface waters and groundwater from treated or untreated wastewater. As two examples, triclosan and silver nanoparticles are antibacterial compounds that are now commonly found in many consumer products, such as personal care products including hand soap, detergents, and healthcare products, and even in clothing (Islam et al. 2021). These chemicals are regulated in many countries, but are still common in many products, and are released into freshwaters with municipal wastewaters. These chemicals can kill bacteria in the natural environment. Triclosan is not only toxic to bacteria, but also to algae, invertebrates, and fish (Carey and McNamara 2015). Triclosan and silver nanoparticles are known to contribute to antimicrobial resistance, and both of those chemicals can harm other organisms (Clarke et al. 2019). Measurable concentrations of the antimicrobial triclosan are found in surface waters around the world since it is not effectively removed by wastewater treatment (see Box 5.2). In one experiment using flow-through flumes, bacteria in stream biofilms were killed when environmentally relevant concentrations of triclosan were present, and there were toxic effects on algae at those concentrations (Ricart et al. 2010). Other studies have shown toxic impacts of triclosan on fish, amphibians, and other organisms (Clarke et al. 2019). While there was a no-observable-effect level of 0.21 µg/l estimated for triclosan (Ricart et al. 2010), even concentrations as low as 1 ng/l can shift the microbial community structure in streams (Clarke et al. 2019). Development of antimicrobial resistance and shifts in community structure due to triclosan, silver nanoparticles, and other antimicrobial agents in freshwater can affect disease transmission and alter aquatic food webs (Marti et al. 2014). Likewise, pesticides of all sorts can cause both direct (toxic) and indirect impacts in freshwater food webs.

Livestock

Runoff of manure from farms can be a major point source of pollution. Manure from cattle in western countries may be about two billion tons per year, representing about 80% of livestock manure (Lee et al. 2014). Resulting fecal bacteria loads in surface waters can affect water quality, and concentrations may be especially high after rains wash manure into water sources (Lee et al. 2014). This also represents an enormous source of nitrogen (N) and other contaminants, such as antibiotics and Zn (a feed supplement). Agricultural activities are also a major source of fertilizers and pesticides (antibiotics, antifungal, insecticides, herbicides), and sediment, all of which will be expanded upon in Chapter 8. Livestock feedlots and other intensive husbandry sites are considered as point sources of pollution.

Electric Plants

Thermoelectric power production uses a lot of water, especially associated with nuclear and fossil fuel power plants, but also including solar power (Langford 1990). Most of the water used in these plants is for the production of steam in order to power turbines or for cooling, and the majority of such water comes from surface freshwaters. Release of heated water may be detrimental or lethal to aquatic life, and the amount of water used is problematic in water-scarce areas and at warmer times of the year. Moreover, coal-fired thermoelectric plants release considerable quantities of wastewater with high concentrations of heavy metals (for instance, lead, mercury, chromium, and cadmium). Along the Mississippi River, USA, alone there are at least 128 once-through (water taken in and used once before release back to the environment) thermoelectric plants, which are subject to government regulations, releasing heated water back to the river (Miara et al. 2018). The Mississippi River is considered to have the highest thermal loading of any river in the world (Raptis et al. 2016). In the Mississippi, the upper limit is supposed to be 32°C, which can sometimes mean that power plants have to limit their power production rates, but river water temperatures do occasionally exceed regulatory limits. Elevated water temperatures cause thermal stress and also have lower oxygen saturation that can lead to higher pumping rates across gills increasing contaminant exposure, with the result being reductions in some species' populations, increased disease, and shifts in community structure (Hester and Doyle 2011; Raptis et al. 2016). Regulations in most countries are intended to avoid lethal temperatures in surface waters, and many ensure thermal environments are not compromised at all by the release of heated effluent water from thermoelectric power plants.

Landfills

Landfills are ubiquitous as humans produce large amounts of solid waste, and these disposal sites need to be managed to avoid widespread contamination from gas production and leachates. In particular, water seeping through materials in landfills can carry contaminants in a solution referred to as leachate, and if landfills are not well managed, they can contaminate groundwaters and surface waters. *Leachate* is a generic term for water from a landfill and may contain any range of materials that are hazardous. Not every country or town can afford the best technology to treat the effluents, so the impacts of landfill leachates on domestic water supplies varies enormously across the planet. Leachates that affect domestic water supplies or even crops can have adverse health effects for humans and other organisms.

Mining

Various kinds of mining activities, such as mountaintop mining, marble mining, fracking (hydraulic fracturing), and other extraction industries also impact freshwaters (see Box 5.1). Mountaintop mining involves removing the surface rock layers from coal mines, and then filling stream valleys with that rock overburden. Whether mines are surficial, pits, or subterranean, such landscape alterations will affect hydrology and elevate concentrations of sulfides and sulfate (which combined with water forms sulfuric acid), some metals, selenium, and other chemicals, all with negative consequences for downstream ecosystems (Palmer et al. 2010; Lindberg et al. 2011). The most notable downstream effects are reduced reproduction, deformities in fish, and a loss of sensitive macroinvertebrates. Runoff into mines is controlled by pumping water, which may then enter the local watershed, carrying with it acidic water and potential contaminants.

Fracking for oil and gas uses large amounts of water, and affects groundwaters through inputs of solvents under pressure to fracture deep layers of shale in order to release hydrocarbons. The footprint of fracking wells alters local hydrology, and the input of chemicals into deep rock layers is known to affect groundwater (Meng 2017). The rates of fracking have increased rapidly in recent decades and regulations on water quality effects were slow to be put in place.

INTERACTIONS WITH OTHER STRESSORS

Often pollutants interact with sediment particles, particularly organic particles, onto which they adsorb. This may make contaminants more or less available and likely to affect organisms. For instance, organic particles with adsorbed pollutants that are eaten by consumers may have a more direct effect on growth or survival than pollutants dissolved in the water in which the consumer lives. This depends on the pathways of effects, such as direct consumption, absorption through respiratory or osmoregulatory organs, or impacts on their surrounding food web.

Many point sources release complex cocktails of chemicals, even though some might be at low concentrations, and the net impact of the many chemicals is hard to disentangle (see Chapter 13). Calculations using toxic equivalents, cumulative criterion units, and other ways of tallying up the combined effects of contaminant mixtures are being developed, assuming that the impacts of these mixtures are likely at least additive, although there are other models as well (Schafer and Piggott 2018; Clements et al. 2021). As you learned in Chapter 4, there are thousands of chemicals entering the environment, and many/most are not routinely monitored.

We expect that the impacts of point sources of pollution might be exacerbated by higher temperatures (climate change, summer, thermal pollution). Higher water temperatures accelerate biochemical processes, reduce oxygen saturation concentration, interact cumulatively with the many contaminants, and may place some species under extreme stress. Also, warmer seasons are generally when young individuals are born for many species, and youngsters are generally more sensitive to stressors than older individuals. However, warmer temperatures in winter resulting from warm effluents from WWTPs and thermoelectric plants may attract mobile species, such as fish. This may result in an ecological trap whereby fish are attracted to warmer conditions when it is very cold, but then end up being exposed to high concentrations of contaminants from such treatment plants (Mehdi et al. 2021).

SOLUTIONS

For decades *the solution to pollution is dilution* was an acceptable way to handle pollution. In some places that is still a principle of waste management because of economic constraints. Those who are responsible for point sources of pollution are typically regulated by government policies regarding management and permitted discharge rates. Most of the solutions discussed here are engineering in nature, and most engineering programs have courses in water management that are mostly focused on supply, treatment, and wastewater treatment, and there are many books on the topic (see Mihelcic and Zimmerman 2014). In recent decades, industries were increasingly legislated to reduce their emissions of pollutants into waters, and as a result, most are now releasing fewer toxic chemicals than in the past.

One solution that can help is for individuals to be more aware of what they put down their drains, and consumer education can assist with that goal. Instead of flushing pharmaceuticals down the drain, many pharmacies collect unwanted drugs, and thus, they do not end up in the wastewater stream. Hospitals and care facilities often are very concentrated sources of pharmaceuticals and in some locations these sources have their own wastewater treatment to deal with toxicants before entering municipal waste streams (Lienert et al. 2011; Verlicchi et al. 2013). Likewise, there are ways to dispose of other materials that may be less convenient than washing them down a drain, but are better for our freshwaters. Also, products that get washed into storm basins or are purposefully dumped there (such as waste oil) end up in the waste stream and may not be completely treated before the water is released into the natural environment. To avoid this, municipalities sometimes have collection depots for potential contaminants—for instance, paints and paint strippers, oils, etc.

Governance regarding the flow of pollutants is critical. Release of any waste into surface waters, or onto the land, is regulated in most countries. We will not explore this here, but policies and laws are enacted to control discharge of any potential contaminant. While there may be laws to safeguard freshwaters, enforcement is essential and may be insufficient in some places, especially less wealthy countries. As discussed in Chapter 4, safe levels are largely determined through ecotoxicology, and there are often different concentrations allowed for drinking water standards versus for the protection of aquatic life. However, as you will learn in Chapter 13, there are often mixtures of contaminants released and their combined impacts are not well understood.

Municipal Wastewater Treatment Plants and Stormwater

WWTPs are a concentrated source of contaminants, but are intended to reduce the concentrations of substances that are being discharged into receiving waters by treating the wastewater. However, many municipalities still have CSO, and in the case of extreme storm events, water may end up overflowing into sewer systems and untreated sewage can be released directly to surface waters. Many cities are working to completely separate sewage and stormflow networks to avoid this issue.

WWTPs, sometimes called *sewage treatment plants*, are important for reducing the chances for contamination of our surface waters, but they cannot remove everything, especially the growing array of pharmaceuticals and other chemicals (see Chapter 4). WWTPs are primarily the domain of engineers, but biologists also get involved in monitoring any impacts of discharges in the receiving waters (see Box 5.2). Nevertheless, it is worth getting an understanding of how toxicants are potentially removed from wastewater. There are different levels of treatment—primary, secondary, and tertiary (also referred to as advanced). The first level is a simple screening and settling of material, resulting in *sludge* or biosolids, which are sometimes used as fertilizer in some applications. The second level uses filtration (such as sand filters), aeration, activated sludge processes (engaging bacterial degradation and mineralization of waste to CO_2), and oxidation (similar to aeration, but longer). Before discharge from secondary or tertiary treatment, the remaining water is often chlorinated or treated with ozone or ultraviolet (UV) light, but chlorine must be removed before the water can be released into the environment. Tertiary treatment can involve sterilization to remove pathogens, biological methods to take up N and P (phosphorus), finer filtration, etc. There has been biological work to search for and select new strains of bacteria that may do a better job of processing certain chemicals in wastewater treatment. One can also invest in membrane bioreactors to increase removal of some pollutants remaining in wastewater (Gavrilescu et al. 2015). Additional treatment for specific contaminants is further possible through activated carbon filters and other technologies, but is not widespread.

The result of primary treatment is referred to as *sludge*, and this material is sometimes used as fertilizer. Application of sludge needs to be done carefully since there can be high concentrations of heavy metals, organic contaminants, microplastics, etc. Sludge can even be used as a biofuel for thermoelectric power production. In many places around the world, primary treatment is the final step before release into the environment (in some places, there is barely primary treatment).

Before the treated water is released into the environment, it is often treated with chlorine, UV radiation, or ozone. However, given the dangers of chlorine and chlorinated byproducts, chlorine-treated water is typically dechlorinated before release. Secondary treatment does little to remove heavy metals, nutrients, some pharmaceuticals, etc., therefore, tertiary treatments are used. In municipalities that can afford it, tertiary treatment of water using activated carbon, biochar, reverse osmosis, or other methods are used to remove some of the major nutrients and other targeted contaminants. Other methods used in less affluent regions include wastewater stabilization ponds, overland treatment, macrophyte treatment, and nutrient film removal (a variation on the macrophyte treatment).

One kind of improvement in WWTPs can be to move from carbonaceous activated sludge treatment to nitrifying activated sludge treatment, which will reduce the toxic ammonium compounds being released. Photooxidation (photocatalysis) of pharmaceuticals using nano-particles—titanium dioxide nanoparticles (P25; particles of approximately 25 nano-meters) and P25 modified by silver nanoparticles—has shown promise at removing pharmaceutical and personal care products in wastewater (Kanakaraju et al. 2014). Titanium dioxide microparticles are used to bind to other organic contaminants as well, promoting photolysis of organic molecules through solar radiation. Recent work has shown some effectiveness of using fungi as an agent to process some of the organic chemicals in the waste process. In some cases, at the final end of the WWTP processes, or as intake to municipal water supplies, reverse osmosis may be necessary. Upgrades to WWTP processes to incorporate advanced treatments like those mentioned can be effective in reducing the concentrations of some PPCPs and estrogenic compounds in effluents (Hicks et al. 2017).

In small-scale WWTPs, such as small towns, one can use *bioballs* to help remove N by increasing the surface area for denitrifying bacteria (under anaerobic conditions). By increasing the surface area, the use of plastics or ceramics can allow biofilms to develop and help oxidize even more of the chemicals, such as pharmaceuticals and N. Osmosis membranes have also been used—although these are useful primarily in small scale. This remains an active area of research to make treatment of wastewater more effective, while keeping the cost low for small municipalities.

On very small scales, homes (most rural homes and cabins) and farms may rely on septic tanks that store waste and allow for very slow seepage of the water into groundwaters. However, the accumulated sludge in the tanks still needs to be pumped and removed to a central treatment plant like the larger municipal plants that were previously described.

On a small scale, stormwater that runs off of streets, parking lots, buildings, sports fields, etc., also needs to be treated. Bioswales have helped with this, but they mostly remove some nutrients; contaminants still may accumulate within their sediments, as mentioned in Chapter 3. These contaminants are typically small, but constructed wetlands are sometimes built to serve a similar function, sequestering nutrients and oxidizing organic contaminants before the water flows to more natural surface waters or into the groundwater.

Industrial Treatment

Thermal pollution from thermoelectric plants is generally treated by ensuring that there is either cooling prior to the water re-entering surface waters or by setting targets for flux rates relative to the dilution in receiving waters (Miara et al. 2018). Cooling can occur in ponds or cooling towers prior to flowing back to surface waters. Projections for increased demand for electricity suggest around a 25% increase by 2035, and the large need for water for these plants means that placement of these power projects needs to be situated where water is not scarce (Tidwell et al. 2012). Projects go through a regulatory review to determine potential impacts on surface waters, and proposed power plants can be refused if it will excessively impact surface waters. Many thermoelectric plants use *once-through* cooling technology (direct intake from surface or groundwaters, and single use before release back to the environment), and more recirculating technologies are being tried. Technologies are developing to make more efficient use of water in generating thermoelectric energy.

One means by which industry deals with its pollution is to modify practices to reduce contaminants in the waste stream. For instance, when pulp and paper mills used a chlorine bleach process in kraft pulp production, several by-products, such as dioxins and chlorinated phenolics, were released at high concentrations in their waste process into receiving waters. These waste materials had high biological oxygen demand (organic wastes), had endocrine disrupting properties (Ussery et al. 2021), and even had neurological effects due to high concentrations of chlorinated organics and other contaminants in their effluents. In the 1980s, the impacts of mill effluents were demonstrated to cause endocrine disruption of several fish species near mills, with consequences for fish reproduction, growth, and population sizes. Global changes in operations during the 1980s and 1990s resulted in shifts away from elemental chlorine to other bleaching agents (for instance, chlorine dioxide and hydrogen peroxide), bans on release of polychlorinated organics, and better treatment of waste (through oxidization and stabilization ponds) prior to release into the freshwater environment. Some of the methods used include processes to recapture chemicals in order to remove them. Other processes filter wastes through osmosis, flocculate them to filter out waste, or oxidize them to a nontoxic form.

Livestock

Livestock operations, especially feedlots where enormous numbers of livestock are kept, produce large volumes of manure that contain nutrients, but possibly also antibiotics, growth hormones, etc. In most cases, there are rules about the storage and disposal of manure. Often manure is spread as fertilizer in some farming operations, but there are regulations about how much and at what time of year in order to avoid excess runoff of nutrients to surface waters and groundwaters.

Landfills

Landfills can be localized sources of groundwater contamination, and the choice of location for these can affect the intensity of the impact. The primary solution is to design these sites well as a means to minimize leakage to groundwater—typically lining landfills with plastic or clay liners. However, these liners can fail, and monitoring of groundwaters is needed. There can be considerable amounts of leachate (contaminants in solution) that percolate through landfills and into groundwater, or even directly into nearby surface waters. Landfill operations are regulated by governments to control contamination by leachates. Leachate is often collected, and processes to treat it can include biological, physical, or chemical methods. Some of these include various means to use electro-oxidization, flocculation, or filtration of materials in leachates (Reshadi et al. 2021), but can also include cleaning of water in constructed wetlands (*polishing*). However, there are many places in the world where garbage dumps and landfills are poorly regulated, and better controls would likely improve regional water quality. The main roles of biologists that we are discussing here are the monitoring of water quality impacts on freshwater communities and help in designing wetlands for contaminant processing.

Mining

Tailings ponds at mine sites are intended to hold wastewater so particulates can settle. In almost all mine tailings pond sites, there is some amount of water permitted to be released each year, allowing for a generous dilution rate in receiving waters. Ponds are intended to allow materials to settle out and eventually be treated to reduce acidity. One solution that is being tried is to use advanced methods to extract residual metals from tailings. Most of these solutions are for engineers, but biologists are responsible for monitoring for any potential downstream impacts.

Because of the economic benefits of mining, many governments are reluctant to impose strong regulations on water use and discharge that would help to protect the environment.

Tailings waters from mine processes are impounded behind dams to precipitate materials containing metals and other contaminants. The highly contaminated water in these tailings ponds is often discharged by permit, with a low rate of release based on estimates of rates of dilution. Retention dams for tailings pond water are typically earthen and rock fill. However, there are hundreds of tailings pond failures around the world, and there is even a Wikipedia page devoted to those catastrophic and sometimes fatal occurrences. Unfortunately, an occasional catastrophic failure of such a tailings pond releases a century's worth of contaminated water and sediment in a matter of moments (for instance, see Padcal 1992, Philippines; Payne Creek Mine 1994, USA; Mount Polly Mine 2014, Canada; Brumadinho 2019, Brazil). Ensuring that tailings ponds do not fail is the responsibility of the engineers who design and monitor the conditions of the retaining walls that hold tailings waters. More diligent inspection routines and evaluations of such sites are needed. We should anticipate that rare events will happen and we need to embrace greater safety factors.

There are millions of contaminated sites around the world. Often they are left over from past industrial activities such as mining, nuclear plants, toxic spills, oil wells, and refineries—which are usually left for governments to deal with (due to failed companies). In 1980, the United States created the Comprehensive Environmental Response, Compensation, and Liability Act (CERCLA), also known as the *Superfund*. There are over 1,300 Superfund sites in the United States (see Box 5.1), and it is estimated that there are over a half million abandoned mines there. Some of the activities used to clean up such sites include installing water treatment plants to remove contaminants, removal of mine tailings to a site where they can be isolated or treated, and revegetation of catchments, especially riparian areas (Clements et al. 2010). Some estimates suggest U.S. taxpayers will need to expend $32 to $72 billion to clean up such sites.

Monitoring

Biologists are primarily involved in the monitoring of freshwaters that are potentially affected by WWTPs and other point sources (see Chapter 16). The allowable concentrations of chemicals in effluent are based on requirements determined by biological studies. Biologists can also provide design suggestions to augment the biological roles of microbes and macrophytes when being used to clean water. There is also a need for better sensors for real-time detection of pollutants.

Monitoring water quality and its impacts on freshwater ecosystems requires good sampling designs to rigorously isolate the specific effects. One such design is the before-after-control-impact (BACI) study approach, which you will learn more about in a later chapter. We often know in advance when a project will occur since most projects that affect freshwaters have to go through a number of impact assessment and regulatory steps first. This means that we can often measure the ecological condition of a site before a project begins—and then again afterward. That gives us the before-after comparison, and we need to have some idea of variation (over time or around the site) in both time periods. However, we also need to have a reference (or control) site because we must be able to show that any changes in our *impact* site were not just due to year-to-year changes in weather, or a particularly large recruitment of a species, or even some invasion of a species that is not related to the project. Having replicate reference sites is especially advantageous in this design. This is useful to keep in mind as we go through subsequent chapters in terms of how we can be certain what the cause of an impact actually is attributable to.

PERSPECTIVES

These releases of contaminants and allowable concentrations in surface waters are generally regulated by legislation and inspection by governments. There are many emerging impacts—with some examples being microplastics, hormone mimics, and illicit pharmaceuticals, as well as other chemicals—that are not removed entirely by wastewater treatment. The continued contribution of all these chemicals to freshwaters adds up in ways we are uncertain about, and we will look at some cumulative effects in Chapter 13.

Box 5.1   Superfund Sites

Many places in the world have sad legacies from resource exploitation (mines, smelting), spills of toxic substances, or abandoned waste sites. In the United States, some of these sites have been part of the Superfund program (technically called the CERCLA), administered by the U.S. Environmental Protection Agency since 1980. There are more than 1,300 sites that fall within the purview of the Superfund program. One example is California Gulch, Colorado, which includes the legacy impacts of lead and zinc mines from a century of mining activities, and where tailings and soils are exceedingly contaminated with lead, arsenic, and other toxic materials. This area is about 46 km^2, and the town of Leadville is within its borders. Extensive piles of waste rock and slag leach heavy metals and wash suspended solids into groundwater and surface waters, creating extremely contaminated conditions.

   Over the course of almost 30 years, teams of scientists, and in particular Dr. William Clements at Colorado State University, have studied the effectiveness of clean-up efforts to rehabilitate streams flowing from California Gulch in the upper part of the Arkansas River basin near Leadville. Their work used BACI (see Chapter 8 for more explanation) designs to compare contaminated streams with suitable reference streams in the area (Clements et al. 2010, 2021). They also used laboratory mesocosms, i.e., flow-through, replicated flumes where scientists could add water of different sources and concentrations of contaminants to communities of stream invertebrates seeded in the mesocosms (Clements et al. 2019) (see Figure 5.1). The mesocosms allow for greater control and replication to test observations in the field. Field studies are typically more complex because of a wider range of natural variation and absence of replication except for the reference sites they are compared with (see Chapter 4, Figure 4.5). Their findings from many years of mesocosm studies showed that sensitivity of aquatic invertebrates was highest in small-bodied species and was influenced by other traits such as gills, body shape, and life history patterns. These mesocosms showed that aquatic invertebrates are more sensitive to some contaminants than observed in single-species, laboratory ecotoxicological studies that were described in Chapter 4 (Clements et al. 2019). They also found that the responses to exposure to metals in mesocosms were also dependent on the specific community structure used, indicating food web effects operating within the community.

   The long-term experience with monitoring four streams in four different states of the western United States after remediation for heavy metal contamination from mine waste has shown that there was a lot of similarity in their recovery patterns (Clements et al. 2021). All four sites

continued

recovered the abundances of invertebrates relatively rapidly after remediation, but it took 10 to 15 years before the community structure began to converge on that of reference sites in each region (see Figure 5.2). It is essential to provide evidence when remedial practices are successful and over what time frame, otherwise it is difficult to convince the government and taxpayers that the efforts are worthwhile. Moreover, evidence allows for an adjustment to practices if the outcomes do not come out as expected.

Figure 5.1 Picture of mesocosms. Baskets of rock sediment are colonized in study streams, then moved to the laboratory and placed in flowing water flumes. The communities of organisms (bacteria, algae, and invertebrates) that develop in these baskets (mesocosms) while in the field can then be subject to different kinds of contaminants in the laboratory—in this case, heavy metals—and then compared with control mesocosms (no treatment) to determine the impacts of those contaminants. Photos courtesy of Dr. W. Clements, Colorado State University.

continued

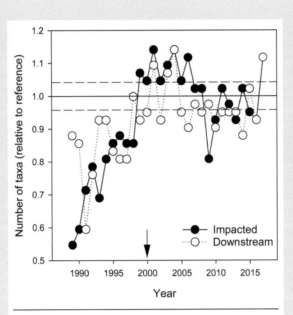

Figure 5.2 Data for long-term, proportional differences in taxon richness of stream benthos at two sites downstream of a Superfund site in Colorado, USA, relative to the average from an upstream reference site. The average for the reference site over time is shown as the zero line on the y-axis, with its 95% statistical confidence interval. The impacted stream, California Gulch, flows directly from the large mine site, and the downstream site is about 10 km downstream of the mine and with several tributaries entering between the mine and the sample site. The arrow indicates the year when restoration activities were completed. Redrawn with data courtesy of Dr. W. Clements (Clements et al. 2021).

Box 5.2   Wastewater Treatment Plant Effluents

Because WWTPs deal mostly with domestic sewage, certain chemicals can be used as tracers (representative compounds) from these plants. Two chemicals that are interesting to consider this way are caffeine and sucralose, which are the results of human consumption and excretion (Oppenheimer et al. 2011; Cantwell et al. 2018). Artificial sweeteners, such as sucralose, saccharin, cyclamate, and others can reach high concentrations in surface waters (Spoelstra et al. 2013), begging the question of how they might affect the ability of fish and other organisms to sense their environment by smells. Sucralose is a reliable indicator of human wastes as it is resistant to digestion in the body and degradation in WWTPs (Oppenheimer et al. 2011). These tracers can also be helpful to detect leaks or unauthorized discharges of sewage. There are many kinds of pharmaceuticals that enter surface waters from WWTPs and more distributed sources, but many are expensive to monitor closely, and so tracers help identify sites with higher exposures/inputs, and perhaps in general, the concentrations of other co-occurring pharmaceuticals (see Figure 5.3).

Oxygen depletion can be an important issue because some wastewater effluents, depending on the level of treatment, can contain elevated nutrient concentrations and a high biological and chemical (such as ammonium from treatment plants) oxygen demand (Venkiteswaran et al. 2015). This is particularly problematic at night when there is no compensating oxygen production by primary producers, and also during warmer times when oxygen saturation is already low and respiration rates of most species are higher. The hypoxia produced can limit what kinds of species may live in reaches near and downstream of WWTPs. A nitrification step in removing N during water treatment can greatly help reduce one of the causes of hypoxia. However, it is difficult without experimentation to isolate causation of impacts given the many simultaneous agents affecting water quality below WWTPs.

The outfall of the WWTPs in the Grand River, Ontario, has been studied for decades, and some of that was to determine the effectiveness of upgrades to their treatment procedures. It was known that the estrogens in effluents of the WWTPs were responsible for the feminization of male rainbow darters (*Etheostoma caeruleum*), and that species has become an indicator for effects from the myriad of contaminants, some of which are endocrine disruptors (Hicks et al. 2017). Within the effluent released into the river, endocrine-disrupting compounds can be found—including estradiol (see Box 4.2) and many other PPCPs (Arlos et al. 2015) (see Figure 5.4)—which is an issue that has been identified globally in rivers as a result of wastewater treatment (Wilkinson et al. 2022). Of two major WWTPs on the river, the one in Kitchener was upgraded in 2012–2013, while the one in Waterloo was not, allowing a test of the effectiveness of the upgrade on downstream fish responses (Hicks et al. 2017). They found that the upgrades reduced estrogens in the effluent (along with other contaminants), and that the male darters downstream of Kitchener had reduced feminization as a result. Lab studies also showed effects of other PPCPs on fish. For example, the opioid codeine altered their hormone systems and could cause a reduction in the numbers of eggs produced by female Japanese Medaka (*Oryzias latipes*) at concentrations detected in the Grand River (Fischer et al. 2021). Caging studies were also done to test whether the water that was released could affect male fish. Rainbow trout and fathead minnows (*Pimephales promelas*) were placed in cages near the WWTP outfall and upstream, but no increase in plasma vitellogenin—a signal of estrogen exposure and feminization—was found (Tetreault et al. 2021). This is one way to determine directly if discharge is harmful to aquatic life, and not just through using laboratory assays.

continued

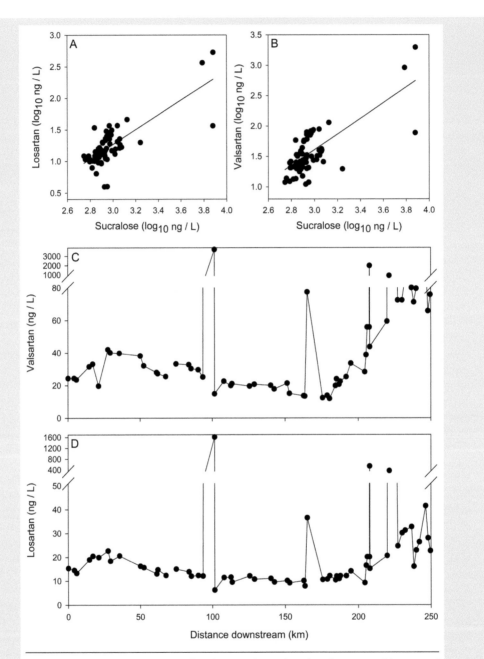

Figure 5.3   Trends in antihypertension drugs and sucralose (used as a tracer) in water of the Hudson River, New York City (New York), USA. A and B show relationships between sucralose concentrations and concentrations of two hypertension drugs, Losartan and Valsartan. Note that the regression is highly weighted by three high points; but even in the absence of those points, the positive relationship between the two medications is evident. C and D show concentrations of these two antihypertension drugs in the Hudson River from Troy, New York (about 10 km north of Albany, New York), to downstream of New York City, New York (The Battery, New York City). Note that the large peaks in concentrations in panels C and D are associated with WWTP outfalls, but there are still considerable concentrations along the length of the river. Redrawn from Cantwell et al. (2018).

continued

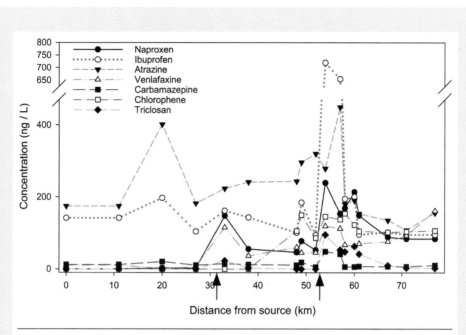

Figure 5.4 Trends in PPCPs in the Grand River, Ontario, Canada. Two WWTP plant locations shown with arrows. Plot redrawn from data in Arlos et al. (2015). Note, lines connecting points are for visual convenience, and not intended to imply continuous changes in concentration along the stream gradient. Points from stream km 33 and 53–55 were averaged as there were several values within a short distance. Data courtesy of Dr. Maricor Arlos (Arlos et al. 2015).

ACTIVITIES

1. Where is your community's wastewater treatment plant?
2. Are there any possible impacts to downstream freshwaters and other communities downriver?

6

WATER WITHDRAWALS AND TRANSFERS

INTRODUCTION

People have greatly modified the flow of water around the planet. Globally, water withdrawals (also called *abstraction*) were estimated to be about 4,000 km³/y in 2019 for all human uses (including consumption, agriculture, industry) (see Figure 6.1). This amount represents approximately 10% of the global, annual freshwater supply (a volume between that of Lake Huron and Lake Michigan), of which more than 50% was consumed and not returned to its original watershed (Gleick 2003; Foley et al. 2005). Moreover, Jaramillo and Destouni (2015) estimate that evaporation from reservoirs to store water may increase human impacts on water supplies by an additional 18%. There is uncertainty in the total global volume of freshwater used by humans annually, and other estimates suggest it could be as much as 30%, which is therefore not available to natural ecosystems (Jaramillo and Destouni 2015; Albert et al. 2021) (see Figure 6.2). Despite this, nearly two-thirds of the human population experiences some degree of water insecurity (reduced supply, need for deeper wells, more costs for pumping, water contamination, bigger dams, etc.), and along current trends this will only get worse (Vörösmarty et al. 2010). These aspects of human water security obviously also pose problems for conservation of our freshwater ecosystems and is not a simple challenge (see Box 6.1).

Most people get water from surface waters (streams, lakes, springs), but billions of people depend on pumping from groundwater (including aquifers) as their direct source of water for all domestic needs. This represents water that is lost to freshwater ecosystems or degraded by human use. In some countries almost all agriculture depends on irrigation (Postel 1999), and one estimate suggests about 85% of global consumptive use of water is for agriculture (Gleick 2003). Europe, a relatively water-rich region, still uses 44 to 67% of its water on irrigation of crops (Antonelli et al. 2017), and that number is much higher in arid regions. In the United States, estimates for water withdrawals for all uses were 445 km³/y of freshwater (around 10% of the world's withdrawals; 2019 estimate), with a further 57 km³/y of brackish water (Konar and Marston 2020) (see maps at https://ourworldindata.org/water-use-stress). Extraction of water—or redirection or diversions of water between catchments for such purposes—changes the hydrology of both receiving and losing catchments.

Many projects move water between various watersheds to supply public needs for drinking water, irrigation supplies, power production, transportation, and other uses. A very large

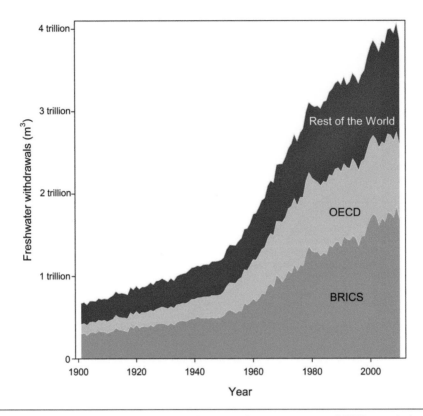

Figure 6.1 Trends in human water consumption over the past century. OECD stands for the Organisation for Economic Co-operation and Development, which is a unique forum where the governments of 37 democracies with market-based economies collaborate to develop policy standards to promote sustainable economic growth. BRICS is an acronym for the grouping of the world's leading emerging market economies, namely Brazil, Russia, India, China, and South Africa, which are not member countries of the OECD. Data from Global International Geosphere-Biosphere Programme (IGB) https://OurWorldInData.org/water-access-resources-sanitation.

amount of surface water is taken for irrigation and for human consumption, often resulting in water moving between watersheds (catchments). A good example is the Colorado River Aqueduct, a canal that carries water from the Colorado River draining the Rocky Mountains to Los Angeles and southern California, the latter of which is a desert region. This 389 kilometer water transfer system moves about 1.5 km³/y from the Colorado River to southern California through canals, pumps, and tunnels.

Human use of water has been quantified as the *water footprint* (Hoekstra and Mekonnen 2012). The water footprint captures water withdrawals, including rainwater use, water for waste assimilation and water used elsewhere to produce imported goods. Trade of goods using water in their production is often referred to as trade in *virtual water*, the latter of which the United States is the global leader in contributing virtual water (Hoekstra and Mekonnen 2012). There is a huge international market that transfers water as virtual water, that is, the products of water, from one place to other regions or countries lacking water, especially production of crops, such as grains. Water withdrawals account for the proportion of the total water that is not returned by flows to within the watershed it started in (Hoekstra and Mekonnen 2012). Countries with

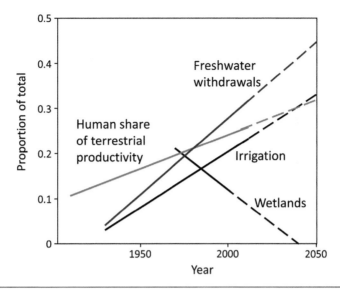

Figure 6.2 Proportional changes in human demands on freshwater supplies over recent decades for irrigation and total water use of annual global freshwater supplies. Dashed lines indicate projections into the future. Note that predictions suggest wetlands will effectively disappear this century. Redrawn after Albert et al. (2021).

considerable water produce crops, manufactured goods, and other products, and then ship them to places with little water, who effectively pay for the services of water in the goods they buy, but do not actually pay for the water itself (Antonelli et al. 2017). One estimate of the international virtual water transfer for rice alone is 31.1 billion m³/y of water (Chapagain and Hoekstra 2011). As a minor example, some countries bottle water to be shipped globally, especially waters from mineral springs, but this is a small volume in comparison to virtual water amounts. Trade in virtual water can also take place within countries, where products (for instance, agricultural and industrial goods and hydroelectric power) from part of a country are transported to regions with less water availability. The economics of water and how water is valued monetarily is an enormous topic by itself and outside the scope of this book (for a text on water economics, see Young and Loomis 2014), but water use and pricing is hotly debated almost everywhere.

Global water demand has increased at least six-fold in the past century (Boretti and Rosa 2019) (see Figure 6.1). Water is extracted for irrigation and domestic use around the world. Many industries consume large amounts of water, and may contaminate it in the process (such as mining). Water is lost through evaporation from reservoirs and canals. There are also losses from leaking pipes and other infrastructure, which can be substantial in some regions. All of these human uses reduce the water availability for ecosystems.

Some commentators suggest that water will become a resource that wars are fought over (Postel 1999), if it is not already. On the other hand, water may be considered too critical to fight over and negotiations could be more likely. Certainly, the world court hears many issues about international water rights. One can compare the price of a volume of bottled water (often filtered groundwater) versus gasoline to get an idea of its commercial value. Some aspects of water use are regulated by governments at different levels (municipalities, states, countries, international) (Eckstein 2009). However, within jurisdictions a variety of agencies control disparate aspects of water use, and may be acting across purposes from each other. Water law, governance, and allocation of water is complicated, and it is an important challenge to balance needs for humans and our ecosystems.

IMPACTS

Impoundments (any kind of dam that restrains water) can have a large impact on the natural flows of rivers, especially when those rivers have been enormously modified (see Chapter 7), including being diverted to other watersheds for drinking water, irrigation, power production, or other uses. These diversions or transfers are generally done with canals or pipes. Streams that are downstream of the diversions and impoundments can be left nearly dry. For instance, during periods with little precipitation or snowmelt, all the water in a reservoir might be retained for human purposes, leaving little for downstream ecosystems. Flow regulation has also eliminated flushing flows and generally regulate water release rates. The elimination of flushing flows from peak discharges means that fine sediments are not washed away, and this can exacerbate the problems of fine sediments for aquatic ecosystems covered in Chapter 3. The absence of normal peak flows results in fewer newly created gravel bars and areas of fine sediments along stream margins that are needed for colonization by many riparian plants, such as cottonwoods (*Populus* spp.). Peak flows are also an important cue for some organisms that are used to time life-cycle events, and the absence of high discharge can result in the mistiming of such activities (Poff et al. 1997). Impoundments and diversions can also misdirect anadromous species of fish, and other fish that are seeking spawning areas based on smell. Global estimates are that only about 37% of the world's rivers that are longer than 1,000 kilometers are free flowing (Grill et al. 2019). Moreover, an estimated 57% of the variability in the world's seasonal water storage is in reservoirs, and the elevation of the surface of reservoirs varies almost four times more annually than natural waterbodies (Cooley et al. 2021). In the next chapter we will learn more about *environmental flows* that are intended to sustain freshwater ecosystems downstream of diversions and dams.

There are many diversions of water between basins for power production and other purposes. For instance, the Nechako Reservoir in British Columbia, Canada (53° 30′ N, 126° 30′ W), transfers water that would have flowed eastward as part of the Fraser River system and stores it in a reservoir at about 850 meters above sea level, before it plunges down 780 meters through two 16-kilometer tunnels to the town of Kemano at sea level on the Pacific Coast near Kitimat. This water is then used to produce power for aluminum smelting and other industrial activities, such as new liquid natural gas plants. There are many enormous projects proposed for water diversions in Brazil, India, China (Dudgeon 2011), and also North America. Some of the water in these redistribution projects are channels open to the surface and will be subject to heating and evaporation. We will discuss the link to movements of invasive species through such channels in Chapter 11. These all remove water, without returning it, from historical receiving areas and from the ecosystems they supported.

Water security, or rather insecurity, for humans often results in large projects to collect and divert water. Some of these water transfer megaprojects may be seen as solutions, at least for humans, but these projects in turn reduce the water available for the ecosystems from which they were diverted. Proposals in the 1950s in North America, such as the North American Power and Water Alliance, would have created diversions of many of the largest rivers of the continent toward drier or more populous regions. The International Joint Commission between the United States and Canada was created to consider such projects, and they stopped most of those proposals. Currently, about 110 projects globally may be considered megaprojects (projects costing more than $1 billion each, diversions of long distances, etc.), which could add up to 1,910 km^3 of water per year being translocated out of their original watersheds (Shumilova et al. 2018). The loss of water from whole regions and new surface connections to other watersheds would imperil enormous areas of many ecosystems, result in invasive species, and probably cause the local (or global) extinction of many species (Dudgeon 2011). Nevertheless, proponents of megaprojects and smaller initiatives, largely funded by governments (see Jeuland 2020 on economics of dams), continue to push their proposals forward, and some parts of these potential diversion networks have been built.

Water withdrawals for human use from surface waters or groundwaters (including aquifers) reduces the amount of water that is available to freshwater ecosystems. Water use can lead to water tables being lowered to the point where there is little to no water supply to sustain wetlands, lakes, and streams (Foley et al. 2005). Lowered water tables can result in streams and wetlands losing water into the groundwater, that is, recharging groundwaters (and aquifers) which in a stream is called a *losing reach*. In many places, springs, even culturally significant springs, have dried up due to abstraction of groundwaters for human use (Cantonati et al. 2021).

Surface water that recharges groundwater can carry contaminants that impair the quality of the groundwater. These contaminants will cause problems as they are carried to other ecosystems via the groundwater and will also end up in wells for human use. Groundwater depletion has also resulted in subsidence in many regions, damaging infrastructure and reducing aquifer capacity. Reduction of the volume of freshwater habitats, or reductions of their quality (such as from pollutants, increased concentrations of salts), pose some of the major threats to freshwaters (Dudgeon et al. 2006; Vörösmarty et al. 2010).

Water supplies are often overcommitted, especially in arid regions. Depletion of water supplies by direct extraction from the surface or from groundwater pumping, regardless of the use of the water (drinking water, irrigation, etc.), results in less water available to native ecosystems. In dry regions of the world, groundwater depletion rates have more than doubled since the 1950s (Wada et al. 2010). There are many examples where groundwater and other aquifers are overexploited (exploitation greater than recharge, also known as overdraft) for irrigation, drinking water supplies, and other uses. This overdraft has detrimental impacts on natural ecosystems and reduces availability (quantity and quality) to humans (Jaramillo and Destouni 2015). As one example, spring ecosystems of all types are disappearing in many places. There are many examples where springs that were historically important (at temples, towns, etc.) have vanished due to the lowering of water tables or have been piped away (Cantonati et al. 2021). Many companies that bottle groundwater (at almost no cost) overdraft the water supply, which leads to springs drying up, wetlands decreasing, and stream flows diminishing. Hot springs (geothermal) have also been extensively exploited around the world for recreation and therapeutic purposes, and there are global lists of examples of these online. Springs are also tremendously important sites for biodiversity. Many endemic species of algae, invertebrates, fish, and other taxa are found in only one or a few springs (see Cantonati et al. 2021 for examples). The loss or conversion of springs has resulted in losses of many species, and we probably are not aware of some species that may have been lost.

Salinization of soils and water by concentration of the salts dissolved in water can be caused by irrigation in arid locations. Irrigation leads to much higher evaporation rates, and losses of water that will not contribute to flow generation or groundwater recharge (Albert et al. 2021). High rates of evaporation of irrigation water concentrate the salts in the water and accumulates in soils since there is not enough water to rinse these away. Even very low conductivity water (soft water) has some salts, and over time will contribute to the salinization of agricultural soils and groundwaters. Salinization of soils and surface waters has become a major threat in many places, and Australia stands out as one country that has had to deal with this issue. The continual application of irrigation water adds to the salt concentrations and can result in soils that are too salty to support crops. Foley et al. (2005) estimates that globally 1.5 million ha of arable land is lost annually due to salinization attributable to irrigation. This is a particular problem with arid landscapes that are dependent on irrigation to support agriculture. Eventually such areas become incapable of supporting agriculture and the actions lead to desertification. More efficient irrigation practices (see Figure 6.4 later in Box 6.1), including drip irrigation, can reduce this problem, but does not eliminate it entirely.

Saline soils can also occur near coastal areas where the withdrawal of water from surface or groundwater lowers the water table and allows seawater to penetrate the ground displacing the fresh water. This form of saltwater intrusion is affecting drinking water wells near the coast of Florida, where municipalities use different forms of reverse osmosis to remove the salts from drinking water (at great energy costs).

Abstraction of water from natural waterways results in the constriction of the amount and quality of water that is available for natural habitats and their supporting ecosystems. As previously mentioned, for springs, entire habitats may be lost. The reduction in habitats that are available, and in some cases seasonal reductions to accommodate human demands, will set upper limits on population sizes of freshwater species that can be sustained, especially those with life cycles of a year or more. Some species are able to tolerate intermittent flows (Datry et al. 2017), but there is a limit to that tolerance, and most freshwater species lack such abilities.

INTERACTIONS WITH OTHER STRESSORS

Lower flows that are due to water exploitation also lead to other problems. Less water means less dilution of nutrients and contaminants (see Chapters 4 and 5) and harmful salinization (as previously noted). Smaller volumes of water also result in the same amount of solar radiation heating up less mass of water, resulting in higher water temperatures. As volume decreases during periods of low flows, the ratio of surface area to volume increases, and those are usually times with warmer temperatures (more evaporation and greater heat exchanges warming the water). Reduced flows also decrease the amount of habitat and/or ability of migratory species to pass barriers such as small waterfalls, thereby restricting individuals to deteriorating habitats (see Chapter 7).

Many endangered species are found in ecosystems with small volumes of water, especially those of vernal pools (spring snowmelt pools) and other temporary pools, springs (Cantonati et al. 2021), and other small bodies of water. One example is the critically endangered Devils Hole pupfish, *Cyprinodon diabolis*, which lives in a tiny, isolated rock pool in Nevada. Pumping of groundwater for agriculture reduced the remaining water volume, and hence interacted with the already small range extent of the species. This situation, combined with climate change and other direct human manipulations, further imperils it (Hausner et al. 2016).

Persistent withdrawals can lead to loss of wetlands, and even large waterbodies may disappear. One stark example is the Aral Sea, at the border between Khazakstan and Uzbekistan. It was once the fourth largest lake in the world by surface area (around 68,000 km^2). Withdrawals for irrigation resulted in most of the lake disappearing by about 2010. As an endorheic lake (no outflow) it was also saline, which got more concentrated (and more polluted) as water volumes diminished. Khazakstan has built a dike and tried to restore some flows into the lake in recent years with some success, but the dry eastern basin is now considered a desert. Unfortunately, the drying of most of the lake has likely resulted in the extinction, or at least extirpation, of a number of species. Another example of an endorheic lake is Mono Lake, California, where abstraction of water that would have flowed into Mono Lake resulted in the lake nearly drying up. Fortunately, intervention by a nongovernment agency resulted in court decisions to reduce the magnitude of water withdrawals, but the lake's volume remains well below historic levels and damage to the lake ecosystem has been done.

SOLUTIONS

This chapter is mostly about how water supplies are managed, which is most often the domain of engineers and policy people. Other aspects of water use by humans is covered in other chapters, but here it is useful for freshwater biologists to be informed about how these activities impact aquatic habitats. Water allocation is largely a policy issue, but in many parts of the world, water rights are hotly contested. Overallocation of water rights to different users often prioritizes water for human consumption, irrigation, and power production, over the protection of freshwater ecosystems. Water rights and access to water can be hotly contested between countries. This can also occur between local jurisdictions within countries, and between various user groups. In Chapter 7, we will discuss more about dams and flow management. There are many opportunities for biologists to be involved in debates about water use.

Governance, Management, and the Surface-to-Groundwater Connection

Management of water withdrawals and allocations needs to better incorporate the connection between groundwater and surface waters, and how these components of water budgets interact. In many parts of the world, groundwater flow rates and flow paths are unknown, thus making it difficult to determine recharge (replacement) rates. Groundwater sustains surface waters and vice versa and are not independent of each other. In the absence of appropriate recognition, there are agencies that still count them as two separate resources, and thereby overestimate supplies and sustainable limits to their use. Consequently, both groundwater and surface water resources become overallocated.

Water allocations are sometimes based on historical periods (scale of decades) that were wetter than millennial or century-scale averages and may allocate groundwater supplies to the detriment of surface waters. For example, in California water allocations have exceeded supplies for many years. These allocations address the water demands of growing cities, expanding agriculture, the increasing need for hydropower, and so forth. Unfortunately, what that usually means is less water for aquatic life. Globally, relatively smaller ecosystems such as springs and wetlands are especially hard hit by groundwater pumping and unsustainable rates of water use. An important element of all of this is better mapping and estimation of subsurface flows, and the surface return flows (groundwater returns to the surface) that produce springs, wetlands, and streams (Cantonati et al. 2021). Included in these calculations should be the changing contributions of glacier melt in mountainous regions, often thought of as nature's *water towers*. Planners need to account for the increasing (and eventually decreasing) magnitude of this addition of *fossil* water to stream flows (Schindler and Donahue 2006). As we will see in the next chapter, many jurisdictions require that environmental flows for natural ecosystems be maintained. This aspect of modern flow regulation (see Chapter 7) and diversions is intended to ensure particular species and the ecosystems that support them are provided with enough water. Related to this are interests in removing the many dams, large and small, that impound water, increase evaporation, and which no longer produce the power that they once did (Hart et al. 2002).

Even when the interaction of surface water and groundwater is better understood, management of water use is challenging due to the myriad of water rights allocated, and priorities given to different sets of rights-holders. Access to water (water rights) is often tremendously overallocated, and that is a topic beyond the scope of this book. Alteration of access and allocations depends on government intervention in order to mitigate emerging problems. For example, the Central Valley of California is an important food production region for North America, and a large portion of the irrigation water comes from surface supplies. However, intensive irrigation there has resulted in groundwater pumping and overdraft of groundwater supplies in the shallow aquifer, which has reduced the aquifer and allowed for saltwater intrusion into the groundwater close to the coast. *Managed aquifer recharge* has been proposed as a remedy, which would take advantage of excess seasonal flows (floods; in excess of 90th percentiles of flows). These excessive seasonal flows are found in northern and eastern California, where the extreme seasonal discharges that reservoirs are unable to store (mostly storage of water for summer irrigation), can be diverted to distant parts of the state and allow that water to recharge the Central Valley aquifer (although that still potentially impacts surface waters; see Chapter 7). Nevertheless, groundwater pumping rates are predicted to remain too high for such diversion projects to work across the entire aquifer. These ideas for California's central valley are still large infrastructure projects dependent on identifying suitable infiltration areas (such as areas of high soil permeability, deep percolation, property availability and suitability) for recharging groundwater and constructing diversion channels to convey water to those infiltration sites (Alam et al. 2020). However, such projects are the slippery slope to larger and permanent water diversion projects mentioned under the previous section titled *Impacts*.

Evaporation must be considered in the connection between surface water and groundwater. In recent years the amount of evaporation from diversion channels (aqueducts) and reservoirs has been shown to be a big component of this overall water balance (see more in Chapter 14). Los Angeles and other jurisdictions have experimented with ways to reduce evaporation, such as floating, plastic balls on the surface, or canopies of solar panels (for instance, India). This can also be accomplished by creating deeper reservoirs to reduce the surface area available to evaporation relative to the volume in the reservoir.

Water supplies and use vary over time. There can be enormous variation interannually and throughout the seasons, and even on decadal and century scales. Water management and allocation is often insufficient to cope with extremes, and yet we know they are increasingly likely to occur due to climate change and increased human demands for water. This variation is difficult to address given current laws around rights to water access, and this is particularly problematic for low precipitation years and seasons. When these events do occur, it is challenging to ensure that there will be enough water for natural ecosystems.

Irrigation

Irrigated agriculture provides a large percentage of the food people eat. The majority of these crops are grown in warm, arid areas with lots of sunlight. In arid regions, the largest proportion of the water supply comes from pumped groundwaters. In addition, irrigation in many places includes spraying water into the air, often during the hottest part of the day, onto the crops. The problem with this is that a large amount of the water evaporates before contacting the ground. Movements toward more efficient irrigation practices have become widespread (see Box 6.1), and involve drip irrigation instead of spraying, better timing, and avoiding overwatering, which contributes to the salinization of groundwaters. Many crops, such as fruit trees and grapevines do very well with drip irrigation systems. Irrigation of crops, including food for livestock (for example, alfalfa), can be altered to use much less water.

Water Supply Costs

Another solution to address the more efficient use of water is to charge more money for water use. In some places, the pricing on municipal water supplies is very inexpensive; for instance, paid for by taxes, but no direct, volume-based cost to individual consumers. Under such an arrangement, there is little incentive for individual households to be efficient. Using water efficiently could mean more water for natural ecosystems. Water tends to be priced at *cost* for processing and delivery in places like California, so there is room to incentivize the more efficient use of water. Water losses through leaks in the distribution systems also need to be addressed to increase the efficiency of delivery.

Landscaping that is appropriate to location can bring water use down. For example, in arid regions of the world and where water is expensive, it makes little sense to have the typical suburban lawns and gardens (see Box 6.2). In some cities, such as Tucson, Arizona, USA, the high price of water and social movements to support *xeriscaping* resulted in gardens full of native plants—mainly cacti, succulents, and other water-efficient plants. Once such conventions become common, a *standard* green lawn is seen as wasteful of water (and space). Such social movements toward gardens with more native plants not only typically conserves water, but it is also a benefit to other native biodiversity, such as birds and insects that cannot thrive on green lawns with a few flowerbeds.

By governing or managing themselves, individuals can do a lot for water conservation, reducing overall per capita requirements. This includes reducing losses from leaking pipes, installing more water efficient toilets, taking shorter showers, turning off taps, creating more water-appropriate gardens (see Box 6.2), and others. In locations where people pay for the amount of water used, individuals benefit from water conservation in cost savings, and other users pay. However, there are many jurisdictions where every citizen pays taxes for water, but there is no distinction in costs between those who conserve water and those who do not. Many local agencies have advocated for individuals to do their part.

PERSPECTIVES

Laws that deal with water rights and usage are complicated in most countries. Rights to water use are sometimes based on whoever claims the water first and puts it to *beneficial use*, although this does not mean ownership (in most cases). The initial intent of these rights was to encourage certainty about water supplies to those investing in farmland or other land uses. While human populations were still relatively small (the population is now about four times what it was in the early 1900s), water was considered abundant and was often overallocated from a modern perspective. One complication regarding water withdrawals is that the source of surface waters from groundwater was largely ignored and the water rights were often allocated as if these two components of water were independent—despite obvious knowledge that the surface waters are mostly supplied from the groundwater. Withdrawals of water by humans have large effects on freshwater ecosystems, and balancing these two needs for water is complicated, but is a necessary challenge. Many governments have legal mechanisms that restrict water use to protect freshwater ecosystems, provided there is the political will.

Box 6.1   Aquifer Loss

Aquifers can be unconfined (water table) or confined by rock or sand layers below the typical water table, and there are situations between these two extremes. Withdrawals by pumping wells can exceed recharge rates, leading to a drawdown of aquifer levels, also called overdraft. Lowering the level of aquifers can result in dry wells and the reduced resurfacing of water elsewhere downslope to support streams and wetlands. Reduced amounts of recharge water may carry pollutants from the surface that concentrate in the aquifer. In coastal areas, a reduction in the recharge of aquifers with surface freshwater may cause seawater to flow into these zones, making the water saline. In karst landscapes (calcareous sedimentary materials), subsurface water (remember carbonic acid from Chapter 1) dissolves limestone, which can create tunnels and, effectively, underground streams (similar to preferential flow paths). The depletion of such aquifers can also result in the development of sinkholes.

The Ogallala Aquifer in the United States Midwest (also known as the High Plains Aquifer) is estimated to support about 20% of agricultural production in the United States (see Figure 6.3). Pumping of water from this enormous (about 451,000 km^2), unconfined aquifer has exceeded recharge, and the aquifer is seriously depleted, although more so in some parts of the aquifer than in other areas (remember that water moving from recharge areas moves slowly through the ground). Federal government involvement has created irrigation districts with more efficient irrigation and nutrient management. Research efforts to improve use of limited water supplies are ongoing, and will save water, reduce rates of salinization of soils, potentially reduce costs of water pumping, and provide for surface return flow to the freshwater ecosystems they support. Solutions include drip irrigation systems or spray close to the soil (see Figure 6.4) for crops, which can save 30 to 60% of the water used by conventional spray irrigation.

There are a surprising number of specialized organisms (known as stygobionts) that live in groundwaters around the world. In karst landscapes, water flowing through subterranean *pipes* are part of aquifers and can host another set of specialized organisms based on organic matter (mostly dissolved organic matter) that makes its way below the surface to fuel these food webs. Bacteria serve as the bottom trophic level in these lightless environments. There are species of fish and amphibians (often sightless) and invertebrates that are found only in these aquifers and are

continued

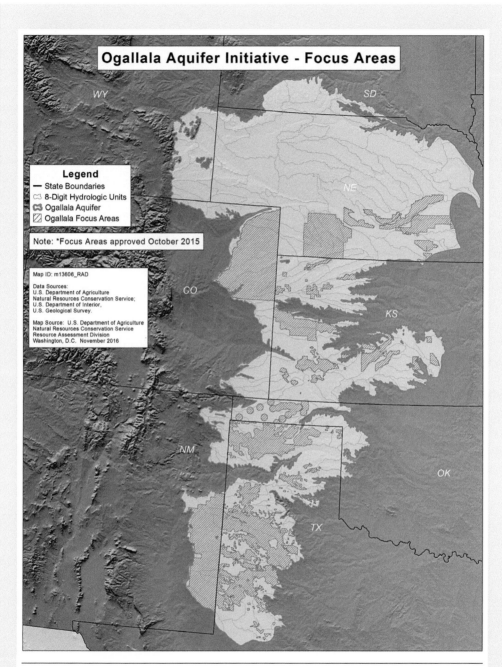

Figure 6.3   The Ogallala aquifer covers a huge area of the western United States and provides irrigation water to an estimated 20% of the agricultural production. *Source*: U.S. Department of Agriculture.

continued

Figure 6.4   A modified center pivot sprinkler irrigation of soybeans using drop nozzles that help minimize evaporation losses. Photo Courtesy of Kansas State University and Dr. Freddie Lamm.

challenged by reduced flows and pollution; many are listed as species at risk or are candidates for future listings. For instance, the Edwards Aquifer in Texas supports several species, such as the widemouth blindcat (*Satan eurystomus*), the toothless blindcat (*Trogloglanis pattersoni*) (catfish, family Ictaluridae), and the robust blind salamander (*Eurycea robusta*), as well as crustaceans, worms, and other species. Isolation of populations from no (or few) connections between aquifers results in specialization and speciation of taxa that can maintain their populations in such ecosystems. Better management of groundwater extraction will allow these aquifers to maintain water levels and support these specialized freshwater ecosystems.

Box 6.2   Urban Water Use

Many cities are built in arid and semi-arid locations where water from distant mountains (*water towers*) provide water at the surface or through groundwater. Diversion of water for human exploitation has resulted in diminished flows of water for freshwater ecosystems and has severely degraded freshwater and riparian ecosystems in such regions. Expansion of urban areas in dry regions has mostly exceeded the capacity of water supplies and in many cases, the solution is to pump even deeper groundwater or divert water from other distant locations. For instance, the city of Tucson in Arizona gets some water through a 540-km-long aqueduct from the Colorado River via the Central Arizona Project (heavily subsidized by federal funds), which is now mostly used to recharge groundwater—an idea known as water banking. However, another dimension to the supply issue is to find ways to reduce water use and increase its efficiency. In a number of cities, a solution is xeriscaping. Xeriscaping encourages the use of native plants throughout a yard,

continued

instead of a monoculture of green (turf) grass. So, rather than well-watered green lawns (usually also supported with fertilizers and pesticides), one can have gardens of native, dry-adapted species requiring little to no watering (see Figure 6.5). Another solution is an emphasis on domestic rainwater capture in cisterns, and even greywater use (there are strict regulations for this practice), for garden applications. Some well documented examples come from the western United States, for instance, Tucson, Arizona, and Denver, Colorado, which were early adopters of such practices.

A.

B.

Figure 6.5   An example of xeriscaping of (A) homes and (B) streetscapes in Arizona, USA. The planting of desert plants in this arid region eliminates requirements for watering gardens and dramatically reduces water demands on a limited water supply. Photos courtesy of David Ramey.

continued

Scarce water supplies in the Sonoran Desert made water expensive for residents of Tucson. The city made a dramatic shift to xeriscaping. High water prices, coupled with an effort to supply appropriate garden plants, created a societal shift in acceptability of desert gardens, and saved enormous amounts (and costs) of water that would have been used to water lawns. In many parts of the world, the traditional expectation of an urban home is a turf grass lawn with a few flower beds (and maybe a tree), and that societal norm is difficult to escape due to pressures from neighbors, and possible loss of property values.

A study of Phoenix showed that despite the successes in nearby Tucson, the social conformity and momentum of turf grass lawns (and watering to sustain them in a desert) has been difficult to escape (Larson et al. 2017). Atmospheric moisture resulting from outdoor water use for parks, golf courses, commercial spaces, and residential lawns can be detected above Phoenix (>4 million people), where typical urban, grassed lawns are still common (Templeton et al. 2018). This represents a large evaporative loss of water resources. As in Box 6.1, irrigation of lawns and crops comes with a huge loss of water, which limits the amount of water available for surface waters.

Tucson also uses treated wastewater for municipal purposes such as watering parks and golf courses, fire suppression, public toilets, and others. These methods allow cities to limit *per capita* water demand, even while most urban centers are expanding and requiring greater supply. Tucson has generally been able to reduce the overall use of water at the same time that its population is growing (Zuniga-Teran and Tortajada 2021). Finally, Tucson sends treated wastewater into the Santa Cruz River, which with renewed, regular flows has become a functional stream, riparian, and wetland ecosystem again. Around Tucson, similar efforts have been able to increase aquifer levels and return surface waters of the San Pedro River system to a better state for conservation of ecosystems (Richter et al. 2014).

Increasingly, municipalities are promoting more efficient use of water. In some places, that includes rainwater harvest (at the household and municipal scale) and greywater (not permitted in many jurisdictions) to be used for watering and other processes. Outside of areas lacking water distribution infrastructure, the use of cisterns to hold rainwater can provide sufficient water for all uses, and thus avoid the need for wells; this is very common on islands and in tropical countries. Other tools include recommending low-flush toilets, shorter showers, and minimum amounts of water for laundry, along with restrictions on times and days for lawn watering (or even car washing). Other municipalities have also used groundwater recharging with their treated wastewater to restore nearby surface waters. Water use efficiencies and the reuse of treated water can be important methods of supporting freshwater ecosystems.

ACTIVITIES

1. Where does your water come from and how is it priced in your state or province?
2. Is the price paid by households in your municipalities the same as for agricultural users?
3. Which crops are the most water intensive?
4. Do you think that water scarcity might shift the type of agricultural production near you?
5. Are there any annual restrictions on use, such as prohibitions on watering lawns or washing cars?

7

FLOW MANAGEMENT AND CHANNELIZATION

INTRODUCTION

For thousands of years humans have been modifying streams and lakes for various uses. These modifications included channelization (straightening and deepening channels) to allow transportation, stabilizing harbors for shipping, diversions for irrigation, flood protection, drainage, and other purposes. Similarly, humans have created impoundments of streams (dams and reservoirs) for energy production and water storage. Dikes (also called artificial levees) and other engineering works built to protect infrastructure have created problems, as indicated in Chapter 6. Early civilizations depended upon their ability to sustain water supplies through flow management to irrigate crops, as do modern societies, which we covered in the last chapter.

In many countries, streams were dredged and straightened (channelized) to provide canals for transportation. Often these canals had paths on either side so that horses (or oxen) could pull boats or barges along (see Figure 7.1A). Canals were also dug to connect different watersheds for transportation by boat, such as the Rhine and Danube rivers in Europe, and these are often responsible for novel species being able to invade new habitats, as we will explore in Chapter 11. The water to supply these canals and lock systems usually were based on the damming of wetlands and lakes nearby the canal system. Many of these connections were for transportation, but there are also many canals designed to provide water supplies to distant urban centers, such as Los Angeles' water canals or China's Grand Canal. In forestry landscapes, streams were channelized in the days before truck transportation to allow for log floating from the forests to lakes, rivers, or estuaries where they supplied mills. Sometimes streams were channelized and diked to provide for construction of infrastructure or expansion of arable land area (see Figure 7.1B). In some places, streams were diked with banks or berms of heavy materials to protect against flood damage to property and infrastructure (see Figure 7.2A).

Rivers have been managed for centuries with engineering works in places to provide flood protection using channelization, dikes, or both, to prevent flood damage. Estimates for the United States indicate that at least 228,000 km of artificial dikes exist along streams of the country (Knox et al. 2022) and that number for the European Union is at least 60,000 km (van Woerkom et al. 2021). Dikes are intended to protect infrastructure and human life from floods (from storms or even ice jams in northern regions), and these are often referred to as flood defenses. However, in some cases, river management may actually lead to more flooding by

A.

B.

Figure 7.1   (A) A canal intended for transportation of goods, with locks—and along the sides there are paths for horses or oxen to pull the boats and barges (near Birmingham, United Kingdom). (B) The Okanagan River in the southern part of British Columbia, channelized and diked to provide for agricultural land—in this case, grazing and orchards.

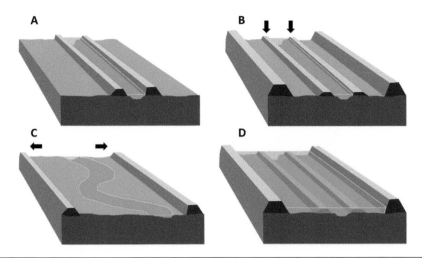

Figure 7.2 (A) A straightened and channelized stream lined with dikes on either side, as is often the situation in urban and agricultural areas. (B) Dikes adjacent to the stream reduced in elevation above the stream, with bigger dikes constructed further back to accommodate large floods. An example of such a scheme is the Netherlands' *Room for the River* program. (C) Dikes moved away from the stream edge to create space for the stream channel to exhibit natural dynamics, including the ability for floodwaters to spill onto floodplains. This design allows for reconnection of streams to floodplain wetlands, renewal of the fertility of floodplain soils, and diminished flood risk by dissipating the energy of flood waters. (D) The result of floods in panel B, where water spills over the smaller dikes during big floods. Infrastructure within the area between dikes needs to be built above potential flood height or be in areas such as pasture that can withstand inundation. Figure created by Kasey Moran, UBC.

limiting the space for stream floodwaters to spread out onto floodplains. Because of channelization, erosion during high discharge might be more powerful due to concentration of flows without the ability of the water to spill into floodplains. River reaches where diking has been done may simply move the erosion problem further downstream (due to lack of energy dissipation) where the erosive power can act against more mobile shorelines. Overbank flooding upstream might occur if the flow is constrained through the diked reach and backs up the stream.

In many places, such as the delta of the Rhine River in the Netherlands or the Mississippi River delta, dikes have enabled the use of floodplains for agriculture and other purposes where natural, seasonal flooding would otherwise make such activities impossible. Floodplains are particularly fertile due to the deposition of fine and nutrient-rich sediments with each flood event, making them valuable for agriculture. Of course, these depositional events are now curtailed by dikes, and agriculture is increasingly dependent on synthetic fertilizers. The disconnection of floodplains from their streams due to the construction of dikes reduces water storage and the flux of sediments and nutrients from those floodplains (Knox et al. 2022).

Humans have been damming streams for millennia, and dams now affect most of the world's large rivers (see reviews in Nilsson et al. 2005; Grill et al. 2019). Dams may be any size from small weirs to extremely large structures that are hundreds of meters tall. Weirs (usually stone, wood, or concrete) across streams have been used to divert flows for irrigation or toward mills (such as water wheels) for grinding grain, milling fabrics, or for manufacturing (see Figure 7.3). Larger dams typically are concrete or earth fill, and may impound water for irrigation, power

A.

B.

Figure 7.3   Weirs and small impoundments are common globally to divert water for irrigation, local water supplies, and mills to create power for a variety of small-scale industrial uses. Such small, water-powered mills are mostly obsolete in developed countries, but their weirs, also known as low-head dams, are still common. Even weirs of only a meter can create a barrier to upstream movements of organisms. (A) A low-head weir and mill pond looking downstream in northwest Spain. (B) A weir on the Garonne River, France.

production, drinking supplies, or even mine tailings ponds. The tallest dams in the world are 220 to over 300 meters tall (such as Jinping-I, China, and Nurek, Tajikistan), but dams can also be ranked on the basis of the volumes of water stored upstream (such as the Kariba Dam, Zimbabwe and the Bratsk Dam, Russia), their hydroelectric power production capacity, or even the amount of material used to construct the dam. Some of the largest dams by volume of material used to construct the dam are tailings ponds, and two of those are in Canada's oil sands area. Dams create barriers for many organisms, and even small dams can alter flows. There are many impacts of dams, which we will explore in the next section.

Many species use natural annual discharge cycles for key life-cycle events, such as migration between habitats. Many species evolve phenologies to avoid certain conditions. For example, fish do not normally lay eggs in gravels of stream beds during spring floods, nor do adult stages of aquatic insects typically emerge from streams at those times. Natural discharge patterns have five main elements: magnitude, frequency, duration, timing, and rate of change, all of which stream ecosystems can respond to (Poff et al. 1997)—to that, we can add water quality (Poff et al. 2010). These ideas about how discharge patterns affect biology have come to be known as the *natural flow regime* (Poff et al. 1997), and elements of this have been incorporated into flow management.

Dams enable energy production and water storage, but there are many impacts of dams and reservoirs (see next section). Some countries are working hard to remove older dams that have filled in or become obsolete (Hart et al. 2002). However, other countries are moving toward increasing the numbers and sizes of dams to provide for water storage and hydroelectric power generation, and this will cause enormous consequences for downstream ecosystems—and people dependent on them (Dudgeon 2011). Some of these dam projects are extensive and will modify habitats and water supplies for many tens of millions of people downstream. In the Mekong River, which is considered the world's most productive river, dams in China and further downstream are turning the river into a series of linked reservoirs, thereby altering habitats, changing flow regimes, degrading water quality, and creating barriers to movements of species (Dudgeon 2011).

IMPACTS

Dams and Reservoirs

Dams created to impound water in reservoirs have many impacts (see Table 7.1). An obvious change is the reduction of flowing water (lotic) habitats and the increase in the amount of still water (lentic) habitats. Water leaving reservoirs has different properties than river water, in terms of temperature regimes, the composition of particulates, nutrient concentrations, and flow regime. The still water in reservoirs is exposed to the full input of solar radiation and can result in very warm water being discharged downstream, thus water released in the summer may have elevated temperatures, much warmer than what would have existed in the downstream reaches before the dam. Warmed water in reservoirs can also have much higher rates of water loss through evaporation (see Chapter 14). On the other hand, some dams release water from the bottom of the reservoir, sending overly cold water downstream compared to natural systems (for example, the Glen Canyon dam, Arizona, USA) (see Melis et al. 2015). Reservoirs trap sediments, thereby limiting sediment supply downstream of the reservoir. Consequently, the erosion of any remaining mobile sediment downstream by peak flows will not be replaced from upstream. This phenomenon is called armoring, and leaves the reach downstream with very large piece sizes of overlapping sediment that are typically unsuitable for spawning of fish, or for many other organisms that typically live among finer sediments (sands, gravels, and cobbles). However, alterations to patterns of discharge downstream can be managed.

Table 7.1 Impacts of dams and reservoirs on freshwater ecosystems.

Process Impacted	Alterations Caused by Dams	Solutions or Other Outcomes
Discharge	A wide range of release patterns	Management for environmental flows
Access past dams: connectivity	Populations of many organisms experience barriers to movements	Fish ladders and other passage structures
Loss of habitats	Former streams become lentic environments unsuitable to most lotic species	Smaller storage areas or run-of-river power production
Temperatures	Surface release leads to high temperatures downstream in summer; bottom release results in cold water downstream	Adjustable release points to modify temperatures of release water
Sediment supplies	Stored within reservoir	Additions of medium-sized sediments to downstream (sands to cobble sizes)
Turbidity	Fine particles contributing to turbidity settle out in reservoir	Generally, this is a positive change to clearer water
Water quality	Reservoirs may end up with anoxic sediments at the bottom, transforming some chemicals into soluble forms, such as mercury and phosphorus	Reduce nutrient pollution which can add to the organic loads (and oxygen demand) on the reservoir bottom
Suspended organic particles	Very little detritus, often high-quality plankton (phyto-, zoo-, bacterio-, ichthyo-)	Often good for consumers, although may cause shifts in types of consumers
Evaporation	Large surface area exposed to direct solar radiation and warming leads to higher evaporation rates	On small reservoirs, *bird balls* or covers can reduce heating and thereby reduce evaporative losses

It is entirely possible for dams to store all the water during drier periods (or when filling initially), and the reaches downstream of a dam could effectively dry up for periods long enough to eliminate most aquatic and riparian life (see Figure 7.4). Alteration of natural hydrographs is a feature of almost all dams, affecting an array of processes. These alterations to flow lead to changes in physical habitat, the loss of a set of cues for certain life-history events such as the timing of reproduction, the loss of longitudinal connectivity, and changes to native stream communities (Bunn and Arthington 2002). There is strong evidence that modification of flows from dams that typically have less variability favor increases in the occurrence and abundances of nonnative species (Mims and Olden 2013) (see Chapter 11). Another impact is hydropeaking, where water is released during the day to produce hydroelectricity and stored at night when power demand is lower. In some places, water is dropped to a lower reservoir and at night any excess power is used to pump water back to the upper reservoir. In some situations, water from a reservoir is diverted to another watershed to contribute flows to power production elsewhere. Any of these situations affect flows downstream. Even when flows are provided, the rate of change in the flow magnitude—referred to as ramping (such as from releasing flows quickly or not at all to accommodate instantaneous needs for hydroelectric power)—may be too rapid for organisms to move in response.

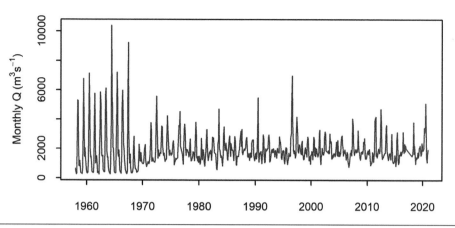

Figure 7.4   Hydrograph of monthly average discharge in the Peace River, British Columbia, below the W.A.C. Bennett Dam (Hydrometric station 07HA001) from before, during, and after dam construction (hydroelectric power dam). Notice the extreme suppression of flows downstream of the dam during the filling of the Williston Reservoir starting in 1968 until 1971. It is obvious that annual peak flows after damming were significantly altered compared to the native flow rates before dam construction. Note that there are still occasional peak flows after the dam construction, but they are not evident in monthly averages. Figure courtesy of Dr. R. Dan Moore, University of British Columbia.

Among the multiple impacts of dams on freshwater systems, fragmentation can eliminate the longitudinal connectivity through the drainage network. This lack of connectivity may mean organisms can no longer reach part of the habitats they formerly occupied, or it may reduce gene flow between populations by isolation (Staentzel et al. 2020). The Columbia River in western North America is the largest hydroelectricity producing river in the United States, and has 14 large dams on its mainstem, with many others on tributaries (see Box 7.2). This series of dams has impacted fish and other organisms in many ways, but especially in terms of blocking access to former habitats upstream. Dams can prevent species from reaching their spawning grounds—such as the Mekong giant catfish (*Pangasianodon gigas*) that is critically endangered and on the International Union for Conservation of Nature's *Red List of Threatened Species*—and dams can be a cause for many other species to become endangered. The prevalence of small dams is generally underappreciated, but may cause large impacts on flows and connectivity of habitats in many smaller catchments (Morden et al. 2022). In the United States, it is estimated that there may be between 2.9 and 9 million small impoundments (most <1 ha), mostly on streams, which affect hydrology, and since many of these are also stocked with species not found in that drainage, it can also result in the expansion of invasive species (Pfaff et al. 2023). Dams can even restrict dispersal of plant seeds (hydrochory) because they get trapped in reservoirs and are not transported by water to downstream sites.

The passage of fish through dams, either going upstream or downstream, can be hazardous. Dams can act as absolute barriers leaving no way for fish or other species to move past them. When passage is available, it can be dangerous. For example, fish going downstream may be swept into turbines where they are often killed outright, disoriented, or subject to nitrogen narcosis from dropping elevation so quickly. This provides a food source to some predators that do very well below reservoir releases. Most of the research on the impacts of dams has been on Pacific salmon, eels (*Anguilla* spp.), and river herring (*Alosa* spp.). To provide access for fish heading upstream, some reservoirs have fish ladders (see Figure 7.5), and in some cases, there is transport by trucks or barges, and even elevators to get fish upstream (Roscoe and Hinch 2010). Some of these access methods do not help particular species, such as white sturgeon, which can be too large for the design of fish ladders for salmon (other methods have been tried for sturgeon). However, complex currents downstream of a dam caused by water released through turbines or overspill, combined with the large volume of water, can make it difficult for fish to discover entrances to fish ladders. When fish find the fish ladders, it is still an energetically challenging swim up and some fish *fall back* and have to make additional attempts to move upstream in these fishways, which can lead to exertion with fitness consequences—including mortality. In the Columbia and Snake Rivers, out-migrating juvenile salmon may be transported downstream by barge to reduce mortality from turbine passage (Keefer et al. 2008). However, transport by barge or truck may also cause some stress and might alter their timing for physiological adaptation to the marine environment. For some populations of fish, there may be multiple dams to pass, thereby causing additional exertion and risk. There can also be delayed mortality of fish moving upstream or downstream through dams, which can be difficult to estimate.

For lacustrine species that live in reservoirs, there may be other challenges. Most dams are operated to store water when it is plentiful and release it as needed for irrigation, drinking water supplies, or hydroelectric power production. This *drawdown* can leave reservoirs with very little volume, and can sometimes drop by tens of meters annually, or even more over longer time periods. Some reservoirs operated for peak power production can experience rapid drops over the span of hours, which is related to ramping rates for stream flows downstream from a reservoir as noted previously. Conversely, reservoirs may fill very rapidly if water is pumped back up at night (in some power production schemes there is an upper and lower reservoir) or if water is being held back when it is not needed for power or water supplies. This habitat

Figure 7.5 A fish ladder around the Schiffmuehle hydropower plant on the Limmat River, Switzerland. The lower part on the left is designed to appear more nature-like, and the part to the right is a typical, vertical slot fishway. Photo courtesy of Dr. Armin Peter, Fish Consulting GmbH.

compression (and expansion) affects the amount and quality of habitat and also alters chemistry through exposure of reservoir sediments to oxygen.

Channelization and Diking

The channelization of a stream (straightening and deepening of streams) removes structural complexity (large wood, large boulders, variability in depth and velocities) and habitat, and alienates riparian areas. Channelization is typically done to provide a transportation corridor or for more space on land for development, and to increase the rate of discharge to avoid flooding. There are often dikes (artificial levees) along the sides of channelized streams to reduce flood risk and possible erosion from streams overtopping their banks during high flows. Dikes are also frequently built along large and small streams, as are many of our roads, irrespective of channelization, and most often means there are no trees retained along the stream margins. Dikes might be constructed close to the water's edge which usually then requires heavy armoring to avoid dikes being eroded. Dikes can be built further back from the normal stream margin to provide room for the stream's flows to expand laterally without getting substantially deeper (see Figure 7.2C; also see Chapter 15). Typically, such engineering simplifies channels in a way that organisms cannot easily evade floods by moving to off-channel habitats such as floodplain wetlands or tributaries, or to shallower and lower-flow environments. Dikes eliminate interactions with floodplains for species that are adapted to using those areas and the seasonal or perennial wetlands refreshed by flood waters (Knox et al. 2022), as in the flood-pulse concept. Seasonal flooding of floodplains is one of the contributors to the productivity of such areas for agriculture. The idea of the flood pulse concept derives from this annual reliance on flooding of areas beyond the average bankfull channel (Junk et al. 1989). In places like the Amazon River and many others, whole suites of freshwater species depend on seasonal access to floodplains for food and areas for reproduction (see Tockner and Stanford 2002). The losses of these annual flooding events are detrimental to many species and biological communities.

INTERACTIONS WITH OTHER STRESSORS

Water stored in reservoirs can heat up at the surface with the direct inputs of solar radiation, sometimes leading to stratification as occurs in natural lakes. If dams release surface water, it can create long reaches where temperatures are elevated above optimal temperatures for native species. This may slow migration rates of fish returning to spawn, and may put them beyond their aerobic scope, leading to mortality. Likewise, if a dam releases bottom water, it may be too cool for some species. As previously discussed, the modification of timing and magnitude of flows have many impacts related to other stressors such as pollution, lack of sediments of appropriate sizes, change in food supplies, and the alteration of habitats for at-risk species. All of these alterations will be further influenced by climate changes.

Dams can fail, and when they do, it can cause catastrophic impacts. There are many such examples around the world, often with large loss of life—human and otherwise (especially terrible are the failures of tailings dams, as discussed in Chapter 5). However, dams sometimes have to release large rates of flow at times and magnitudes that downstream ecological communities may not withstand. For instance, in anticipation of large storm systems, reservoir management may release water so that it has capacity to moderate storm flow releases to an extent. This management intervention may become more common as climate changes lead to more intense storms.

Reservoirs store sediments that deposit in the basin, and this may include toxic substances from upstream (particularly in the case of tailings ponds discussed in Chapter 5). Since the bottoms of reservoirs may become anoxic, this can change the form of some chemicals, thereby making them more readily transferred into food webs. For instance, mercury can become methylated under anaerobic (anoxic) conditions, and methyl mercury can then be bioaccumulated in consumers within freshwater food webs. Likewise, phosphorus stored in bottom sediments may become biologically available under anoxic conditions that prevail at the bottom of deep reservoirs.

Dikes (levees) typically remove possible slow-water areas of refuge, such as floodplain wetlands or shallower (and lower velocity) habitats during floods. Dikes can fail catastrophically by floodwaters spilling over dikes or washing them away (such as floods on the Mississippi River in June 2008) during large or extended storm periods (such as hurricanes, for example Katrina in 2005). The flooding of infrastructure and other property when dikes fail costs billions, if not trillions of dollars in damage, as well as lives lost. The typical response by local governments and engineers is most often to build bigger dikes, instead of dealing with the impacts that the loss of floodplains has on water storage and flows. As climate change raises sea levels, dikes are being augmented and heightened in coastal areas, continuing the impacts of dikes on stream ecosystems (see Chapter 14). Demand for more land for development also often leads governments to favor more and bigger dikes to enable such expansion at the expense of freshwater ecosystems. In some countries such as Japan, the shortage of relatively flat land for development and agriculture, coupled with intense storms at certain times of the year, has resulted in large dike works, usually concrete, and are sometimes referred to as three-sided streams (a bottom and two almost vertical sides).

SOLUTIONS

Finding solutions to support freshwater ecosystems in the face of flow regulation is tricky since we need to balance human needs with the needs of nature. For instance, as humanity tries to solve the atmospheric carbon (greenhouse gases) causes of global warming, we look to more production of hydroelectricity; but while that is *greener* than burning hydrocarbons, it still has major environmental consequences. We also need dikes as flood defenses where extensive infrastructure and homes are vulnerable. The next section considers some of the options available, while also knowing that there are places where these solutions may not be possible. The key is to be creative within the constraints of our built, human environments.

Environmental Flows

Hydrology can be modified in many ways by diversions, withdrawals, land use, flow regulation, and climate change. Thus, one of the first solutions is to ensure that flows released from dams take into account the environmental needs downstream, known as *environmental flows*. Some of the tools for environmental flows are primarily aimed at providing minimum flows to ensure that some small amount of flow exists at all times. This is especially true of dams that divert water from one reservoir to another or during low-flow periods where dam operators could store all water. However, as previously noted, some species need flood peaks to arrange habitats in ways that provide appropriate habitats, or to provide life-cycle cues (Poff et al. 1997). Environmental flows now include not only minimum flows, but flow peaks to mimic natural seasonal patterns and flows to arrange habitat features that require high flows to sustain.

The need to provide for *environmental flows* has resulted in a set of metrics referred to as the Environmental Limits of Hydrological Alteration (ELOHA) (see Figure 7.6). These metrics are used by some agencies to consider the range of flow conditions necessary to maintain some level of ecosystem function and resilience (Poff et al. 2010; Poff and Zimmerman 2010). Releases of flushing flows to clean out fine sediments or medium duration peaks to stimulate some life history events usually result in less water available for power production or irrigation. Sometimes these trade-offs are contrasted as humans versus ecosystems, but the benefits of nature to humans (*ecosystem services*) are enormously valuable. This management approach to environmental flows considers ecological needs linked to a particular river type, and also considers the management framework (social aspects) around flow regulations.

As dams manage flows for producing electricity, the demand may change through the day, and the rate of change in the increases and decreases of discharge from these impoundments is referred to as ramping or ramping rate (usually in the cm/h range). In general, mobile species such as fish, can move in response to change in flow if they are not too rapid. However, rapid decreases in discharge can leave fish stranded in pools or out of water and trapped among cobbles or gravel (Irvine et al. 2009) (see Figure 7.7), and rapid increases might flush them

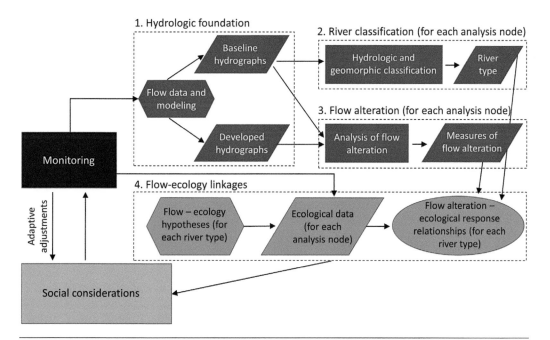

Figure 7.6 Schematic of the ELOHA process for evaluating hydrological patterns to sustain functional ecosystems (redrawn after Poff et al. 2010). A lot of ecological and hydrological information goes into the decision matrix, as do social and economic considerations. Balancing of ecological and social values is vital to this process. Note that monitoring is a central feature of this process, which allows for revision of actions as information is gained about the success (or failure) of outcomes.

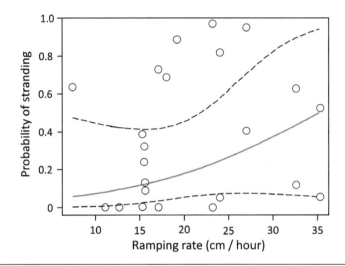

Figure 7.7   Relationship between ramping rate (rate of increase of water depth from releases from reservoirs) and stranding rate of fish. The red line represents the mean value from the statistical model and the dashed lines are the upper and lower calculated 95% confidence intervals around the estimated line. Figure redrawn after Irvine et al. (2009).

downstream. Many dams are operated in a way to minimize the ramping rates to protect fish and other organisms. These rates are determined empirically, as different target species and age classes vary in their abilities to avoid altered flows, and this is further affected by time of day, season, channel geometry, overall flow rates, and fish density.

There are tools for estimating the amount of suitable habitat available to an aquatic organism at specific discharge rates, which can assist in defining required environmental flows or flows for particular life stages. Some of these include the Instream Flow Incremental Method, Physical HABitat SIMulation (PHABSIM), River2D, and other methods for predicting habitat availability. PHABSIM evolved in the 1970s and is still in use (for example, see Reiser and Hilgert 2018), although some find the method does not adequately reflect ecology and there are now other methods available (Railsback 2016). PHABSIM is relatively simple, using measures of velocity, depth, and substrate, but assumes that density of fish or other organisms is a reliable predictor of habitat suitability, along with other assumptions. Densities of target organisms are mapped relative to those measures to estimate what combinations will support what density of that organism when flows are altered. However, there is evidence that such models do not address all life stages adequately, even for highly valued species, such as salmonids (Rosenfeld and Naman 2021). Many environmental flow schemes target single species, and then the flow regimes proposed may be detrimental to many of the other freshwater species, including prey of the target species. It will be important to take a broader ecosystem perspective to balance the flow needs of entire freshwater ecosystems (see examples in Tonkin et al. 2021).

More recent habitat prediction models are more ecologically based, that is, considering all life stages (not all of which will be within the same reach). Models such as occupancy modeling (hierarchical multispecies occupancy models) or suitability models incorporate more details about habitat features needed beyond flow velocity and depth. Such models also need to integrate the distribution of predators, prey, competitors and other actors, water quality, and distribution of other features (such as subsidy resources, shading, etc.). Another class of models are individually based models that build in ecological interactions of the target species (Railsback et al. 2013). Models are useful for exploring options, but no model is ever 100% correct and will not replace proper monitoring of outcomes (see Chapter 16).

Connecting Habitat

Early dam projects did not always consider fish passage, and sometimes projects had to be retrofitted for bypass structures or processes, although many remain as barriers. In the European Union and elsewhere, free passage of fish past such barriers is mandated by law. A very common method used around the world are fish ladders or fishways, where a series of step pools are built of concrete and steel that allow fish to make their way up a long series of small steps and thereby pass up and past a dam. Fishways may be concrete basins in a series allowing passage by some species (most often for a target species, such as adult Pacific salmon), but can be constructed of more natural materials with a lower gradient. One challenge is that migrating fish may have difficulty finding the beginning of the fishway, often because they seek out faster flows, which fishways do not provide. One solution has been to release additional *attraction* flows of water near the start of a fishway. The extra water will provide the faster flows that will draw the fish toward the entry into fishways (Burnett et al. 2017).

Even though fishways facilitate passage, not all individuals that move upstream will survive since this is such an energetically costly activity that can put fish at their aerobic limits, and may have fitness consequences (Burnett et al. 2017). This can even literally result in fish having heart attacks. Some survival impacts may be immediate, but measuring the delayed mortality of fish migrating upstream or downstream requires following those fish for some days (or weeks) after passage. Studies of Chinook salmon have suggested that passing multiple dams on the way downstream may incur higher, delayed mortality, but estimating the effectiveness of fish passage (upstream or downstream) requires more research. In some locations it is not possible to construct or retrofit dams for fishways. One creative but expensive means of providing for fish passage is to transport salmon around dams (upstream or downstream) on the Columbia River (see Box 7.2); this method has reduced direct mortality. Such trap and transport methods have been used for other migrating fish and other species to get them past dams. Engineers have even developed a *fish cannon* that uses hydraulic pressure to send fish through a tube, up and over obstacles like dams.

A solution for downstream passage to keep fish from passing through turbines is accomplished by nets or screens that direct fish that are migrating downstream to the fishway, other spillway, or some other bypass devices. The mesh sizes used may be insufficient to hold back small fish, or can be challenging to maintain if there is a lot of debris, such as wood, leaves, or trash, floating in the water of a reservoir. Fishways are most typically designed for upstream movements, and spillways for downstream passage are still being refined. Other solutions include changing schedules of power production (turbine use) during peak periods of migration by reducing flow rates or stopping the turbines. For instance, stopping the turbines could occur at night when electricity demand is typically lower. There are also turbines that are less hazardous—such as Kaplan turbines as opposed to Francis turbines—but that depends on the powerplant design and hydraulic head (Algera et al. 2020).

Temperatures of outflows from reservoirs depend on whether water is drawn from the top (epilimnion) or bottom (hypolimnion) and those temperature differences might severely affect fish and other organisms downstream. Based on location, temperatures could be too hot or too cold, depending on whether they are top or bottom releases and on the river system that the water is being released into. The temperatures of released water can be adjusted if the dam has releases from both levels that can be mixed to modify temperatures to suit species downstream. This is possible with some dams. There are designs to mix water from the top and the bottom of a reservoir to meet a target temperature, and there are other mechanisms to ensure that temperature pollution is mitigated (Michie et al. 2020).

Restoration of Habitats Downstream of Dams

Due to the loss of sediment inputs to a reach downstream of existing dams along with the flushing of mobile sediments from those areas, two solutions have been used (see also Chapters 3 and 15). One is to augment gravel of a size that is appropriate for the spawning of particular species. This was used in the Sacramento and San Joaquin rivers in California for Chinook salmon and rainbow trout (or steelhead trout, as the sea-run individuals are called) and where the limited supply of appropriately sized gravels is eroded away. In these sites, mounds of appropriately sized gravels were piled along the stream margins and enter the streams as high flows wash them into the channel, providing spawning areas. Most data available indicate that the ecological effects of these additions of gravel are local and transient; that is, these gravels are often transported away downstream with the next seasonal flood. These approaches to fixing habitat (see Chapter 15) require sustained efforts and input costs to replace gravels that are displaced further downstream (Kondolf et al. 2008; Staentzel et al. 2020).

A second idea is to restore functional riparian ecosystems downstream of dams by washing remnant, mobile sediments out of the channel to create gravel and sand bars. This has been tested through releases of very large peak flows from a reservoir over a short duration to rework downstream sediments (Staentzel et al. 2020). One such example is the Glen Canyon dam project on the Colorado River where large flows were released from the dam in an experimental approach to create flood disturbance downstream (Melis et al. 2015; Korman et al. 2023). It is becoming clearer that environmental flows alone are insufficient to address downstream impacts of dams (see Table 7.1). Many dam management projects where environmental flows are mandated are also combined with physical restoration actions—for example, the gravel augmentation projects that were previously mentioned.

Removal of Dams

The removal of dams can have great environmental benefits. Dam building—especially low-head dams as were typical of 18th- and 19th-century mills for grains, fabrics, and other uses—has resulted in large numbers of inefficient (or useless) dams (Hart et al. 2002). Some dams were removed over a century ago to allow salmon passage, such as the first dam on the Wallowa River, Oregon, which was blown up in 1914. Dam removal in some regions is increasing, particularly the removal of low-head dams (weirs) (Hart et al. 2002). Even large dams, such as the Elwha and Glines Canyon dams (31 and 61 meters tall, respectively) in Washington State (removed 2011–2014; see Box 3.2), were removed because their benefits no longer outweighed the environmental costs that were displayed by downstream impacts and loss of habitat through impoundment. Removal of these two dams in Washington resulted in a 67% increase in the numbers of spawning Chinook salmon from 2014 to 2020, relative to the period before the dam removal (1986–2010) (Pess et al. 2023). In the eastern United States, removal of low-head dams has been effective in restoring access to spawning sites for anadromous and catadromous species, including American eels (*Anguilla rostrata*), and has resulted in increases in fluvial species (Watson et al. 2018). Removal of dams often happens when the reservoir fills with sediments and the costs to increase the ponding volume are too high to justify (Warrick et al. 2015). Dam removals are often a component of freshwater restoration projects (see Chapter 15).

Of course, it is possible to find alternatives to the construction of new dam projects. This means finding new ways to generate electric power. A recent example from Albania protected the Vjosa River from the construction of 45 hydropower stations, which was the successful result of public pressure. The river will be given park status and remain one of the rare wild rivers of Europe. A similar outcome of public resistance was the banning of plans for dams on the Mary River, near Brisbane, Australia.

Flow Management for Environmental Restoration

One solution to impounding water behind dams and affecting flows is to use run-of-river power installations. These can be scaled from micro-projects for individual homes or small villages (micro hydropower plants <100 kW) to industrial-sized power producers (typically <25 MW). The idea is to divert some portion of the flow around a stream reach through a tunnel (or pipe) to the turbine in order to produce hydroelectric energy downstream. This usually requires steep topography that provides sufficient hydraulic head (elevation difference) to produce a reasonable power output over a short distance.

In the case of a micro hydropower installation, the small turbine is attached to the pipe and can be done with as little as 3 to 20 meters of hydraulic head. In larger industrial installations, the distance from intake weir to turbines is usually less than five kilometers. In larger systems, there are requirements to maintain minimum environmental flows, as we discussed earlier. This ensures some flow is available to sustain the stream ecosystem below the intake weir in the reach before the water is reintroduced (instream flow requirement). Thus, the reach downstream of the power turbines has its natural flow regime. These installations require a weir that can potentially block some species' movements, and need a small headpond, but no large impoundment.

In western countries (for example, United Kingdom, United States, Canada, France, Czech Republic) these projects are typically, but not always, built above the upstream limit of fish (or at least some higher-value fish). Studies of impacts on salmonid populations mostly show slightly negative effects within the diversion reaches, but the results are quite variable among projects, and may depend on the proportion of flow diverted and other local characteristics. The impacts of such projects depend on whether the diversion weirs create barriers to movements of organisms or cause impoundments more similar to typical reservoirs. One other limitation is the landscape disturbance due to the large numbers of installations, as well as the service roads and transmission lines necessary to connect these facilities to the power grid.

As channelization is usually accompanied by dikes, it can alienate large areas of habitat that cannot be accessed beyond the dikes, such as floodplains and wetlands. These habitats are important seasonally for organisms to escape floods, and they may provide essential habitats for certain life stages. In some cases, water is pumped from diked areas back into rivers, but such pumps are lethal to most organisms. Pumps based on Archimedes screws can alleviate that problem, although that is still mostly in one direction—that is, from floodplains to a receiving waterbody. Another way this is solved is sometimes with gates that swing open when water pressure inside the dikes is higher than along the channelized river, but that is rarely successful, except in some tidally influenced areas.

There has been movement toward giving streams more space by moving dikes further away (Figure 7.2C), creating multiple stage (flow height) dikes (see Figure 7.2B, D), or recreating the meandering form of streams (see Box 7.1). Removal of dikes to reconnect streams to their floodplains is possible, but often is costly if there is infrastructure within the floodplain. Knox et al. (2022) give examples of the environmental, economic, and societal benefits of reconnection of streams with their floodplains, but estimate much less than 1% of dikes in the United States have been altered for such renewed connections. In some places, it has been possible to reconnect off-channel areas where they remain—for example, oxbows and former high-flow channels. This offers at least two benefits, one being access for organisms to lower-flow environments to escape floods or to utilize resources there, and a second being floodwater storage to reduce the instantaneous discharge in a stream. Another option is to have dikes that are further from the stream margin to give a stream room to move or dikes of different heights (see Figure 7.2B, D). In the Netherlands, this has been called the *Room for the River Project*, which allows for more space for floodwaters to spill over and reduces floodwater depths. Dikes are often built with accompanying groins, installations of rock roughly perpendicular to the edges of dikes to dissipate some of the flow energy and reduce erosive forces against the dikes during floods. One related activity is construction of roads along our streams since these are often a consistent gradient in valley bottoms, but also require heavy armoring as with dikes. We should resist building roads so close to our streams when possible. Allowing more room for the river also allows for expansion of a river's width and better expression of natural geomorphic channel evolution, creating more physically complex ecosystems.

MONITORING

An important element for developing solutions to protect freshwater ecosystems is *monitoring*. Many agencies monitor water quality and quantity, and this can be used to evaluate outcomes of management activities. There are also programs to monitor biological communities; some of these are project specific and some are ongoing monitoring programs to assess freshwater ecosystems. Measuring the outcomes of management is the only way to learn and advance our ability to do better at protecting freshwaters (see Chapter 16).

PERSPECTIVES

Although hydroelectricity is much greener than burning fossil fuels, it is not without its own environmental consequences. As we rush to increase our dependence on electricity from renewable sources, especially water, we will need more dams. We will also need more dams for water storage, irrigation, and drinking in a warmer world. In this chapter you have learned that dams and flow regulation have their own impacts, and we need to consider how to balance impacts against benefits.

Box 7.1   Reconnecting Floodplains with Streams

The European Union (EU) has the goal of having all its streams in good condition (based on hydrogeomorphic criteria) by 2027, and one means to achieve this is stream restoration activities (EU's Water Framework Directive 2000/60/EC). Many alluvial streams have had their floodplains alienated by dikes (60,000 km at least in the EU), armoring, channelization, and infrastructure. Even small streams have been straightened to allow for faster drainage and more intensive use of land for agriculture and other purposes. It is also common for large wood to be removed from stream channels, simplifying structure. Moving dikes back further from streams, or removing them entirely, can do a lot to re-establish floodplain wetlands, which have often been lost by infilling or loss of water. Restoring floodplain and stream habitats takes a landscape perspective, and one example is the work that has been done in the streams of Germany's lowlands (typically low-gradient streams). Much of this work has involved re-meandering straightened reaches (see Figure 7.8A, B) and reconnection

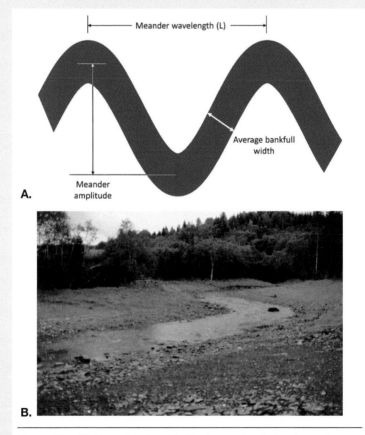

Figure 7.8   (A) Example of an idealized meander design for a stream—useful in a general context but application depends on more details and requires engineering expertise. As a rule of thumb, the wavelength of meanders is about seven times the bankfull width, but in reality, nature is rarely so orderly, so it is best to have room for the stream to move around after re-meandering. (B) A re-meandered stream segment, Gråelva River, Norway.

continued

with former floodplains, along with additions of large wood and rocks. Similarly, agencies in the United States are looking for opportunities to reconnect floodplains by modifying some of the 228,000 kilometers of dikes in that country (Knox et al. 2022).

Studies of re-meandering and lowering of floodplains along the reaches of two large streams (1 and 1.9 km long, respectively) in Germany were each compared with a paired straightened reach through time (see also Chapter 15, Box 15.1). The study showed improvements in biological diversity in the re-meandered reaches, including many taxa not present in the straightened streams, as well as greater macrophyte and wood abundance (Lorenz et al. 2009). In that study they found only small increases in invertebrate abundance through time, consistent with many other studies (Louhi et al. 2016).

The consequences of re-meandering and other changes created more heterogeneity in the channel and gravel bar areas, as well as side channels and off channel wetted areas, and reduced sand coverage. In these projects, they found greater diversity and abundances of macroinvertebrates (Lorenz et al. 2009). Recolonization of sites by any organism depends on there being a source population of potential colonists available, so recovery of species diversity may be limited in areas where such colonists are not present.

Restoration of these lowland streams by re-meandering and other actions has also resulted in a higher diversity of organisms in their riparian areas. Based on stable isotope analyses, the improved stream ecosystem led to tighter linkages of riparian area organisms that use stream resources, such as prey from streams (Kupilas et al. 2020).

Box 7.2   Columbia River Dams

The Columbia River is among the 40 largest rivers in the world based on average annual discharge (about 7,500 m³/s) and was once a major producer of Pacific salmon and steelhead trout. At the beginning of industrial-scale logging in the late 1800s, splash dams were common in Washington on the Columbia River and its tributaries to store logs before flushing them downstream to mills using the impounded water. The massive geomorphic changes from the sudden floods triggered by the releases from these splash-dam impoundments—along with large amounts of moving logs—disturbed habitats and caused long-lived changes to streams. In the early 1900s, the first of many dams was built on the Spokane River (a tributary), largely for hydroelectric power, but without any means of fish passage. There are now at least 39 large dams on the Columbia River. It was not until the 1930s that the U.S. government passed a law requiring most dam projects to consider fish passage. However, a great deal of damage to salmon (five anadromous Pacific salmon species) and steelhead populations over the decades has resulted in large decreases in population sizes and many stocks of those species being listed as endangered (see Figure 7.9).

Some, but not all, modern dams are equipped with fishways or fish ladders to facilitate the passage of fish. Many passage devices have since been built retroactively around older dams, including vertical slot fishways (see Figure 7.5). In some cases, the costs of retrofitting dams with passageways were prohibitive. At some of the dams in the Columbia River system, the solutions have included fish elevators, or transport of trapped fish by barge or even truck to move them upstream past dams (and sometimes downstream for the young)—some examples of which can be found in the Columbia River in the Pacific Northwest (Roscoe and Hinch 2010).

continued

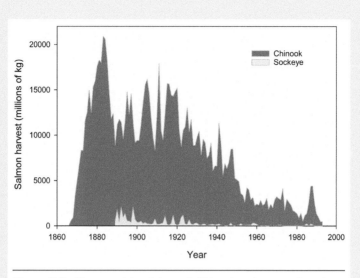

Figure 7.9 Commercial catches of Chinook salmon and sockeye salmon in the Columbia River over the past century from when reasonable estimates began in the 1880s. Note the dramatic declines in numbers associated with the period of dam building on the river beginning in the late 1930s. Numbers include wild and hatchery fish. *Source:* Institute for Fisheries Resources (1996).

Estimates are that dams on the Columbia River now block access to about 40% of the original spawning and rearing habitats of salmonids in the basin. One solution to this was to build hatcheries, but these create their own problems of reduced genetic diversity and maladapted behavior, which can affect wild fish (see Chapter 10). There are many other restoration activities that are intended to sustain and increase salmon populations in the Columbia River watershed. Some of these activities include fertilization with nutrients and salmon carcasses to replace nutrients (and protein and fatty acids) that historically would have augmented productivity at spawning sites as adult salmon returned to freshwaters.

An additional complication for managing water flows of the Columbia River is that it is a transboundary river—that is, part is in Canada and part is in the United States. Thus, there is a *Columbia River Treaty* between the two countries to regulate management. Four of the large dams on the Columbia River are in Canada, but these are managed in order to provide water to United States users downstream for irrigation and power production in the summer. Flow regulation frequently affects other jurisdictions, and legal remedies can be complex and controversial.

ACTIVITIES

1. How many dams are located in your state or province?
2. What is the primary purpose of the biggest dams near you?
3. Are they for irrigation, drinking supplies, power production, or flood control?
4. Are there any other purposes for dams?

8

FOREST AND AGRICULTURE MANAGEMENT

INTRODUCTION

Forestry and agricultural activities occupy a large portion of the earth's land surface. Thirty-one percent of the earth's land surface is forested, with about 28% of forests dedicated to forestry (Global Forest Resource Assessment), that is, about 9% of the land surface of the planet. Agricultural use accounts for about 37% of the earth's terrestrial surface (crops and livestock). Both forestry and agriculture can have widespread impacts on freshwaters (see Tilman 1999; Smith et al. 2016). Land use for these two activities has resulted in a net loss of forest over the past 300 years, estimated to be in the range of 7 to 11 million km^2 globally (Foley et al. 2005). Both land uses cover large areas and have somewhat similar effects on freshwaters and the solutions have similarities, so we will explore these land uses in this chapter.

These activities can vary in scale from subsistence to large-scale industrial uses. The effects of forest harvest and agricultural practices on freshwaters can be considered at different spatial scales, such as catchment-scale (watersheds) impacts (such as hydrological alterations) and reach-scale (local) impacts (such as shading), although these scales are not truly discrete or independent. The impacts affect hydrology, sediment fluxes, habitat structure, nutrients, amounts and types of organic matter, temperatures, and pesticide concentrations (Malmqvist and Rundle 2002; Danehy et al. 2022) (see Figure 8.1). Because of all the physical and chemical changes, biodiversity is also altered, and many species are negatively affected. The literature about impacts on streams from forestry and agriculture is much larger in comparison to the studies of the effects on lakes and wetlands, which may be due to the much more apparent effects resulting from the erosive power of running water. However, there are still impacts on wetlands and lakes, even if those might appear to be of a lesser magnitude (Becu et al. 2023). There is extensive literature on the impacts of forestry and agriculture on freshwater ecosystems, so this chapter will provide a summary, while some topics will be omitted. Note that the effects of agriculture and forestry will differ according to geography, climate, kind of activities, and politics.

In the United States, annual water withdrawals are in the range of 455 km^3 (2019 estimate), of which 74 to 93% is for agriculture (crops and livestock). About half of withdrawals for agriculture are from surface waters and just over half are from groundwater sources (Konar and Marston 2020). The water that is used to produce agricultural products is part of the virtual

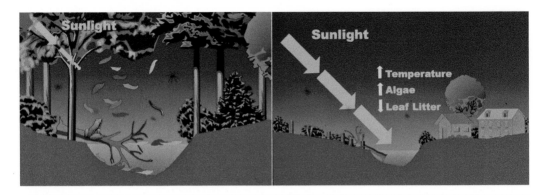

Figure 8.1 Schematic of the effects of land-use, such as forestry or ubanization, on some freshwater processes and structures. On the left, the intact forest provides shade, organic matter and invertebrate inputs to the stream, a source of large wood (LW) as physical structure, a complex riparian habitat, and nutrient uptake. On the right, removal or alteration of the streamside forest leads to increased light (heating and increased algal production), loss of organic matter inputs, reduced nutrient uptake by riparian trees, and loss of LW supplies from the streamside area. Figure credit: University of British Columbia.

trade in water, and water is transported away from watersheds in such products (see Chapter 6). For instance, grain production in the United States represents about 728 km^3/y (2012 estimate) of water (this includes rain), equivalent to about 60% of water stored in reservoirs in the United States (Ruess and Konar 2019). Note that this number exceeds the annual water withdrawals that were previously mentioned, indicating the tremendous uncertainty in some of these estimates.

One form of agriculture that directly affects freshwaters is farm ponds and large-scale aquaculture operations. Rice cultivation in many parts of the world requires large amounts of water, and that water can be affected by pesticides and the accumulation of heavy metals, while at the same time, removing a large volume of water from other freshwater ecosystems (Chapter 10). Rice paddies may be small subsistence farms or enormous industrial operations. Farm ponds for the production of ducks and geese and the subsistence production of fish occur in many parts of the world as part of local agricultural production. There are also extensive aquaculture production areas for fish, crayfish, prawns, and other species that are intended for human consumption (Chapter 10).

IMPACTS

The impacts of land use, whether forestry or agriculture, can occur at different scales. It is useful to consider watershed scales (affecting hydrology, nutrients, etc.) versus reach scales (affecting direct shading, inputs of litter and wood, etc.). General impacts of forestry and agriculture can include alterations in hydrology, light regime (shading), organic matter input, reduced bank stability, less LW, downcutting of channels through higher peak flows coupled with less bank stability, elevated pesticide loads, higher nutrient flux, and changes to biodiversity (see Sweeney and Newbold 2014; Richardson et al. 2022) (see Table 8.1).

Table 8.1 Forestry and agricultural impacts.

Impact	Mechanism	Moderated by
Hydrology	Higher peak flows, especially rain-on-snow Greater yield Less vegetation to take up water, little resistance to flows	Soil depth; drainage system.
Temperatures	Higher peak summer temperatures	Relative exposure to short- and long-wave radiation (shading); volume of water; degree of shading; area of the clearing relative to the watershed area; relative amounts of groundwater versus surface water contribution to flow generation.
Light	Affect thermal inputs; potentially increases primary production; may alter the behavior of some species	Height and density of vegetation; width of buffer; sun angles (latitude).
Bank stability	Tree harvesting and fields sometimes to edge of streams and wetlands; particularly with livestock	Affected by channel gradient (steeper has more stream power) and stream size (bigger is more powerful). Also, windthrow of trees that are not windfirm in the buffer may expose bank sediments to erosion.
Erosion	Soils and streambanks contribute to sediment supplies	Floodplains often store sediments, whereas steeper banks supply more for a given stream size. This is strongly linked to bank stability.
Reduced wood inputs	Removal of large trees	Retention of trees; size of remaining trees.
Reduced organic material inputs	Removal of vegetation or shift in species composition	Retention of trees; size and species of remaining trees.
Nutrient flux	Removing vegetation (trees) reduces nutrient uptake in the short term; addition of nutrients flushes into waterways	Depends on which nutrient(s) are most limiting and retained by vegetation and soils.
Pesticides	Insecticides, fungicides, herbicides, etc. Sprayed to kill pest insects or unwanted plant species or fungi (rusts, etc.)	Depends on solubility of chemical and rate of dilution in receiving waters.

Hydrological Impacts

Several aspects of hydrology (see Chapter 1) are influenced by land use. We have already addressed the impacts of water withdrawal for irrigation in Chapter 6, which is another influence of agricultural practices on water systems. The large amount of water diverted from surface waters for irrigation and the pumping of groundwater both lead to lowered water tables and reduced surface water flows. This is particularly problematic in arid regions, which also tend to be good places for crop agriculture.

Flows are altered by changes to the balance of water inputs and losses. With the removal of extensive vegetation cover, the transpiration component of water balance may be greatly reduced (Moore and Wondzell 2005), leaving only evaporation, so more water may be available to contribute to flow generation. The particular tree species may also vary in how much water they take up, which will affect water yield (Lopez et al. 2021). However, the direction and magnitude of changes to flow timing and amounts also depend on how rates of evaporation change with soils exposed to direct solar radiation. Removal of forest canopy can also result in greater peak flows during periods of precipitation. Forest harvesting in temperate regions can result in greater snow accumulations in cleared areas and faster melting in areas without forest cover, increasing the rates of floods from snowmelt. This can be exacerbated by rain-on-snow where warm rain can add to melting causing even higher flows than the rain alone would predict. In many landscapes, small streams are converted to ditches or passed through culverts with all obstructions removed to ensure rapid passage of water during rain events, which contributes to flooding in receiving streams. These changes due to forestry and agriculture alter the timing and magnitude of flows and, depending on conditions, might lead to greater or lower annual water yield than similar undisturbed sites.

Often, soils are compacted by land use and roads across the landscape, which reduces infiltration rates, and thereby increases rates of flow reaching streams and wetlands and can contribute to higher or more frequent peak flows. Roadside ditches and culverts are designed to convey water quickly to streams. Peak flows are even more extreme in urban settings where water flows along roads, ditches, catch basins, and storm drains to local streams (see Chapter 10). Roads and ditches can be major sources of inorganic sediments to freshwaters. Culverts are installed to connect ditches to larger drainage systems to carry water away from roads, but these culverts can concentrate flows causing downcutting on the downhill side. Culverts in streams can also be clogged or left *hanging* where erosion on the downstream, flow-focused side can leave a culvert beyond the ability of a fish or other organism to move into (see Figure 8.2).

In agricultural landscapes, particularly crop agriculture, small streams are mechanically straightened (using excavators) and turned into ditches to promote drainage of fields, and in some cases, streams are piped into tile drains and become buried streams (Stammler et al. 2013). Ditches often become severely downcut and even undercut due to erosion of soils washed downstream, and can be a large contributor to sediment supply in streams further down in the network (Hanrahan et al. 2018). Ditches have also been used in forestry situations to improve drainage within peatlands and other areas (Rajakallio et al. 2021), which decreases residence time and speeds water on its way downstream, leading to more rapid peak flows. Ditches are constructed along forestry roads for drainage (to prevent roads from being washed out) in most landscapes, and these ditches can be a source of sediments, alter local hydrology, and may become clogged, requiring maintenance by scooping out accumulated materials.

Figure 8.2 A hanging culvert under a road in coastal British Columbia where the downstream side is downcut and creates a barrier to upstream movements of fish and other organisms.

Shade Removal

Shading of freshwaters by trees can reduce thermal inputs, restrain temperature increases, and limit algal growth (Kaylor et al. 2017). The clearing of land for crop agriculture or pastures can include the harvesting of tree cover close to freshwater bodies. This decrease in shade coverage increases the solar radiation reaching these waters. This results in water temperature increases, increased algal production, and reduced organic matter inputs, including loss of LW inputs. Temperatures are known to increase above what they would have been in a shaded forest, especially in summer when radiation is highest and often when volumes of water are least (Gomi et al. 2006). Removing shade can also result in greater daily fluctuations in temperature and even greater cooling in winter. In most parts of the world, removal of tree cover can result in increases in water temperatures that may exceed the thermal capacity of some freshwater species, leading to their loss locally. Finally, more light generally leads to higher rates of primary productivity in streams, albeit with a shift in composition of the algal community.

Sediments and Organic Matter Inputs

With large areas of land cleared for forestry and agriculture, along with the roads and ditches that run through them, the percentage of exposed soils is greatly increased. They are also exposed by the destabilization and failure of slopes (landslides). When water moves through these areas via precipitation and irrigation, the soil particles (inorganic and organic) are easily transported away—eventually to receiving waters. Note that the word *soil* here refers to any loose natural material on the earth's surface, which includes gravel, sand, mud, organic matter, etc. As a result, when soil is exposed and starts to become transported through the force of moving water, it contributes to the flux of sediment through freshwaters (see Chapter 3).

All sediment does not move at the same rate, and following forest harvesting, the rate of transport might be increased, especially for fine particles, until sources of sediment supply are stabilized. On the other hand, logging slash (branches and leaves) may clog streams and actually store sediments in the shorter term, so there can be time lags in sediment transport rates after the disturbance (Gomi et al. 2001; Jackson et al. 2001). The amounts of sediment (including organic materials) moved through rivers such as the Mississippi River or Yangtze River have created estuaries that are considered dead zones.

Large Wood and Shoreline Vegetation

An important element of streams and lakes that is affected by land management is the removal of the long-term supply of LW since dead trees and branches can contribute to freshwater ecosystems. LW (defined as pieces >10 cm diameter and >1 m length) provides hiding places (security cover) for fish, invertebrates, and other species. LW also creates a structural element that can create back eddies in streams with deeper, slower water; it can help trap organic materials and store sediments. Wood also provides another energy source to food webs because fungi and bacteria convert cellulose into biomass (Eggert and Wallace 2007). However, forestry and other riparian clearing removes the future supply of LW, and any remaining wood breaks down over time, leaving streams with a reduced amount of LW. Moreover, stream *cleaning* (removal of LW from streams and shorelines) has left a legacy of reduced amounts of LW and structure that is still detectable a century later (Mellina and Hinch 2009; Zhang et al. 2009).

Alteration of the types of vegetation surrounding freshwaters can have a variety of impacts. For instance, crops may be planted up to the edge of water, or plantations of trees—some of which are nonnative—may replace native forest (see Figure 8.3A, B). Some crops have been genetically engineered (often called genetically modified organisms) to have resistance to pesticides, such as glyphosate. Or crops may have been modified to have constituent chemicals such as *Bt*-modified corn (corn with genes that produce the insecticidal protein normally produced by *Bacillus thuringiensis* against Lepidoptera pests). Leaves from these corn plants can get deposited into freshwaters and may be toxic to freshwater insects (Tank et al. 2010). Plantation trees, such as eucalyptus, rubber trees, spruce, and pine, are often planted in tight spacing, reducing understory and terrestrial habitat, heavily shading water, and their leaf litter might be of very low quality. Studies of litter decomposition indicate that eucalyptus leaves and pine needles have decomposition rates that are 22% (or more) lower than native plant tissue that had been tested in the same streams—and these nonnative plantation trees contribute less to the productivity of freshwater food webs (Martinez et al. 2013; Ferreira et al. 2016).

Figure 8.3   (A) Eucalyptus plantation alongside a stream in coastal Chile. (B) Spruce plantation in northern Sweden (photo courtesy of Dr. Lenka Kuglerová, Swedish Agricultural University). Around the world, plantations of many sorts—most of nonnative species—are often planted right up to streamside and suppress or eliminate native plants and other biodiversity.

Nutrients and Pesticides

Nutrients, such as nitrate and potassium, are well-known for being mobilized from soils after disturbance from land uses such as forestry and agriculture; and without a lot of plant cover those nutrients are flushed to streams, wetlands, and lakes. Crop agriculture applies large amounts of inorganic and organic fertilizers, especially of nitrogen and phosphorus, and those nutrients frequently exceed uptake by plants, particularly at certain times of the year, contributing to eutrophication (Bennett et al. 2001). As nonpoint sources, it is harder to regulate these nutrient inputs. Fertilizer application has increased at least seven-fold in the last half century (Foley et al. 2005). A typical practice in pastoral agriculture is for animal wastes to be applied to fields as a source of nutrients, which can result in excessive flows into freshwaters if too much is added (Belsky et al. 1999). Moreover, cattle are known to preferentially defecate in water if they have access (Oudshoorn et al. 2008).

Pesticides, for agricultural uses and sometimes for controlling forest pest insects, can get into freshwaters easily. Pesticides from agriculture and forestry can occur at high concentrations in groundwater and surface waters (see Chapter 4). There have been many ecotoxicological studies of the impacts of chemicals like common pesticides, as well as glyphosate and neonicotinoids, on freshwater ecosystems. Herbicides are used in agriculture and forestry as well. In forestry, herbicides such as Glyphosate (commonly known as Roundup), and 2,4-D (dichlorophenoxyacetic acid) and 2,4,5-T (trichlorophenoxyacetic acid) (known as *Agent Orange*) are, or were, used extensively to reduce the survival of noncrop trees; for example, alders, cottonwoods, or willows that might overgrow the conifer seedlings that were planted to replace the commercial forest (Östlund et al. 2022). Herbicides were also used to control shrub growth that could interfere with the crop trees. Many agricultural crops have been engineered to be resistant to glyphosate so that *weeds* can be controlled by spraying the chemical, which frequently ends up in freshwaters. In addition, fungicides are used frequently in a variety of agricultural contexts and are often overlooked in freshwater monitoring, but can have dramatic impacts (Knabel et al. 2014; Zubrod et al. 2019). One of the most common agricultural applications of fungicides is in vineyards (Zubrod et al. 2019).

INTERACTIONS WITH OTHER STRESSORS

Given that forestry and agriculture affect so many processes, they can all interact with each other to cause impacts greater than any one alone. Of course, climate change will exacerbate most of these effects by changing the timing and amounts of discharge, warming the air temperatures, and causing increasing droughts that will affect bank stability, and so forth. Climate change predictions include more extreme precipitation events and longer periods of low flows during dry seasons—and the impacts from forestry and agriculture can intensify these extremes.

The combined effects of forest harvest on streams have been referred to as a forestry syndrome (Richardson 2008), similar to the many simultaneous impacts of urbanization, called the *urban stream syndrome* (Chapter 10). Via stream networks, the impacts from upstream are transmitted downstream. However, in many jurisdictions, it is rare to provide much protection to smaller streams and wetlands (Kuglerová et al. 2020), which can result in cumulative impacts carried downstream (more in Chapter 13).

In earlier times (and still a practice in some places), streams were used as a way to transport logs, which generally meant straightening and clearing out obstructions, such as LW and boulders. Since streamside trees were rarely retained, it was relatively simple to use horses (or men) to help drag logs through small streams to larger streams where they could continue to be floated down to mills or splash dams. Almost a century later, the lack of complex channel structures and LW accumulations that were caused by stream cleaning (removal of LW), removal of the supply of LW by clearing of riparian areas, and big floods from splash dams is still measurable (Mellina and Hinch 2009).

The waste from livestock (manure) can run off into nearby surface waters (over land or through groundwater). Livestock, in many landscapes, have direct access to surface waters. With this freedom, the animals can degrade banks, thereby exacerbating erosion and sediment influx. Livestock can sometimes contribute nutrients directly by defecating in water, impairing water quality. Livestock also eats much of the vegetation found in riparian areas, which may reduce bank stability and shading that is provided by such vegetation. In the western United States, it is estimated that 80% of rangeland streams have been negatively impacted by cattle (Herbst et al. 2012). These impacts have been widespread across many parts of the world (Conroy et al 2016).

Habitat alterations for freshwater (and riparian) organisms come about from the physical, chemical, and biological changes caused by forestry and agricultural land use. Barriers to the dispersal movements of organisms can occur from the use of small head dams, log jams, culverts, irrigation water diversions, and such. Changes to habitats cause large shifts in freshwater communities. It is not possible to explore all of the solutions to these problems here, but in the following sections, we will explore some of them.

SOLUTIONS

Finding solutions to the impacts that were previously described is ongoing; and while there have been decades of intensive research, there is no single solution that is applicable to every situation. Solutions will depend on the particular ecosystems involved, the kinds of land use, and other geographic differences. One difficulty with solutions to protecting freshwater ecosystems from both forestry and agricultural land use is the obvious need for food production and for fiber for all sorts of uses. Thus, there are trade-offs between human economies and societal needs, and the ability to adequately protect freshwaters. Here we discuss a number of solutions, remembering that these may not completely protect freshwaters, but they do reduce the intensity of impacts.

Pesticide Exclusion Zones

One of the relatively simple measures one can take is to avoid pesticides, fertilizers, and other chemicals near water. These pesticide exclusion zones typically forbid overspraying (by plane or helicopter) or ground spraying within a specific distance of water. The distance varies by jurisdiction and land use, but keeping pesticides (insecticides, herbicides, fungicides, etc.) away from our surface waters can be very helpful. This is not so simple, as it is not easy to avoid small water bodies, including ditches, which can carry these chemicals downstream. Some of these pesticide exclusion zones are merely a few meters from the edges of water (see Louch et al. 2016). This usually also applies to nutrient applications as well. There is an enormous amount of variation in rules depending on jurisdiction, land use, chemicals involved, and other considerations.

Silt Fences and Landscape Cloths

The banks of freshwater ecosystems can become eroded, or sediments may be washed to the edge by overland flow. Silt fences, or landscape cloth intended to capture fine sediment, are often used to reduce transport of fine sediments from roads, fields, and other open sources of particles. This is often done near active forest operations, especially in road ditches. In some situations, sediment retention ponds are dug out in ditches to reduce sediment transport, which can also be seen in urban situations (Chapter 9). Such methods are also used in many agricultural and industrial activities near water to reduce the impacts of sediments.

Riparian Areas and Buffers

One of the most common ways of protecting freshwaters from forestry and agriculture is by the retention of a band of vegetation, often referred to as a riparian buffer or riparian management zone (Castelle et al. 1994; Lee et al. 2004; Richardson et al. 2012). These buffers can be native (preferably) or nonnative vegetation, with the aim of stabilizing stream banks, providing shade, ensuring a long-term supply of LW, taking up excess nutrients, supplying organic matter to food webs, and protecting biodiversity (see Figure 8.4A, B). However, the width of this buffer

Figure 8.4   (A) A retained buffer around a point of initiation of perennial flow, which can be critical habitats for some species. (B) A treed buffer around a small stream. Both photos from the state of Washington.

and the management within the buffer can vary tremendously (Blinn and Kilgore 2001). The widths may be based on local objectives for wildlife, fish, water quality, or other management targets. In some jurisdictions, much of the harvestable timber may be removed from a buffer, whereas in others there may be no removal of vegetation. These best management practices may be rules or recommendations, depending on who is responsible (Danehy et al. 2022).

Another advantage to leaving buffers is that it may reduce the overall rate of stream heating and help moderate rates of change in water temperatures due to climate change (see Chapter 15). The big challenge is always to determine how much vegetation to leave (buffer widths, amount of partial harvesting, vegetation composition) since this varies with location, latitude, forest type, and slope, as well as by what specific objectives and targets are considered important for the responsible government agency.

Often riparian buffers are left on larger streams, without concern for protecting the source streams that supply them. However, once natural vegetation cover is removed from alongside these small streams due to forestry or agriculture, sediments can erode, nutrient flows increase, organic matter supplies are reduced, and water can be warmed by exposure to direct radiation—and all of those impacts are likely to propagate downstream (Richardson 2019). In some jurisdictions these smaller streams get some degree of protection, recognizing their contribution to integrity of downstream ecosystems and the cumulative effects of altering source areas (McIntyre et al. 2018; Kampf et al. 2021).

The widths and management of riparian buffers continue to be debated. The United States' Forest Ecosystem Management Assessment Team's (FEMAT 1993) approach was to use potential tree height as a scalar for width of the buffer, therefore, taller (potential) forests would have wider buffers. As a rationale for these widths, it would be advised to see the original publication or Figure 10 in Naiman et al. (2000).

Where land had previously been cleared to the water's edge, planting as part of a restoration action can achieve similar objectives but will take time to meet those objectives (see Chapter 15). While the idea of planting riparian areas around streams and lakes provides for many of the key functions that link land and water, the choice of plants matters. In plantation forestry, trees are often planted outside of their natural range or in monocultures that do not provide for diversity effects that are important to freshwater ecosystems. For example, eucalyptus and pine species are often planted up to the banks where they are nonnative species and contribute little to the freshwater food web. Even if native species are planted, they might be to the exclusion of other native species, for instance, in Nordic countries or the UK where native, deciduous species are typically excluded from riparian areas as part of plantations. In agricultural settings, the planting of transgenic crop plants, such as Bt-corn, can contribute potentially toxic leaf litter to adjacent freshwaters. One solution is to ensure that the diversity of native tree and shrub species are maintained along the edges of freshwater systems.

Forestry

Often authorities at the state or federal level determine a set of best management practices for land-use activities in order to protect freshwaters. These include a number of actions, which we will expand on later. One of the simplest is that most jurisdictions do not allow streams to be used for dragging logs along, and moreover, most places no longer allow logs to be dragged across streams.

Road crossings, culverts, and ditches create problems as previously noted. One of the simplest solutions is to minimize the number of road crossings of streams, which requires planning, but it can also save money since carefully constructed bridges are expensive. Likewise, it is important to anticipate the largest peak flows (often 100- or 200-year return flows) to determine the appropriate size of culverts so they are not overwhelmed during storms. Culverts may be metal pipes or concrete. There are culverts designed as half-pipes (open part on the bottom) to allow a stream bed to remain unaltered, which can avoid downcutting and the barrier effects of culverts. Other culverts, such as smaller ones that run under roads, can be installed such that the rate of erosion on the downstream side is minimized by ensuring there is no large drop off of water from the culvert to the stream channel below (in contrast to Figure 8.2). Well-designed culverts that are at the same level as the downstream stream or channel can allow access for organisms to swim to upstream areas and avoid concentrated flows that can create higher rates of erosion.

Since changes to hydrology are often considered one of the driving effects of forestry impacts on streams (Moore and Wondzell 2005) (see Figure 8.5A and Box 8.1), the interrelated concepts of hydrological recovery and equivalent clearcut area (ECA, i.e., the area of forest that has been disturbed) are sometimes used. This considers the percentage area of a catchment harvested (large catchments), but also how long ago it was harvested (Moore et al. 2016). For an example, let us assume that the time to hydrological recovery is 10 years. Consider that two harvest areas, one area of 15% of the catchment was *just* harvested, and another area of 5% of the catchment was harvested *eight years prior*. The most recent harvest would obviously represent 15% of the area, but the second harvest block is nearly recovered hydrologically. To calculate that, we would consider the impact to be 5% of the harvest block times 0.2 (that is, 8 years of recovery would equal 80% recovered, with 20% *not* fully recovered, thus the *times 0.2*). Therefore, the ECA for this older harvest block would be only 1% of the second catchment area, which, added to the 15% of new harvest, would mean 16% of the catchment is still affected hydrologically by harvesting.

Often ECAs are used as guidelines to limit watershed-wide harvest rates, although this is a very crude measure that deals poorly with details of a watershed, harvest methods, and different aspects of the hydrograph such as seen in summer low flows. Studies often conclude that there is no hydrological effect when harvest areas represent less than a specific ECA of a catchment, or the effect may not be detectable against other background sources of variation (Moore and Wondzell 2005). This control on harvest areas is frequently used as (weak) guidance to ensure that there will be no large hydrological effect. Often a catchment with regenerating forest is considered to have recovered hydrologically within 10 to 20 years, but this can vary greatly. However, the time to hydrological recovery varies due to alterations in canopy interception, transpiration, snow accumulation, alteration of flowpaths from roads and compaction, and changed forest composition. In general, agencies have focused on peak flows, but growing evidence shows that land use also affects seasonal low-flows in a detrimental way (Moore et al. 2020)—a period that can be very stressful for stream fish and other organisms (Penaluna et al. 2015). Most sites that were studied over the longer term show residual alterations beyond a decade or two (see Figure 8.5B and Box 8.1).

A common practice for foresters and others developing areas in forests is to avoid steep slopes and saturated areas. Often, this is referred to as terrain stability (or instability) and forest engineers are trained to consider this to avoid causing slope failures (landslides). Wet areas, such as focused groundwater discharge zones along streams, lakes, and wetlands are likely to be very vulnerable to damage from any kind of operations, and foresters avoid these areas for that reason, and also because machinery can often get stuck and the sites are unlikely to return to productive condition in the short term. Riparian management schemes that take such variation into account are now regularly considered (Kuglerová et al. 2014).

In many forestry situations up until the mid-1970s, practices typically left no standing trees beside streams, and this removed the intermediate-term supply of LW. Planting of streambanks

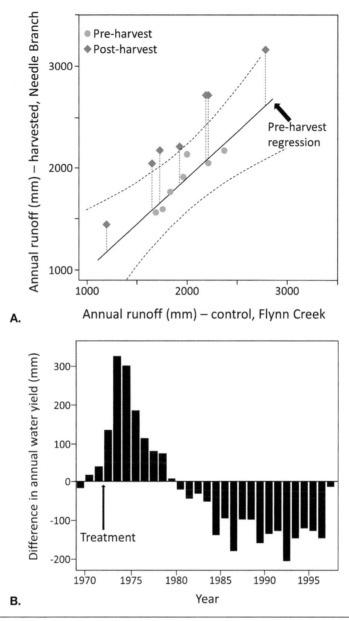

Figure 8.5 (A) Illustration of the paired catchment comparison from the Alsea Watershed Study (Stednick 2008), with a regression relating pretreatment annual flows from the two streams (green circles). The deviations from that regression relation are shown with red diamonds for years after the clear-cut forest harvesting in Needle Branch, using Flynn Creek as a reference. The seven years post-harvest show significant increases in annual flows from the harvested watershed. (B) Long-term flow patterns in a stream in a Mountain Ash Forest in Australia after forest harvesting of 78% of the watershed relative to an annual rainfall of 1,195 mm/y (redrawn from Vertessy et al. 2001).

and restoration (see Chapter 15) are common activities along damaged freshwaters, but of course, this takes anywhere from decades to centuries to provide sufficient LW. Note also the challenge noted before about the planting of nonnative species. One practice that is common in stream restoration is adding LW to streams to replace the wood that may have been in channels and has since decomposed. This replenishes the lack of natural supply that has been removed from riparian areas; however, most studies show little to marginal value to this activity (more in Chapter 15).

Crop Agriculture

Two of the simplest measures that are applied to agriculture are: (1) to ensure that the furrows (for crops) are parallel to hillslopes, rather than creating pathways that will direct fine soil downhill to streams, and (2) to leave ground cover of some sort close to streamsides to reduce erosion. Riparian buffers, as previously mentioned, for forestry are also used for agriculture, although the details are often slightly different and widths generally narrower because farmers do not want to give up a lot of productive land (Castelle et al. 1994). Most often, a small margin along larger streams is left to reduce erosion. Leaving a treed buffer in agricultural landscapes (where permitted) provides for shading, reduces water temperature increases, and mitigates some of the diffuse pollution from agricultural pesticide and fertilizer applications (Turunen et al. 2021). Furthermore, in crop agriculture, a buffer can provide benefits to the crops by supporting pollinators and perhaps predators of pest insects (Kovács-Hostyánszki et al. 2017).

In general, applying only the fertilizer amounts that are useful would be a start. However, nutrients in excess of what crops can take up are wasted as it is washed into groundwater and surface waters. The timing of application of fertilizers in order to avoid wet periods and low growth periods (autumn, very early spring) can also reduce excess nutrients. Most nutrient applications are on the soil surface and are very easily dissolved into the groundwater with precipitation. Injecting fertilizers directly into the soil would help reduce these loss rates.

In agricultural areas, ditches can continue to downcut with high flows and cause problems with erosion of adjacent soils. One solution has been to create an artificial *floodplain*, or two-stage ditches (see Figure 8.6), along agricultural ditches that allow floodwaters to spread out reducing their erosive force (Hanrahan et al. 2018). This, accompanied by robust vegetation along the edges, can reduce erosion and downcutting that lead to additional soil loss. There is also evidence of such ditch construction providing phosphorus uptake and sequestration (Kindervater and Steinman 2019).

Livestock Management

Pasture for livestock may account for around 67% of the global area used for agriculture. Fencing to keep livestock from natural freshwaters, while ensuring that the animals have another water supply, has done much to reduce erosion and other degradation from livestock. Fencing has been effective in reducing erosion and direct defecation, while enabling riparian vegetation to re-establish (Belsky et al. 1999; Riley et al. 2018) (see Figure 8.7). Studies show that when excluding cattle from streams, water quality (nutrients and turbidity) has recovered rapidly and eventually, other characteristics (benthic communities) improve over time (Scrimgeour and Kendall 2003; Conroy et al. 2016). In addition to benefits to water quality in general, eliminating direct animal waste run-off can reduce loads of pathogens such as *Cryptosporidium*, *E. coli*, and *Salmonella*.

Livestock can produce a lot of manure, and many jurisdictions allow farmers to apply animal waste to fields as fertilizer, but set standards for the amount applied. This needs to be carefully

Figure 8.6 A two-stage ditch in an agricultural setting, the Shatto Ditch, Indiana, United States. The wider level provides for water to expand laterally and reduces the erosive force somewhat, while also providing a greater surface area for nutrient mineralization and less nutrient flux downstream. Photo courtesy of Dr. Jennifer Tank, University of Notre Dame.

Figure 8.7 Riparian area of a stream fenced against cattle in New Zealand's Canterbury Plains. The wire fence is not obvious here, but note the vegetation within about 10 meters on either side of the stream, and then the grazed grass beyond (upper left of the photo).

managed as to the amount applied and the time of year it is applied in order to ensure that it does not exceed the field's capacity to use the nutrients. Otherwise, it contributes to nutrient issues in nearby freshwaters. Managing nutrient applications is also tremendously helpful. Excess nutrients from fields (or forests) that are not taken up by plants are easily washed into nearby waters. This can also be an easy *win* since excess fertilizers are wasted money as well. Along with the application of inorganic nutrients in most crop agriculture, there can also be problems with the storage and application of livestock manure causing nutrient problems. Agencies typically have rules for how waste is to be stored and limits for the timing and intensity of application (USDA 2021).

Farm ponds and other small aquatic wetlands can make up a very large cumulative area globally, and their contribution to things like global carbon budgets is now being appreciated (Downing et al. 2006). These are often designed for fish, crayfish, and duck cultivation, but can also provide habitats for a number of freshwater species. These are valuable to individual farmers and can also contribute to regional biodiversity conservation. However, these ponds also need to avoid oversupply of nutrients, and particularly contamination from waste from livestock, so management is still relevant to the health of the human users and the aquatic life therein.

PERSPECTIVES

Many solutions that have been adopted for protecting freshwaters from forestry and agricultural activities have only been tested under limited ranges of methods and in a limited number of places. There is a broad opportunity for innovation and testing of new practices, complete with proper evaluation. For instance, the actual width of riparian buffers to meet specific targets is rarely tested, and most current guidelines are applied without knowing if they actually work, or whether or not they work in all situations. Often practices are broadly adopted to landscapes where they may be inappropriate or insufficient (Kreutzweiser et al. 2012). As resource use for agriculture and forestry intensifies to meet global demands, we need to ensure it is not at the expense of water, our most precious resource.

Box 8.1   Paired Catchment and BACI Studies

Field studies can be difficult when trying to control for every detail, as we know there is enormous natural variation among streams and lakes, particularly when trying to account for size and location. Fortunately, there are a number of study designs that are available to solve that. Paired catchment and before-after-control-impact (BACI) designs have been developed to deal with the challenge of large-scale experiments, particularly where it is difficult to replicate treatments on a number of different watersheds. Often it is difficult to get a pair of similar watersheds and a long enough funding period to carry out studies at a scale that is appropriate to most land uses and to address temporal variation. Paired catchment studies have been in use at least since the 1930s. Most often, this approach has been used for watershed-scale studies of hydrology and nutrient flux from streams.

continued

The statistical analysis of this paired-catchment design is similar to what is generally called BACI (as previously mentioned), which is especially useful for a wide range of studies (Green 1979; Underwood 1992; Stewart-Oaten and Bence 2001). This paired-catchment design has two (rarely more, but can be replicated) units (lakes, wetlands, streams, etc.), both of which are initially relatively similar. The relative difference between the units is studied for a duration that is sufficiently long enough to account for natural variation, and then one of the two units is disturbed (such as urban development, mines, or forestry) and the difference, compared to the pretreatment measured relationship, is studied for several more years, with the expectation that the difference between units is greater after the disturbance than it was before (see Figure 8.5A). BACI is also similar to *intervention analysis* (Stewart-Oaten and Bence 2001). While these studies are often paired, they are even more powerful when replicated. While there are controversies about how best to use such designs, mostly from a statistical perspective, this remains one of the best available designs for estimating the effects of environmental perturbations.

There are many good examples of paired catchment studies in the literature (for instance, Burt et al. 2015; Moore et al. 2020), but one example is the H.J. Andrews Experimental Forest in Oregon where stream-flow monitoring has a long history. Analyses of long-term patterns in flow show the expected increase in water yield in the years following harvesting, but also reveal clear patterns in the longer term, consistent with the complications in the idea of hydrological recovery (Moore et al. 2016). Some of these include summer low-flow impacts and other long-lasting effects. This can also be seen in other long-term hydrological studies following forest harvesting, as in Figure 8.5B.

One example of the use of the paired catchment approach for biological measures is the study from Coweeta Hydrological Laboratory in North Carolina where one stream was used as a reference and one stream had a net built over it to exclude leaf litter inputs (Wallace et al. 1999). The biological measures in the streams during the leaf litter exclusion period were compared with the period prior to building the net and the differences between those two periods were contrasted with the much smaller differences over the same period for the reference stream. That study demonstrated large reductions in stream production and reduction in many consumer populations, relative to the reference stream, from eliminating leaf litter over the course of several years.

Box 8.2   Testing Forestry Streamside Buffers

Decades of study of the impacts of forest harvesting, forest road construction, and related practices such as getting trees to mill, eventually resulted in requirements to protect freshwaters. For instance, rules were implemented to stop dragging trees through streams, while also protecting streamsides, often with a strip of mature trees, referred to as a riparian buffer or riparian management zone. A great deal of research about how best to protect streams from forestry has come from the Pacific Northwest of the United States and has been very influential for setting rules for forest practices around streams as a global model. One of the earliest studies with a clear experimental design was at the Alsea River in southwest Oregon (Hall et al. 1987; Bisson et al. 2008). In that study, there were three stands, one each of clearcut, patch cut, and reference, and therefore unreplicated, but it was a detailed BACI design (see Box 8.1) using the pretreatment data to determine the effect size of the manipulation (clearcut or patch cut versus the reference). Several other

continued

studies with a similar *before-versus-after design* were also done around the same time in different parts of the United States and Canada, such as Hubbard Brook (New Hampshire), White River and Turkey Lakes (Ontario), the Forest Watershed and Riparian Disturbance project (Alberta), studies at the Coweeta Hydrological Laboratory (North Carolina), the Trask River (Oregon), and others.

A number of *retrospective* studies around the world have looked at stream measures based on time since forest harvesting—what is often called a *chronosequence*—which is then compared to sites that are considered the appropriate reference condition. This approach is also a type of *space-for-time* substitution, and we assume sites that were altered at different times in the past have recovered appropriately to the time since disturbance. This kind of study can incorporate many more sites into the analysis and can provide insights into much longer-term effects, sometimes centuries, which are unlikely with experimental studies, which tend to be shorter-term. For instance, Zhang et al. (2009) showed that negative impacts of forest harvesting were still detectable in streams a century after harvesting, and Mellina and Hinch (2009) showed the deficit of LW in streams from wood removal and that the lack of supply from harvested riparian areas was still clear many decades later.

Since the early days of studies showing the effects of forestry and the value of riparian buffer zones, rules around streamside management have been imposed in many jurisdictions. After the expert opinion exercise from the United States FEMAT 1993, rules about buffer widths were broadly adopted, even without explicit testing of the actual designs. The question about how wide these buffers should be is regularly posed, but rarely answered based on field experiments with a limited range of treatments. Using general relationships about the amount of shading, the distance from a stream edge that a fallen tree might become instream LW, and such has resulted in some relatively standard widths applied in many parts of the world, generally 30 meters (100 feet), although this is probably also insufficient, depending on the targets for protection.

There have been many descriptive studies of sites (mostly streams) having experienced different cutting amounts and widths of buffers (Murphy et al. 1981; Danehy et al. 2007). In recent decades, there have been several watershed-scale, replicated experimental tests of the outcomes of riparian forest management, including the Trask Creek study (Oregon), Malcolm Knapp Research Forest (British Columbia), Hinkle Creek (Oregon), western Washington nonfish streams project, the Balsjö catchment study (Sweden), the Manomet Forest (Maine), and others. There are surprisingly few studies of forestry impacts on lakes and wetlands in comparison to those on streams, but there have been some (Semlitsch et al. 2009; Carignan et al. 2011; Becu et al. 2023).

There is still a need for greater clarity of the targets these protection measures are intended to meet. Questions about the appropriate reference state for these targets remain, such as whether there should be no change following forest harvesting, and what that is relative to. Should the reference be the old-growth state, or a highly modified state as found in managed landscapes? Moreover, more innovation is needed to ensure protection meets the long-term targets of protecting our freshwaters (see Kuglerová et al. 2017).

ACTIVITIES

1. What are the laws or general best management practices for forestry or agricultural activities around water in your state or province?
2. Do practices vary for different kinds of water bodies (for instance, by size, slope, kind of fish) or other landscape features?
3. Why should those rules vary within a state or province?

9

URBAN WATERBODIES

INTRODUCTION

More than half of the world's population now lives in cities, and therefore it is relevant to examine how cities and suburbs affect freshwater ecosystems. Many urban centers are built in areas near water for reasons that are obvious—for supplies of domestic water, transportation, irrigation water, industrial supplies, power, and other uses. Even small villages are typically near water for the same reasons (see Figure 9.1A). Flowing waters were also efficient means of flushing away wastewater. Moreover, settlements often began on or near floodplains, as these zones are usually very fertile for agriculture and flat for building. Estimates are that 15% of Europeans live in floodplain areas. These areas were historically some of the most productive and biodiverse areas, another reason people settled there. However, the consequences of this encroachment on functional freshwater ecosystems and the biodiversity they support have been dire.

The effects of cities on freshwater ecosystems are similar to the ways forestry and agriculture affect streams, lakes, and wetlands in rural settings (see Chapter 8), except that these effects happen in urban settings. These effects often produce a suite of changes, that in the context of streams have been coined the *urban stream syndrome* (Paul and Meyer 2001; Booth et al. 2016), which will be discussed in more detail later in this chapter. Urban waterbodies often suffer from a full onslaught of the stressors we have examined in earlier chapters, including pollution, fast response of flow to storms, bank erosion, heating, and others (see Figure 9.1B). The large number of impervious surfaces (pavements for roads and parking lots, roofs, tennis courts, etc.) found throughout urban areas prevents water from percolating through the ground and leads to fast runoff to streams and wetlands. This runoff may exceed the capacity of the receiving waterbody, which results in flooding. This typically leads to construction of dikes (artificial levees), which can create further problems for aquatic life since the dike *walls off* the main water stream, leaving aquatic life at the mercy of faster water flows with no places, such as slower-flow floodplains, in which to escape (Knox et al. 2022) (see Chapter 7, Figure 7.2B). Moreover, higher flows can cause more severe erosion. Dikes can also cause streams to back up to upstream areas or transmit the flows to downstream reaches, either of which might be less protected and cause flooding of urban areas. The excessive amount of impervious surfaces and the overwhelming of flood passage capacity of the affected stream have created more and more flooding problems in urban settings.

There are many changes to the physical, chemical, and biological features of freshwater systems in urban and urbanizing areas. In general, urban areas create changes to microclimate,

Figure 9.1   Urban streams. (A) Village in southwest China built around a small stream. (B) Santiago, Chile, with the Mapocho River running through its center.

one aspect of which is the urban heat island effect, which results in warmer overall conditions, and higher extremes of temperature. Alterations to hydrology come from loss of groundwater inflow due to pavement (roads, parking lots, etc.), buildings, and compressed soils, all of which leads to greater and faster flows of precipitation directly into streams and lakes, often through storm drains, causing large floods. Lack of groundwater recharge and storage also means streams can experience very low flows during periods of low precipitation. Pollution in the form of nutrients, hydrocarbons, heavy metals, organic compounds, and even artificial light at night is pervasive. Water courses are usually altered or buried (in storm drain networks) or filled in. Wetlands were often filled in to allow for building and to reduce mosquito populations. All these alterations impair or eliminate freshwaters from urban areas.

In public and private parks, there are often *created* freshwater habitats, such as ornamental ponds, but these are often far from natural. Decorative water features, such as fountains, pools, or some other artificial waterbodies are for show and people's enjoyment. However, ornamental water features often use chlorinated municipal water (toxic to most life) and are kept clean of algae and organic materials that would be needed to support food webs (see Figure 9.2). It is not likely that the public would accept such features if they were not clean, so we need to look elsewhere for opportunities to have freshwater ecosystems in urban settings.

Freshwater environments in urban settings provide many benefits. Many ecosystem services, such as flood mitigation, water purification, and recreational benefits, come from urban freshwaters. Frequently parks are situated where there are streams or lakes in urban areas, and these provide tremendous amenity value. Often waterfront homes command higher prices because people enjoy being near water. Walkways and commercial spaces near streams and lakes provide enjoyable areas for public use and many municipalities promote such developments. However, the human imprint on these systems due to loss of riparian vegetation, erosion, heating, pollution, and other impacts, make it a major task to reduce such alterations to maintain functional freshwater ecosystems.

Figure 9.2   Fountain in downtown Lyon, France.

There is significant overlap in the global distribution of water insecurity for humans and the magnitude of endangerment of freshwater biodiversity (Vörösmarty et al. 2010). In Chapter 6, we learned about the large human water footprint and how the magnitude of that impact is proportional to the number of people in an area. Thus, it should not be surprising that urban areas have some of the biggest impacts on freshwater ecosystems and the biodiversity they sustain.

IMPACTS

One of the largest impacts to freshwater ecosystems in urban regions is their disappearance. Streams are frequently incorporated into storm drain networks and disappear from the surface (see Figure 9.3). For instance, in Baltimore, Maryland, estimates are that more than 70% of stream length has been buried in storm drains, and in their downtown area this number is greater than 98% of stream length (Elmore and Kaushal 2008). In many places, these storm-drain systems can spill over into combined sewage and storm flow during high rates of precipitation or snowmelt (see Chapter 5), resulting in release of waste to downstream surface

Figure 9.3 Drain from a small, intermittent stream channel leading to the buried storm drain network, typical of many urban areas. This is the fate of many small streams in urban settings to make way for development of the land surface for built infrastructure.

waters. Ponds, wetlands, and lakes in urban areas are often drained and filled in with soils to provide more land for developers, or even to reduce disease risk in regions with water-borne diseases or disease-vectoring freshwater organisms (including adult aquatic insects, such as mosquitoes). One estimate suggests that approximately 70% of the numbers of wetlands have been lost around the world (Hassall 2014). The water that remains suffers many impacts which are reviewed here.

Urban Stream Syndrome

The term *urban stream syndrome* was coined by Meyer et al. (2005) to describe the suite of changes common to most streams in urban areas. The list of symptoms within this syndrome include hydrology, chemistry (including contaminants), temperature, sediment flux, and others (see Table 9.1) (see reviews in Paul and Meyer 2001; Booth et al. 2016). It is rare to have one component of urban streams change without all the other components changing to some extent, hence the term *syndrome*.

Hydrology in urban and urbanizing areas can be seriously altered by large increases in impervious surfaces, such as pavements, compressed soils, and buildings, hastening runoff and reducing groundwater inputs, as well as burial of streams in storm drain networks. Urban streams often are developed in a way that makes downstream problems even more severe, a consequence of accelerating runoff rates in upstream areas due to roads, compaction, and storm drains instead of streams (see Table 9.2). These move the problem of higher peak flows to downstream areas where flooding and erosion cause property damage, and lead to municipalities using heavy armoring of stream banks for erosion control. Residents downstream incur the impacts caused by development upstream.

A consequence of increased peak flows is the inability of streambanks to withstand such erosive forces, which results in the banks often yielding large amounts of sediment. Sediments also come from exposed soils washed into storm drain systems. The higher amounts of sediment result in increased turbidity, rearrangement of channel geomorphology, and even more erosion downstream. This, too, results in heavy armoring of stream banks, and typically means that as flows increase, there is nowhere that organisms can retreat to with lower velocity areas, such as floodplains or off-channel wetlands.

Table 9.1   The urban stream syndrome (after Meyer et al. 2005; Paul and Meyer 2001; Booth et al. 2016).

• Increases in the frequency and intensity of physical stress
• Increased peak flows
• Decreased channel stability
• Altered sediment regime
• Increases both the diversity and the degree of chemical stress, with increases in the concentrations of nutrients and salts, heavy metals, and pesticides (especially in pulses)
• Modified thermal regimes
• Increased baseflow temperatures or pulses of heated stormwater runoff
• Labile C and nutrient subsidies altered

Table 9.2 Summary of general impacts on urban water bodies.

Impact	Streams	Wetlands and Lakes
Loss of the freshwater ecosystem	Buried in storm drain network; redirected; dammed	Drained or filled in; regulated by dams
Nutrient and contaminant inputs	Runoff from urban and suburban landscapes carries nutrients and many contaminants (pesticides, salts, heavy metals, etc.)	
Altered hydrology	Higher peak flows downstream	Rise in water levels and potential flooding of low-lying properties
Altered hydrology	Lack of regular groundwater inputs to support flow	Lack of regular catchment water inputs to support water renewal and dilution of nutrients
Thermal regimes	Often lacking shading; warm water from storm surface runoff from pavement and storm drains	Water inputs may be warm from surface runoff
Erosion	Higher peak flows contribute to higher erosion; lack of riparian vegetation reduces bank stability; increased sediment movements	Erosion due to human and urban wildlife use; recreational use; fine sediments causing turbidity and other water quality problems
Alteration of basal resource inputs	Reduced riparian vegetation lowers detrital inputs; shifts in the quality of detrital materials; more light and nutrients can lead to algae problems	Shifts in the quality of detrital materials; more nutrients and less water inputs (dilution) can lead to algae problems; reduced detrital inputs from catchment

Water Quality Changes

Temperature regimes in urban freshwaters are seriously altered, mostly warmed. In addition to increased exposure to long- and short-wave radiation due to diminished amounts of shading, water running off pavements can be very hot. In fact, storms in many places can dump precipitation onto warm pavements that drain into storm runoff systems and deliver hot water to downstream locations (Somers et al. 2013). Removal of riparian vegetation along waterbodies is common in order to harden banks, create sight lines for personal security, and to reduce large wood and other *debris* that might *foul* the water.

In temperate regions, salts, primarily sodium chloride, calcium chloride, and magnesium chloride are applied to roads to melt snow and ice. This practice can result in high concentrations of these salts reaching urban waters, and the amount used each year continues to increase (see Figure 9.4) (Kelly et al. 2019; Le et al. 2021; Hintz et al. 2022). Salt, particularly the chloride ion, at moderate to high concentrations in freshwaters can be hazardous and even fatal for many species, and there can be large spikes in concentrations associated with periods of rapid melt. Hintz et al. (2022) show that salt concentrations in lakes and streams can often exceed government guidelines for chronic exposures, and even be beyond acute exposure levels. Moreover, salts getting into groundwaters can cause persistent increases in concentrations in urban freshwaters well beyond cold periods. There are natural sources of some salts, but those are minor and most are from deicing practices.

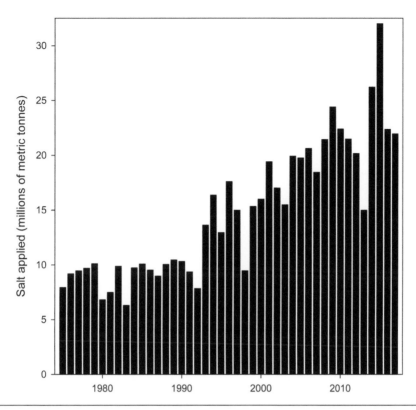

Figure 9.4   Salt application amounts for deicing uses in the United States, showing increasing use through time. The salt, largely sodium chloride, is used on roads, parking lots, sidewalks, driveways, and service roads. Redrawn after Kelly et al. (2019).

As discussed earlier (see Chapter 4), there are many kinds of contaminants that get into surface waters, and particularly in urban areas. These include nutrients and pesticides (for insects, fungi, moss, etc.) from gardens, golf courses, and private properties. Also discussed earlier, there are many kinds of personal care and pharmaceutical products in our sewage. Normally, these chemicals are treated in wastewater treatment plants, but some make their way into surface waters directly through septic systems and leaky sewage pipes. Of course, the higher the human density, the greater the rates of these contaminants entering waterbodies. Since these chemicals are often water soluble and frequently exceed guidelines, these chemicals end up washed into urban waterways (Carpenter et al. 2016). There are also high concentrations of heavy metals from wear of engine parts, and even microscopic fragments of artificial rubber from tires (Johannessen et al. 2021).

Overflows from Combined Sewer Systems

Combined sewer systems are sewers that are intended to collect rainwater runoff, domestic sewage, and industrial wastewater—all in the same pipe. Usually, these combined sewer systems carry their flows to a wastewater treatment plant, where it is treated and then discharged into some surface water. However, during periods of intense rainfall or snowmelt, the wastewater volume in a combined sewer system can overwhelm the capacity of the sewer system or treatment plant. For this reason, combined sewer systems are designed to overflow occasionally and discharge wastewater that is in excess of the system's capacity directly to nearby streams, rivers, or other waterbodies (see Chapter 5). These overflows, called combined sewer overflows (CSOs), carry stormwater, untreated human and industrial waste, toxic materials, and debris. As mentioned in Chapter 5, they constitute a major point source of pollution, but the outflows from these systems are not continuous, rather they are highly pulsed and spread out through time. However, these CSOs remain a major water pollution issue, and over 700 cities in the United States have combined sewer systems. This means that, like sanitary sewer overflows and stormwater discharges, they are point source flows from a municipality's wastewater infrastructure. CSOs provide a bit of a *safety valve* for urban water treatments and conveyance infrastructure, but are hazardous to freshwater ecosystems. Many cities around the world are working toward updating municipal infrastructure to address this problem.

Channelization

Another pervasive alteration to streams in urban areas includes channelization, diking, and bank hardening to limit lateral movements of streams and protect adjacent infrastructure (see Figures 9.1 and 9.5; also see Chapter 7, Figure 7.2). These actions affect the geomorphology, the connections with riparian areas, the flow environments, and usually include a loss of shallow, marginal areas. Straightening and deepening of channels to promote evacuation of floodwaters also increases the erosive power against the banks and reduces slower water areas at the edges that otherwise provide refuge to organisms from powerful flooding. Channelization also typically transmits the power of high flows to further downstream without much local dissipation of the energy of flowing water.

Lakes and Wetlands

Lakes and wetlands in urban areas are impacted by most of the same stressors as streams. Those lentic waters that remain often experience problems from alterations to hydrology, nutrient inputs, contaminant additions, recreational activities, and other problems. In many urban areas, stormwater ponds are increasing in number, and these are intended to ameliorate stormwater flows (short-term storage) and capture nutrients and sediments (Taguchi et al. 2020). Proximity to roads can lead to higher rates of road kill or greater access by predators of some species, such as turtles and amphibians. The urban heat island effect may even lead to higher proportions of female turtles, as sex determination in turtles includes the temperature that eggs develop under. One study of biodiversity of 41 urban ponds in England found that although density remained similar to ponds in floodplains and agricultural areas, the taxonomic diversity was only about half, and the urban ponds supported no species that were considered at risk (threatened or endangered) (Hill et al. 2016).

Ecological Traps

Even if a freshwater ecosystem is in good shape as far as habitat and water quality is concerned, isolation in the landscape can be an issue. For example, amphibians associated with an urban wetland may have nowhere to spend their terrestrial life stage, and may not even colonize the wetland in the first place. Thus, urban freshwaters can become ecological traps, where populations may be isolated, subject to contaminants that affect their fitness, and ultimately do not support replacement of their populations (Bowne et al. 2018; Sievers et al. 2018; Taylor and Paszkowski 2018). For instance, degraded water quality in urban wetlands, and especially constructed wetlands or stormwater treatment wetlands, can end up with frogs laying eggs, but with their young having lower survival rates and lower growth rates than in more natural wetlands (Sievers et al. 2018). While adults may reach urban freshwaters, their offspring may not be able to survive there, or at least not at a rate that would allow the population to survive without further immigrants.

The idea of wetlands and lakes as stepping stones that might enable populations to move through the urban matrix could work for smaller organisms, such as amphibians, reptiles, and small mammals. Results from the New York City region suggest if ponds are relatively near (<500 m apart) and relatively dense (at least 1 per km^2), populations of freshwater organisms may be able to persist in such a landscape (Gibb 2001). Another aspect of urbanization is the high density of roads, and traffic that is responsible for very high rates of road kill of amphibians, reptiles, dragonflies, and other aquatic organisms that have complex life cycles, or can disperse overland at times as seen in some crayfish species.

Artificial light at night is pervasive around urban settings and other areas of human infrastructure (Perkin et al. 2011). Light can alter the behaviors of fish and other aquatic organisms. Juvenile salmon are attracted to lights, increasing chances of predation (Tabor et al. 2017). For instance, light around the bridge over Lake Washington in Seattle makes it possible for seals and sea lions to see migrating salmon at night, increasing their overall predation of salmon stocks. Artificial light also attracts insects and their predators (bats, spiders, and others), and adult aquatic insects may be particularly vulnerable to these ecological traps (Perkin et al. 2014).

INTERACTIONS WITH OTHER STRESSORS

Urbanization creates so many stressors to freshwater ecosystems that sorting out specific interactions can be difficult. For instance, climate change interacts with most aspects of freshwater systems, and in urban settings, the urban heat island and elevated concentrations of most chemicals are obvious issues. Wastewater treatment and sewage outfall are major sources of contaminants to surface waters and groundwaters. In addition, invasive species may be introduced through transportation systems, aquarium trade, aquaculture, restaurants offering freshly caught fish and crustaceans, and other means (see Chapters 7 and 11). All of these are especially concentrated around urban areas. As a result of all these interactions, management solutions appear to be hard to reach. However, attempts with statistical and experimental methods have tried to disentangle some of the effects (Hassett et al. 2018), although not often successfully.

SOLUTIONS

We are not going to be able to remove all the stresses that human settlements impose on freshwaters, but there are ways to reduce some of the impacts. Public awareness of the impacts on freshwaters is a good start, and there are many nongovernmental organizations that contribute to the improvement of surface waters. Protecting the original freshwater systems in urban settings is best done during planning stages of urban development because restoring freshwaters after the fact is expensive and provides marginal returns in most situations (see Chapter 15). This requires controls by municipal governments to restrict land developers, which is possible, but not often done. Many wealthier municipalities have adopted greenway strategies to connect green spaces, and many of these take advantage of stream networks and lakes. Such planning can have positive consequences for the conservation of urban freshwaters and public benefits, especially if areas of native vegetation and shorelines are maintained as part of the network. Restoration of areas near water can help improve water quality and quantities, as well as offer opportunities for the habitat needs of freshwater organisms.

Functional Waterbodies

As mentioned before, public fountains, reflecting pools, swimming pools, etc., are not intended as ecosystems, and they are cleaned in a way to prevent biodiversity, although there are a few tolerant species, such as drain flies (Psychodidae) that can persist. Cities can be contributors to biodiversity conservation, and the Curitiba Declaration on Cities and Biodiversity (2007) states that urban areas should aspire to being part of the solution to the biodiversity crisis. There are other similar agreements by many cities to promote biodiversity conservation within urban areas, such as the C40 Urban Nature Declaration.

It is popular to create water features in public urban spaces and even on private properties such as small ponds (often with some sort of liner to hold water). One estimate from Sheffield, England, a city of just over a half million people, was that there were about 25,200 ponds in the city, and backyard ponds averaged about 2.5 m^2 in area (Gaston et al. 2005). These backyard ponds sustained some freshwater biodiversity, although mostly common species. One challenge is to avoid contaminants since it is tempting to manage such waterbodies using chemical treatment to control nutrient levels and the resulting growth of algae and bacteria with algicides or chlorine. Companies devoted to water features and even natural pools for swimming use biofiltration and uptake of nutrients to keep water clean. Such ponds still often require removal of nutrients by periodic harvest of shoreline vegetation (which then continues to grow). Treatment for nutrients is especially important in closed-loop systems where water is recycled and nutrients concentrate as water evaporates. Many parks, golf courses, cemeteries, and other spaces can also provide for functional ponds, small lakes and streams, and support of freshwater biodiversity, within urban boundaries. Creating or conserving functional waterbodies in urban areas, as opposed to fountains and reflecting pools, requires specific management.

Making Urban Streams and Their Floodplains More Natural

Moving dikes back and avoiding sharp armoring (or concrete walls) would improve urban streams (as discussed in Chapter 7). More shading alongside streams and natural vegetation providing bank stability would help urban stream ecosystems. More room for streams with less hard constructions along the banks could be planned during development, but rarely are. However, there are often few opportunities for these kinds of adjustments in dense urban environments.

An obvious solution to many aspects of urban disruption of surface waters would be to avoid building in floodplains. However, in most cases this is too late, and most of our settlements have occupied areas near water for reasons that are evident. It would improve the condition of freshwater to build roadways and railways away from edges of waterways and wetlands. Floodplains are common places for building roads and rail lines since these areas are flat; however, they are also prone to flooding and cause problems for freshwaters. Other uses of floodplains in urban areas can include public walking and cycling paths. There are often high use rates of such paths which can also lead to exposed soils subject to erosion, bank instability, and damage to roots of trees and shrubs. Paths should be built a specific distance away from the actual water's edge to reduce erosion and loss of vegetation. Well-maintained paths with gravel surfaces (allow water infiltration) or pavements can help reduce damage to areas near stream banks. In high-use areas, the construction of boardwalks can reduce damage to shorelines or sources of exposed sediments, but these are usually expensive to maintain and may incur liability issues.

Rather than building serious infrastructure in floodplains, many cities have used these areas along rivers as recreational zones, including parks, golf courses, and playing fields. These areas are still often protected against flood-caused erosion of their banks, which diminishes habitat quality (see Figures 9.1 and 9.5; also see Chapter 7, Figure 7.2). Most streams could benefit from more natural bank protection, rather than hardscaping with blocks, large rocks, or rock baskets (rip-rap). This would largely be by trees and shrubs that have root strength that can bind shoreline materials and prevent erosion (see Chapter 15 on restoration). However, as discussed earlier, because of the need for unobstructed views for personal security and constructions close to water, this is unlikely in most settings.

The idea of daylighting streams has emerged to return streams that are buried in storm drainage networks to the surface as freshwater ecosystems once more. One famous example comes from Seoul, Korea, where the Cheong Gye Cheon project removed a major highway through the downtown that was built over a former river. The 10.9 kilometer project was very costly, but in the end, it returned the river to a marginally functional ecosystem and became a major amenity as a park in the city's downtown. Creation of this park also increased property

Figure 9.5   Bank hardening along an urban stream to avoid erosion and migration of channel edges. Hiirosenoja (stream name), Oulu, Finland.

values nearby since water features are often appreciated as a space to enjoy (Kang and Cervero 2009). Likewise, smaller streams can be daylighted to enable restoration of streams, contribute to park areas, and reduce downstream flooding by re-establishing a more natural flow regime. Studies demonstrate that daylighting of streams can lead very quickly to the recovery of typical biological communities (Wild et al. 2011; Neale and Moffett 2016) (see also Chapter 15).

Increase Urban Green Spaces and Vegetation

Urban green spaces are most often designed for people, such that there are large shade trees and clear sight lines for a sense of personal security. Often the groundcover is grass or gardens, which offers little to shoreline stability or contributions to shading of water or organic inputs. There are many solutions proposed for urban streams (Cooke et al. 2022). More riparian areas and more shading would be beneficial to urban and suburban freshwaters (Alberts et al. 2018). Parks in floodplains offer some opportunities for freshwaters, but are often still highly modified to avoid erosion of land and other infrastructure (see Chapter 3, Figure 3.6). Opportunities in cities to restore native riparian vegetation around freshwaters have been seized in many cases, reducing typical park vegetation, like grass, and planting of small stands of trees. Some cities have done a good job of protecting their original stream and pond networks in a relatively unmodified form—that is, retention of some or all of their native vegetation along ravines and streamsides. One example is Toronto, Canada, where large parts of their stream networks are in steep ravines as a result of its glacial history, and as such, have escaped development and are now reserved to a degree in parks. Nevertheless, these streams are still altered in terms of their hydrology and pollution. On the edge of Toronto, the Rouge River and its valley retains enough natural features that it became Canada's first urban national park.

Overall, improvement in water quality of our urban freshwaters would come from less use of nutrients, less use of pesticides of all sorts, and more of a separation of human development from aquatic ecosystems. Reduction of chemical use results in a lower burden of chemicals that can be flushed into surface waters during precipitation events.

As mentioned before, urban areas create heat islands, which over and above the removal of tree canopy over water and water draining from warm pavements, can augment the thermal loading of freshwaters. Warmer water, whether short- or long-term exposure, may exceed the thermal tolerances of some freshwater species. Efforts to create greener cities by adding more green spaces, particularly with tree cover, can mitigate this additional source of heating from warm air.

Many cities are exploring ways to increase the amounts of pervious surfaces to allow penetration of precipitation into groundwater, and one is by creating more green space. Some ideas include permeable materials in parking lots (permeable asphalt, open paving stones), storm retention ponds, and bioswales (see the upcoming section titled *Incorporate Constructed Wetlands*; see also Chapter 3). These ideas can help to reduce the extremely rapid runoff that often leads to serious flooding, and may also contribute to better water quality in our surface waters.

A large interest in stormwater retention ponds or detention ponds (or bioswales) has led to more storage of stormwater, more infiltration to groundwater, and slightly lower flood peaks. These can dampen the rate at which stormwater reaches streams. These retention ponds can also be used to capture fine sediments, nutrients, and other contaminants in runoff. Retention ponds can support freshwater communities, as long as they hold water long enough to sustain freshwater organisms. However, the space for such ponds, as well as their potential as drowning hazards to children, has limited the extent of the building of these in some regions.

Protecting Water Quality

Storm runoff usually enters pipes, which may also carry sewage from the drainage network, called combined sewer systems. In some cities, these combined sewage networks can mean that during storm runoff, untreated sewage can overflow and escape to surface waters before reaching a wastewater treatment facility. Or if the combined flows do not reach the surface, the large amounts of stormwater often exceed wastewater treatment plant capacity (or damage infrastructure), and untreated wastewater is allowed to escape (Berland et al. 2017). One solution is to have stormwater runoff systems entirely separate from the sewage network. However, even as a separate pipe network, storm runoff is still likely to carry heavy metals, hydrocarbons, and other contaminants from roadways to surface waters, and create fast storm runoff and peak flows.

Salts applied for deicing can wash into nearby waters through storm drains, from snow piled on lawns, and even mounds created as municipal services pile up snow away from roads. Some solutions include applying salt as a brine solution to streets before snow accumulates, which can reduce the amount of salt used by 75%, and by better estimating the amount of salt actually needed (Hintz et al. 2022). Covering salt piles to reduce being washed away by precipitation can also assist in reducing amounts of salts reaching waterways, as can situating piles away from surface waters (Hintz et al. 2022).

Maintain Urban Wetlands and Lakes

Many urban centers occur along lakes, and so shorelines provide an important amenity to urban dwellers. One goal is to avoid nutrient and contaminant accumulations, and this takes planning. Several decades ago, it was demonstrated that there were high rates of eutrophication from urban situations (see Chapter 4) that even resulted in models predicting limits to development, particularly around recreational areas. Madison, Wisconsin, provides a good example since the city is built on the shores of Lake Mendota, a heavily studied lake (see Box 9.1) that is facing changes that are typical of agricultural and urban landscapes.

Urban ponds and wetlands can provide for conservation of native species (Hassall 2014). In England, a study showed that urban ponds supported rates of productivity and almost as much diversity as ponds in the countryside (Hill et al. 2016). In the New York City region, wetlands around the city provide for conservation of some amphibians, small mammals, reptiles, and invertebrates, especially when there are multiple wetlands in close proximity, and lower human densities as in the suburbs (Gibb 2001). In Portland, Oregon, a block of downtown was used to recreate a wetland that had existed there a century earlier, and Tanner Springs Park has become a major amenity for people, as well as supporting many plants and invertebrates typical of wetlands in the area (see Figure 9.6). However, as is common in many of these systems that have no real inlet and outlet, input costs to maintain water quality and remove accumulated organic material are substantial. In the case of Tanner Springs, the accumulated organic materials in the bottom need to be removed manually every few years to avoid it filling in. Their efforts seem to be paying off since there are reports of many aquatic insects (at least seven dragonfly and damselfly species observed there) and other species living in Tanner Springs Park. Such ponds and wetlands are a typical feature of many gardens and large parks and can contribute to urban biodiversity (see Figure 9.7).

Figure 9.6   Tanner Springs Park, Portland, Oregon, showing the artificial wetland created in the middle of the city. Photo courtesy of Dr. Elizabeth Perkin.

Figure 9.7   Park in Japan showing a pond as part of the space used for recreation and reflection.

Urban areas can support functional freshwater ecosystems, and some cities maintain near-natural freshwater ecosystems. Other ecosystems may be constructed; for instance, stormwater retention ponds are increasingly common and can support a wide component of biodiversity and ecosystem processes. Ponds and lakes in parks, agricultural areas, and even backyards, can work to support functional ecosystems (Davies et al. 2008; Verdonschot et al. 2011). Urban areas may also end up with *accidental waterbodies*, such as ditches and other collections of water that persist long enough to establish freshwater ecosystems (Palta et al. 2017). These features can occur in abandoned areas, lower elevation areas around parking lots and industrial areas, and other locations where they were not established intentionally. These waterbodies can still host some elements of freshwater biodiversity within urban centers.

Natural and artificial wetlands, lakes, and streams can be damaged by recreational use. This can include boat wakes from motorboats damaging shorelines and impacting biodiversity, as was mentioned in Chapter 3. In many high-use areas, such as paths along waterbodies, where protection of shorelines is needed, the construction of boardwalks and other infrastructure can restrict access and erosion of shorelines, while allowing people enjoyment of the feature. A good example is the *Learning Forest* of the Botanic Gardens in Singapore, where boardwalks allow large numbers of people to visit, but minimize damage to waterbodies.

Incorporate Constructed Wetlands and Other Green Infrastructure

Constructed wetlands in urban areas, such as bioswales and storm retention ponds, can help to reduce rates of storm runoff, and also can trap sediments and contaminants, keeping them from entering surface waters (Chaffin et al. 2016; Berland et al. 2017). These blue and green infrastructure solutions, or nature-based solutions (as opposed to grey, like concrete pipes), can help protect surface waters, and may even provide small wetland ecosystems. Bioswales are constructed to capture surface runoff in gently graded and pervious drainages that allow infiltration of water to the ground (see Chapter 3, Figure 3.8). At the same time, bioswales may aid in filtering sediment and surface contaminants before they enter surface waters. These constructed features are usually established with native plants, although they are often managed to restrict the height of the vegetation, and most plants in bioswales need to be tolerant of inundation, potentially for long periods. However, planting of trees throughout urban areas to augment the capacity of water storage, interception, and loss through evapotranspiration of plants can be consistent with the objectives of stormwater management (Berland et al. 2017). Bioswales may be directly connected downstream to surface waters or they may flow into a storm drain. When designing such a water storage and infiltration infrastructure, it is necessary to be cautious about recharging the water tables to a level that could cause neighborhood flooding. Other green infrastructure for urban water storage includes rain gardens, curb cutouts, tree pits (soils exposed around trees in urban settings), and even rain barrels.

Green infrastructure, such as constructed wetlands, can be designed on a larger scale to capture contaminants and sediments from road runoff, but with appropriate design, can also contribute to biodiversity and water purification. Constructed wetlands can trap contaminants in their sediments or plants, and microbial degradation of some pollutants is possible. Generally, such wetlands need biomass removed periodically to dispose of contaminants such as heavy metals, and this biomass could be used as fuel (Hassall 2014). Also, low-impact development and other best management practices can be used to deal with stormwater runoff problems. These ponds can also be useful in removing excess nutrients from surface waters. There are many designs; however, beyond denitrification, other nutrients and contaminants sequestered in the short term can only be depleted if there is periodical removal of vegetation and sediment (Vymazal 2007).

One concern many people have from the creation of small (such as in a private backyard) to large wetlands is the prospect of increasing the numbers of mosquitos and the risk of transmission of mosquito-borne diseases. In more biodiverse wetlands, native predators can keep mosquito larvae in check—though introducing non-native species like mosquitofish should be discouraged. However, mosquitos breed in many small containers of water, including old tires, open paint cans, poorly maintained gutters, etc., and keeping these free of water is probably more helpful in reducing mosquito numbers. Use of a more natural mosquito insecticide formulated of the toxin in the bacterium *Bacillus thuringiensis* var. *israelensis* can provide control and avoid other synthetic pesticides. When constructing such water features, it is useful to ensure that the edges are not vertical, so that animals can leave the wetland when needed (such as metamorphic amphibians).

Increase Connectivity and Reduce Ecological Traps

A feature of urban waterways is that they are often isolated by infrastructure, restricting the dispersal between sites. In some places, there is sufficient protection of stream and wetland networks so that organisms can disperse between areas, providing for gene flow and *rescue effect* (from metapopulation dynamics). For instance, many cities have built connectivity between wetlands and streams into their planning process to enable movement of species (Connery 2009). However, the hard barriers to movement include the ability—or inability—of organisms to pass busy roadways. Constructions such as amphibian tunnels and other underpass structures can aid in reducing roadway mortality and enhancing dispersal. These passage features are often added secondarily and may include fences or barriers along the water side of a road to direct amphibians and reptiles toward the underpass. There is probably little that can be done to reduce the mortality of flying aquatic insects caused by cars, but avoiding building roads immediately alongside freshwater ecosystems could help. However, there are often few options for that.

Artificial light at night is an issue for many ecosystems, and freshwaters are also impacted. Standard lights are often magnets for insects, fish, and other organisms (Perkin et al. 2011) and thus, are a kind of ecological trap. Artificial lights around the world are often high-pressure sodium lights (or sometimes mercury vapor), but these are being replaced with light emitting diodes (LEDs) in wealthier nations. LED lights are more expensive, but also are more energy efficient and last much longer, therefore, they are increasingly being seen in urban centers around the world. LEDs can have their wavelengths tuned so that they are less attractive to animals, thus reducing one more hazard to ecosystems. However, the lower power consumption of LED lights can also result in installation of more lights and with higher intensity than previously, which can result in broader impacts to some organisms.

Public Participation

People and their pets also use surface waters, which can add to problems of turbidity and others. People may swim, wade, or boat in urban waters, and products such as sunscreen lotion can wash off and cause issues for some aquatic species. Dogs running through streams and ponds may also disturb species living in water, especially if they are chasing species near the surface, such as turtles or frogs. Awareness through signs and public outreach (or fencing) can help.

Finally, municipalities of all sizes can encourage conservation of water by citizens, including less water use, more appropriate vegetation, proper disposal of contaminated liquids (rather than down drains), and rain capture in cisterns for garden watering. There are efforts in most municipalities to conserve water through awareness campaigns, domestic water pricing, restrictions on when lawns can be watered and when cars can be washed, and other such limitations. A nice example is altering the common, urban expectations of gardens with grass (usually a non-native grass species) with a border of a few flowers, to a garden made up of native species that require less water and chemical inputs to maintain. For instance, in Tucson, Arizona, in the middle of the Sonoran Desert, extremely high prices for domestic water and the availability of appropriate desert plants enabled the city to develop xeriscaping—that is, growing gardens full of native plants that require little water (see Box 6.2). This also required convincing people to change their habits from green grass lawns to something else, which can work when there are good examples.

PERSPECTIVES

Since these urban waterbodies are the freshwater ecosystems that are most easily encountered by the majority of people, there is value in maintaining them in good condition. Urban waterbodies can be valuable for native species conservation because urban developments are often situated near biodiverse and productive areas. Urban waters can also provide a valuable amenity to people who enjoy being near water or hearing the sound of running water. It is worth noting that urban green spaces are not evenly distributed across neighborhoods, and there can be socioeconomic differences between districts that affect access to freshwater ecosystems. Direct experience with urban waters and their biodiversity can be important when it comes to educating people about the value of their freshwaters and teaching them to care about protecting water.

Box 9.1   Urban Lakes

Many urban areas front onto water, whether that is the sea, rivers, or lakes. The fact that urban areas that were developed near water for reasons of transportation, drinking water supply, and arable floodplains, should not be surprising. Despite enormous degradation of freshwaters and their shorelines in urban areas, many municipalities have embraced their waterfronts as a social amenity; they value their environments and work to clean them up and provide public access. One example of urban lakes is Lake Mendota and Lake Monona in Madison, Wisconsin (see Figure 9.8). These lakes provide an excellent case study, the city of Madison sits between these two lakes, and the Madison campus of the University of Wisconsin is right on Lake Mendota—the lakes have been extensively studied as a consequence. The city of a quarter of a million people is built around the lakes, and the lakes provide an attractive feature for its residents. Both lakes are used extensively for water-based recreational activities, and parks along their shores are popular. However, since the city and surrounding lands have had such an impact on the lakes, they and other lakes around Madison have been studied for decades by limnologists at the University of Wisconsin.

Lake Mendota has the University of Wisconsin's Center for Limnology on its shore. As a global leader in lake research for over a century, there is a tremendous amount of literature on lakes from the institution, and a lot of it focused on the Madison lakes. The North Temperate Lakes

continued

Figure 9.8   Madison, Wisconsin, seen from the air. Madison is built along Lake Mendota and Lake Monona. These lakes are important amenities for the city and important for recreation. These lakes are also vulnerable to other land-use impacts, many of which are well documented by the University of Wisconsin's Center for Limnology. Photo courtesy of Dr. Eric Booth (University of Wisconsin).

Long-Term Ecological Research (https://lter.limnology.wisc.edu/) program has operated as an observatory in Madison's lakes since the 1980s (also a part of the Global Lake Ecological Observation Network).

Residents have an acute awareness of the state of the lakes, and a Clean Lakes Alliance has brought attention to the conditions there. Because of a long period of high nutrient fluxes from the surrounding landscape, including agriculture and the city, the lakes have become eutrophic. For instance, in 2018, their beaches were closed 23% of the season—78 days of which were attributed to excessive amounts of cyanobacteria, *E. coli* (fecal coliform bacteria), or both (https://www.cleanlakesalliance.org/lake-mendota/ accessed June 26, 2019). Controls on nutrient inputs to the lakes are considered to be the primary tool for reducing the water quality problems leading to blooms of microbes.

Zebra mussels (*Dreissena* spp.) and spiny water fleas (*Bythotrephes* spp.)—invasive species that are advancing across North America—are relatively new to the Madison lakes, along with other non-native species (see Chapter 11). *Bythotrephes* spp. are a potential factor in decreasing water quality because they eat other Cladocera that could otherwise consume phytoplankton and enhance water clarity. The wide range of problems and actions taken in these lakes are examples of the challenges facing urban freshwaters.

Box 9.2   Urban Stream Restoration

Many streams and wetlands in urban areas have been adopted by citizen-based groups, such as the Streamkeepers, the Wetlandkeepers, RiverLink, the Catchment Based Approach, and many others, to protect and restore such freshwaters. These groups do a lot of work to conserve, monitor, and restore urban streams and wetlands, and to inform others of the values of these ecosystems. At the campus of the University of California, Berkeley, small streams running through the campus had been severely degraded—most had been buried in pipes and other sections were used effectively as open sewers. In 1987, a group sought to restore Strawberry Creek (see Figure 9.9A). Strawberry Creek has a total watershed of about 4.7 km^2 and a large portion of it flows through campus. As an urban stream, before restoration, a large portion of that watershed had impervious surfaces and storm flow (and sewer) connections, resulting in high peak flows, contaminant inputs, and erosion of the stream. The first step was to identify where the water was coming from (in pipes) and to stop contaminant flows (nutrients, sewage) that were coming from buildings on campus. This also meant daylighting sections of the stream, finding ways to manage storm flows (removing culverts and channel restrictions), and removing concrete or rock walls that were lining the streams. These first steps improved water quality and hydrology. Subsequently, construction to control the erosion of stream banks, channel grading, and the addition of rock check dams helped to re-establish many of the features of the stream. Streambank restoration through the planting of vegetation to stabilize the banks helped stop erosion and returned shading and organic inputs to the stream (see Figure 9.9B). However, much of the riparian vegetation includes non-native species of plants. Riparian vegetation is still relatively sparse. Several native fish species were transferred into the stream beginning in 1989 and these fish are now self-sustaining.

Strawberry Creek is now a valued feature of the University of California, Berkeley campus. As a functioning stream ecosystem, it is also used for teaching purposes as a living laboratory, and many students each year learn about streams and stream biology there. Monitoring of invertebrates has continued and allowed for an evaluation of an accidental diesel oil spill into the stream—the good news is that the stream's fauna recovered fairly quickly after the spill (Peterson et al. 2017), probably because of upstream sources of colonists. The recovery of the streams aligns with the importance of thinking of whole watersheds, not just restored reaches of streams (see Chapter 15).

continued

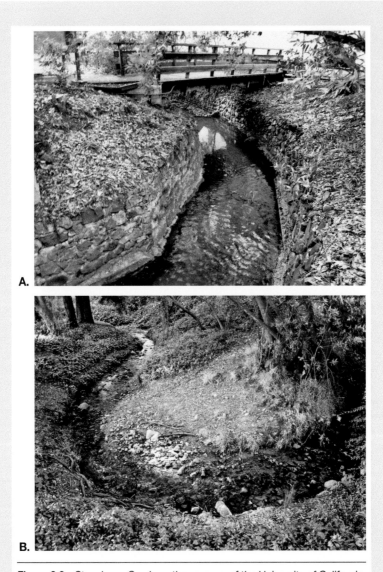

Figure 9.9   Strawberry Creek on the campus of the University of California, Berkeley. (A) This stream was once severely degraded by contamination, sediments, and lack of riparian vegetation. (B) Restoration of this small urban stream, beginning in the mid-1980s, has resulted in a functional stream ecosystem, which is now used as a living laboratory for teaching about freshwater biology. Photos courtesy of Robert Charbonneau (University of California, Berkeley).

ACTIVITIES

1. Many cities and regions have local groups devoted to the protection of freshwaters, such as the Streamkeepers and the Wetlandkeepers. These volunteer groups provide opportunities to get involved in protecting freshwaters in their communities. The people organizing these groups are also extremely well informed about freshwater issues in their respective areas and can be great resources. Are there chapters of these groups near you, and if so, what do they do?

2. The United States Environmental Protection Agency (EPA) and other organizations sometimes make information available about your local bodies of water. For instance, the U.S. EPA has a site for information on your local watersheds (https://www.epa.gov/waterdata/hows-my-waterway), and likewise, the U.S. Geological Survey has information on what science they have underway near you (https://water.usgs.gov/wsc/). Other countries have some similar sources of information. What can you find out about your local watersheds?

10

EXPLOITATION OF SPECIES

INTRODUCTION

There are many freshwater fisheries for fish, crustaceans, and mollusks around the world. These may be commercial, recreational, or subsistence fisheries. To this list we can add harvesting of beaver and other aquatic animals and plants. There are commercial fisheries for many freshwater species, as seen in the Laurentian Great Lakes and the African Great Lakes, and elsewhere (see Box 10.1). The largest freshwater fishery in the world is considered to be the Mekong River in southeast Asia (Hortle 2007). Some fisheries are interested only in certain parts of fish, such as sturgeon harvested in many areas primarily for their eggs, known as caviar (see Chapter 12, Box 12.1). There are also large recreational fisheries for trout, char, largemouth bass, and other species, which has sometimes resulted in the transplanting of such species outside of their original range (see Chapter 11). In many places in the world, local peoples have artisanal and subsistence fisheries to sustain themselves, either directly as food or to sell through local markets. People also harvest other species, such as turtles, frogs, and crayfish, from freshwaters for food.

In addition to harvests of wild-grown species, there are also many aquaculture operations based in freshwaters. The global magnitude of aquatic animal harvest (marine and freshwater) was estimated at about 178,500,000 tons (2018 numbers) and valued at $401 billion (USD in 2020) (see Figure 10.1). Estimates from the UN's Food and Agriculture Organization suggest that global production through aquaculture grew more than five-fold between 1990 and 2018. From these numbers, it is clear that aquaculture in freshwaters is a big contributor to freshwater food production and can help provide high-quality protein for global human populations.

Sometimes harvesting is divided into components depending on the magnitude or scale of the harvest. Industrial (commercial) fisheries (or exploitation of other harvestable species) in marine systems and freshwaters are done on a large scale and remove huge amounts of biomass. This can be contrasted with recreational fisheries, which are typically for people with a rod and reel, and is done for pleasure (and sometimes consumption). While recreational fisheries are more widely distributed throughout different countries, it is still a big industry. Canadian figures suggest that in 2015, 194 million fish were caught by recreational fishers, with direct expenditures of more than $2.5 billion (Canadian)—which does not account for indirect expenditures (for example, meals, accommodations, and travel) (Fisheries and Oceans Canada 2015). Freshwater recreational fisheries have potentially the largest impact on freshwater species in many water bodies since freshwater commercial fisheries are generally restricted to large

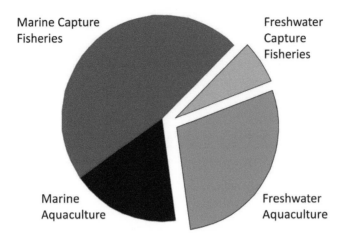

Figure 10.1   The global magnitude of aquatic animal harvest (marine and freshwater) was estimated at about 178,500,000 tons (2018 numbers) and valued at $401 billion (USD in 2020). Capture fisheries (from the wild) accounted for 54% of the total biomass harvested and 46% was from aquaculture (freshwater and marine). Of capture fisheries, freshwater harvests (including fish, crustaceans, shellfish, etc.) contributed 12.4% (about 12,000,000 tons), while marine fisheries produced the rest. Of the total aquaculture production (both marine and freshwater), freshwater aquaculture produced 62.5% (51,300,000 tons) of that value. From freshwater aquaculture production, 91.5% was finfish (especially grass carp [*Ctenopharyngodon idellus*], silver carp [*Hypophthalmichthys molitrix*], and Nile tilapia [*Oreochromis niloticus*]), 7.1% crustaceans, 0.4% mollusks, and another 1% for other groups (including frogs, turtles, crocodiles, etc.). Note that freshwater fisheries are indicated in green and marine in blue. Data from the United Nations Food and Agriculture Organization (2020).

lakes and streams. Another category of harvesting is artisanal fishing, where the fishers use relatively low levels of technology, traditional gear such as spears or hand reels, and small boats rather than larger, more powerful vessels. Artisanal fishing is often for subsistence and/or for cultural reasons, and the number of fish taken is much smaller than industrial scales.

The species targeted in industrial, recreational, and artisanal fishing may be managed as populations, commonly referred to as *stocks*. A stock can be a population (or set of populations) of a fish species returning to a particular lake or river, even if they return with other stocks. Fisheries may be made up of *mixed stocks*. For instance, a particular species of Pacific salmon may return to the major rivers on their way to spawning grounds, and management may need to be focused on protecting some stocks more than others, even though they are difficult to quickly identify. These stocks may require monitoring and management as separate units for a range of reasons—including being unique conservation units. There are also *terminal fisheries*, such as for Pacific salmon, eels that rear in freshwaters (*Anguilla* spp.), or whitebait (New Zealand), where fish (or other species) return to a particular habitat. This is best seen when fish return from the sea to spawn (or rear in the case of eels) in freshwaters.

In many parts of the world, turtles and other species are also taken for food. This has led to many species becoming endangered. Lack of management plans or enforcement for local fisheries may be largely due to the emphasis on larger commercial and recreational fisheries, or to a lack of resources to manage small-scale exploitation. Small and local populations can often be at risk of overexploitation—and one solution is governance by local communities over who gets access to sites.

One of the general methods used to predict allowable fishing pressure is based on a model referred to as maximum sustainable yield (MSY), which is a nice theory, but seriously challenging in application. You have learned about the logistic growth equation in introductory ecology, which has a sigmoid shape for growth in numbers through time (see Figure 10.2A). At the low density beginning of the curve, the rate is accelerating, but the number of new individuals per unit time is small because there are not so many breeders. As the population reaches high density it becomes resource limited, and the number of new individuals added per time unit is also low. The fastest rate of population growth in terms of new numbers is at intermediate density. If we plot this population growth as new individuals being added to the breeding population each year relative to the current density of reproducing individuals, this results in a hump-shaped curve (see Figure 10.2B). Theoretically, if we could keep fishing (harvest) pressure at exactly the steepest part of that curve, we could harvest that sustainably (see Figure 10.2C).

Unfortunately, there are at least four reasons why the MSY model does not work well, with results being mostly theoretical. First, it is almost impossible to even come close to estimating the density or population size—there is a considerable confidence interval around any estimate, meaning that you could easily overestimate or underestimate the actual number. Second, we rarely even know the number actually harvested because of the amount of illegal and unreported harvesting, and there are individuals that die and are not *landed*. Third, the shape of the population growth equation is not likely to be perfectly logistic. Fourth, the carrying capacity (K from Lotka-Volterra equations; as seen in Figure 10.2A), which is a term in the population growth

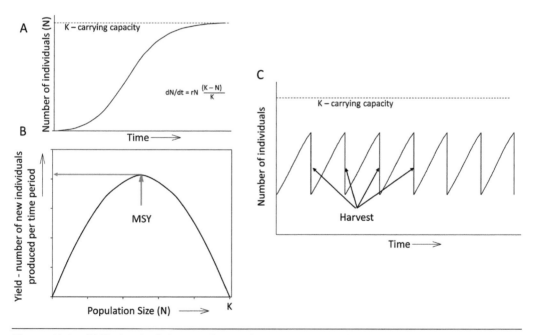

Figure 10.2   (A) Typical logistic growth equation and curve starting from low numbers and reaching carrying capacity. (B) MSY curve from generalized theory about population growth and replacement rates. The idealized, hump-shaped curve of new individuals to a population versus population size. Note that N along the x-axis represents population size or density. (C) Regular harvests (for instance, annual) reduce the numbers of individuals in a population, and ideally continue to sustain high rates of productivity—that is, keeping the population size close to the steepest part of the logistic growth curve.

equation, varies through time based on weather, prey abundance, and other environmental variables that affect the numbers in a population. This means that using this method by itself could result in serious errors that would not be known until a stock is near collapse.

One caution when estimating stock sizes and exploitation rates is that there can be depensatory population reactions, or depensation. For instance, if a population is being exploited and population size is shrinking, the remaining individuals might still aggregate (such as in shoals of fish) for safety or mating. This aggregation means that the *per capita* risk of being caught by a human consumer might increase even as population size goes down. Fishers might also aggregate to the remaining good fishing spots and fish out the last of the population. Depensatory mechanisms can put smaller populations at increasing risk to the point of local collapse of exploited populations (Post 2013).

Many freshwater species are grown in aquaculture to feed human markets. These can be an efficient way to produce high-quality protein, and include prawns, fish, ducks, crayfish (*crawfish*), frogs, etc. However, there are also serious problems with pollution of water supplies, disease transmission to natural populations, escaped animals invading areas to which they are not native, etc. Species such as rainbow trout, Arctic char and others are grown commercially in cooler climates, while tilapia, carp, and catfish are typical of warmer regions. Fish and duck ponds on farms are common and an important source of protein in many countries.

Beavers were a major reason for the race of Europeans across North America as the rodent's pelt was used for top hats worn by Europeans in previous centuries. Trapping precipitously drove down the number of beavers, and in many parts of their former range the beaver is still absent and being purposefully reintroduced for its role as an ecosystem engineer (for instance, in Arizona), which can create habitat for many other species, as well as provide other ecosystem services for water flow regulation.

Freshwater plants are also harvested for human use. Rice (*Oryza sativa* L.) is one of the primary dietary items for half the world's human population. As a semi-aquatic grass, it is dependent on freshwater for its growth, mostly in flooded areas known as rice paddies. Similarly, several other species of wild rice grow in water. About half of the world's rice production of 755,000,000 tons (2019 estimate, Food and Agricultural Organization of the United Nations [FAO]) comes from China and India, and about half is also consumed in those two countries (Muthayya et al. 2014). Many other aquatic plant species, such as watercress, are harvested for consumption. Microalgae, such as *Spirulina* (a cyanobacterium), is grown in relatively small-scale culture for human consumption, and is also one of the species of algae that are consumed by lesser flamingos.

IMPACTS

The exploitation of freshwater organisms provides important sources of protein and carbohydrates to a human population exceeding eight billion people. However, populations of any species that is consumed can be overexploited, and that is obviously not sustainable—for the consumers or for continued biodiversity. Other changes we have already learned about in previous chapters, such as degradation of water quality, can affect the productivity of exploited populations or impact the quality of those food species. Water used for the production of exploited species in aquaculture can reduce or degrade freshwater that is needed by other species in the ecosystem. Sustainable management of exploited populations is necessary to ensure sufficient production and minimize the ensuing adverse environmental effects on our freshwater ecosystems.

Reduction in Wild Numbers

The most obvious impact of any harvest is the reduction in numbers of a species, but sometimes it is the habitat damage that is most critical. It is difficult to manage many species of fish (and other food species) since in most countries there is very little management, monitoring, or enforcement available. Commercial fishing operations in freshwaters are regulated by governments, although enforcement is expensive and often scarce. There is little doubt that the reducing numbers of some species leads to undesired changes in food webs and whole ecosystems. Unfortunately, monitoring efforts in many countries are not up to the task of detecting changes until it is too late.

In the case of harvesting for subsistence, often it is not possible to track the numbers (or biomass) in the fish populations (and other species) or the harvest rate at the species level (McIntyre et al. 2016). Artisanal fishing is even harder to regulate, and access and poverty issues influence and motivate fishers to catch what they can for personal use or sale at market. Estimates of the magnitude of subsistence fishing are hard to make, but estimates based on household consumption rates suggest that these rates of fish harvest are at least 65% higher on average than reported by governments (Fluet-Chouinard et al. 2018).

There are many *terminal* fisheries that typically capture fish or other species as they return to spawn or leave natal areas to migrate to growing areas. For example, Pacific salmon are mostly (but not exclusively) harvested when they return from the sea to freshwater and as they move into estuaries on their way to spawning areas. Other kinds of fisheries along these lines are eels, lampreys, and whitebait (referring to small fish), as they return to or leave freshwater. A difficulty with terminal fisheries is the uncertainty of the actual numbers of individuals of a given population since they cannot be estimated accurately before they arrive at the narrow area where they are fished (such as the estuary of a large stream). This uncertainty can mean more (or fewer) are taken than is sustainable in a given year.

Recreational fisheries can reduce numbers of some desired fish species, and these fisheries are often regulated with catch limits or even restricted access. Catch limits in some places may be a maximum number of fish that one can harvest per day (or per year), or may be strictly catch and release. Such fisheries usually require the fisher to have a license, and conservation officers may check on their activities. Another big impact of recreational fisheries on wild populations is that hatchery fish may be stocked, sometimes even a species that is not naturally occurring there, and that can displace or eliminate some indigenous species.

The selective capture of larger individuals in any fishery may lead to changes in size and age distributions, and in some cases, populations have begun to reproduce at an earlier age. Given that reproductive rate is often positively related to size, this can lead to lower rates of reproduction for a population. There are many studies of such shifts due to harvesting of larger (older) fish (or other species) (Heino et al. 2015), and some that show that such trends can reverse relatively rapidly as size-selective pressures are released (Feiner et al. 2015).

Another form of exploitation of freshwater organisms is the harvest of freshwater mussels for the production of buttons and the collection of freshwater pearls. In the late 19th century and early part of the 20th century, freshwater mussels were extensively harvested for those two commodities (Anthony and Downing 2001). In some locations, mussels have pearls, which resulted in huge harvests to find them. Between these two motivations for harvesting, freshwater mussel populations were severely depleted and is one of the reasons that many mussels are now considered threatened (see Box 12.2).

Aquaculture

The impact of harvesting of species or the introduction of species for harvest potential can each have catastrophic effects. Lake Victoria in eastern Africa was once home to over 500 endemic species of cichlid fish, which were mostly small bodied and not easily exploited. Nile perch (*Lates niloticus*) were introduced in 1954—largely as a way to create a fishery around a fish that was large enough to be exploited (mature fish can be 1 to 2 meters long). Unfortunately, Nile perch like to eat cichlids, and therefore, caused quite the disruption in the food chains. In addition, this introduction is considered to have led to the local extinction and hybridization of hundreds of endemic forms of fish (Hecky et al. 2010). The growth in the fishery for Nile perch and other species from Lake Victoria has expanded by an order of magnitude from the 1960s to the early 2000s (see Figure 10.3). This is just one example of many where small-scale fisheries were not well regulated, or may have been entirely unregulated, leading to unwanted outcomes. Some species are exploited without clear conservation plans.

Most aquaculture operations have a need for inputs of food and nutrients to support production. The consequences of such operations can include pollution (including nutrients, antibiotics, etc.), escape of species, and exploitation of other aquatic species solely as food for aquaculture production. Aquaculture can also increase the rates at which non-native species expand their ranges geographically as they escape from growing areas. This will be addressed further in Chapter 11.

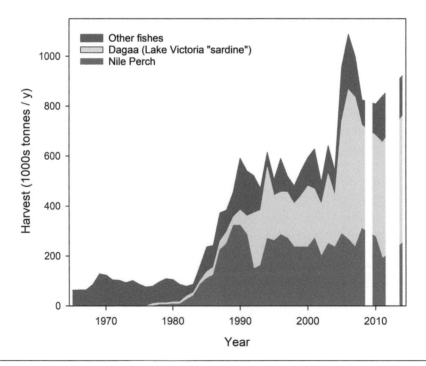

Figure 10.3 Reported catch amounts per year for fish from Lake Victoria (bordered by Tanzania, Uganda, and Kenya), the largest of Africa's Great Lakes. Note the large increase in rates of harvesting of the introduced Nile perch (introduced in 1954) and the asymptote in catches of that species after the European Union (EU) put rules in place regarding import of the species to the EU (1992–1993). Gaps (white vertical bars) represent years without data. Figure redrawn after Taabu-Munyaho et al. (2016).

Rice Production

Rice, a semi-aquatic, annual grass, is the biggest source of calories for half of the world's population. Its production uses the withdrawal of an estimated 34 to 43% of the world's irrigation water, covering approximately 1,670,000 km^2 of the earth's surface, with about 91% of that area in Asia (Surendran et al. 2021). In addition to intensive water use, it is a source of pollution from pesticides and fertilizers, loss of water to evaporation, and greenhouse gas emissions—mostly methane and nitrous oxide (N_2O) from anoxic conditions created at the bottom of flooded fields (anaerobic decomposition of organic materials). Rice seems to accumulate inorganic arsenic more than most other plants, and arsenic from human sources can accumulate in the sediments of rice paddies.

INTERACTIONS WITH OTHER STRESSORS

Disruption of Wild Freshwater Ecosystems

Reduced densities of a particular species due to exploitation (fishing or other harvest) can lead to other changes in rates of community and ecosystem functions. For instance, lower densities and reduced ecological rates can lead to changes in water clarity, nutrient accumulations, and opportunities for other species to increase, including nonindigenous species. Such activities are also often accompanied by reductions in water quality due to increases in nutrients, turbidity, and pesticides. Recreational fishing has many times resulted in nonindigenous species being introduced outside of their range, or of bait fish or crustaceans being released. Fishers also are considered as the main vector for unintentionally moving species from place to place, such as the diatom *Didymosphenia*, mud snails, *Myriophyllum*, zebra mussels, *Bythotrephes*, etc. (see Chapter 11).

Harvesting of large fish can lead to reductions in size at maturity by several mechanisms, but generally, may also reduce reproductive rates of individuals (egg number is positively related to size). Beyond individual species, if the largest individuals are harvested from a population, it can change predator-prey and competitive interactions, with residual effects to community and ecosystem changes. These effects can include shifts in prey in terms of the size ranges and species eaten if consumers are smaller than usual. It can potentially lead to more, but smaller, individuals, and with shifts in prey that are eaten, might alter competitive interactions.

Hatcheries can be important tools in the rescue and increase of population sizes of species at risk. They can also provide recruits to augment populations. However, hatcheries can also be controversial, for instance, hatchery fish that are not wary of predators can mix with wild fish and increase predator densities that affect population sizes of the wild individuals. There may be genetic homogenization as a result of a few adults contributing their genes to the hatchery production. Subsequently, hatchery fish may breed with wild fish affecting genetic variation. Release of hatchery fish can also potentially spread diseases that were contracted while in the hatchery into wild populations.

As aquaculture has grown globally, many species are now grown in places beyond their native range, and escapes are common. As mentioned before, many of the species grown in freshwater aquaculture have become invasive species in the locations to where they have been moved. Invasive species are discussed in more detail in Chapter 11.

Fish and other freshwater organisms can bioaccumulate toxic materials, one of the most severe being mercury contamination. Many contaminants are fat soluble, and these can be bioaccumulated or bioconcentrated from water or food. Other freshwater organisms may concentrate other contaminants, including metals, organics, or pharmaceuticals (see Chapter 4).

Unintended Consequences of Rice and Other Crops

Flooding of areas for aquaculture or production of rice and other aquatic crops can lead to large increases in emissions of methane and N_2Os (especially in areas with nitrogen (N) fertilizers), which are the main greenhouse gases emanating from anaerobic sediments in the bottom of these flooded areas. In particular, rice production using flooded fields produces greater greenhouse gas emissions than any other cereal crop, and because it is so extensive, it is a notable source of gases globally. Increasing the area in cultivation by these methods to support more humans also can lead to more greenhouse gas emissions, and methods are being explored to balance these outcomes. Moreover, with warmer climates, more water will be lost to evaporation from pond surfaces and by transpiration (by plant crops), leading to greater demand for water to sustain current production.

SOLUTIONS

Fishery Population Management

Proper management of fisheries at local and regional scales should avoid the fallacy generated by MSY models—and simply implement good monitoring. Commercial fishing operations in freshwaters are usually subject to regulations, even if those are not entirely or effectively enforced. These rules typically limit the numbers (quotas) and types of fish (or other species) captured, as well as timing and size restrictions. These rules also extend to the gear used, such as the types of nets or fishing lines. The rules are enforced by personnel from state, provincial, or national agencies. Artisanal fishing is harder to regulate. Monitoring of this level of fishing pressure is uncertain and costly to countries with low-income levels. In many countries, there are rules around the number of fish caught, the species, their sizes, and seasons of catch. In wealthier countries, conservation officers or fish and wildlife officials conduct infrequent inspections of the catch of commercial and recreational fishers in order to enforce the rules. However, fisheries of any sort can easily cross the line between sustainable, self-replacing populations and populations becoming locally extirpated. It is difficult to anticipate the trajectory of populations in the absence of independent monitoring.

Another issue is how to estimate the stock size when there are mixed stocks in a terminal fishery, such as Pacific salmon returning to a river mouth on their way to spawn. One solution is rapid genetic testing of fish in a test fishery (a day or so before a fish harvest opening), which has been shown to be feasible with genetic stock identification using single nucleotide polymorphisms (SNPs). This has been demonstrated with Chinook salmon that exist in at least 369 identifiable populations (stocks) across their range, many of which are of conservation concern (Beacham et al. 2021). This highlights the uncertainty in being able to predict any given stock and leaves some populations vulnerable to overexploitation.

Resource management depends on having clear targets for management, but also requires data on the resource, the latter of which is not always available. Another commonly applied method, especially for recreational fisheries, is catch-per-unit-effort (CPUE)—for example, the number of fish caught per angler per day. For some fisheries, CPUE is used as an index to determine resource status. This is a crude tool, but convenient in the absence of other ways to gain data. In recreational fisheries this can be done by a creel census, based on fishers reporting how many hours they spent fishing (effort) and the numbers (also sizes) and species of fish they caught (the catch). If that CPUE number begins to decrease, it generally represents a decrease in the target population and signifies that fishing pressure may be too high. This is generally regulated by limits on the number of fish allowed to be kept per day (including limits on individual fish sizes), areas and seasons where fishing is allowed, and gear restrictions.

With recreational fisheries, there are frequently rules for catch and release as well. Here fish are obliged to be released after capture (often after a picture is taken of the fisher with their catch). This helps sustain fish populations to provide opportunities for others. However, fish that have been exposed to air for a period of time have damage to gills, acute hypoxia, and lower survival rates than if they are kept in the water. The survival rate of fish that have been caught is not 100%—various studies suggest survival of released fish is only 82%—but there is a huge variation between species and across studies (Bartholomew and Bohnsack 2005). Survival of released fish can be significantly enhanced if unhooking is done underwater, and any picture is taken with most of the head still underwater, or just one side above the water surface for an instant (Cook et al. 2015). Some rules can also diminish the satisfaction that recreational fishers get from their outdoor experience, and managers are often sensitive to balancing the protection of fish populations with the good experience that recreational fishers enjoy.

As the number of large fish decline dramatically, a solution to sustaining the production of fish protein for human consumption is to consume smaller fish, such as anchoveta and sardines, although these are marine fish. This has the advantage of eating from lower in the food web, with smaller losses of total biological energy from moving up multiple trophic levels to bigger species. Another advantage is that smaller fish have less opportunity to bioaccumulate toxins.

Avoid Inadvertent Introductions of Nonindigenous Species

Fishers themselves can be a major agent of ecosystem decay. It is best to avoid undesired introductions as a consequence of recreational activities, including intentional or unintentional introductions of species considered as sport fish or species used as bait (see Chapter 11). Bait species (often small fish) can escape or even be discarded into places where they are not native. Some of the worst invasive species are known to have been moved between lakes and rivers by attaching to boats, boots, nets, and other angler and recreational items, thereby being unintentionally released at the next site visited. This is important even for nearby waters, as there are many instances of species being moved and expanding their range in this way. Good practices involve the inspection of boats, boots, or other gear so that no invaders are carried between water bodies. Even for biologists visiting multiple sites, cleaning boots, nets, sampling gear, etc., and using bleach or other decontaminating solutions is important.

Exploitation of Species Other than Fish

Many species of freshwater vertebrates other than fish have been hunted and trapped for their furs, skin, and meat—such as beaver (*Castor* spp.), muskrat (*Ondatra zibethicus*), coypu (*Myocastor coypus*), caimans and alligators (Crocodilia), and turtles (Testudines). Harvesting these species from the wild is generally subject to regulations, but some subsistence and artisanal harvests are allowed. Beavers were reduced in numbers and extirpated from some North American and European landscapes in the 18th century. Lack of experience with beavers over the past century means they can frustratingly disrupt infrastructure, such as blocking culverts and causing flooded roads and fields, or cutting down trees that may have been planted for commercial or ecosystem purposes (such as restoration). Mechanisms to avoid damage from flooding can include water levelers or cages around culvert inflows to reduce flooding impacts (Hood et al. 2018), and this can avoid relocation or culling of beavers. Unfortunately, relocation or culling is still a frequent outcome.

Duck (and other waterfowl) hunting can be considered in this chapter, as they are relevant as a freshwater-based group of species. The conservation of many wetlands around the world are a result of efforts for the protection of duck populations. Several organizations initially grew from interests in sustaining and increasing populations of waterfowl for hunting, which is still one objective, but now their aims extend to broader conservation goals. Organizations, such as Ducks Unlimited and others, are important conservation groups protecting and restoring wetland habitats.

Hatcheries

A hatchery is a place for the breeding, hatching, and rearing of finfish, shellfish, and other species through their early life stages. Hatcheries actually produce the larval and juvenile forms of fish, shellfish, and crustaceans—which can then be transferred to fish farms where they are grown until harvested (see the next section titled Aquaculture). Otherwise, these young stages may be released to the wild, as is common for Pacific salmon and other species. The most typical practice is for eggs and sperm to be stripped from the adults and mixed artificially to fertilize eggs. Propagules are produced in tanks and ponds where the eggs and young are kept in well-oxygenated water that is free of predators. Hatcheries can be valuable tools in conservation efforts for species at risk, including fish, turtles, amphibians, mussels, and others.

Aquaculture

Aquaculture is one solution to supplying high-quality protein to markets (see Box 10.2), but this sometimes has higher costs than wild-caught animals and has its own environmental problems. Of the world's aquaculture production, freshwater fish—such as carp (and other cyprinids), Tilapias (and other cichlids), and catfish (Siluriformes)—make up about 40% of total biomass and make up five of the top ten aquaculture groups by production (FAO 2020). Often, species grown in freshwater aquaculture (fish, crayfish, clams, etc.) are fed high-protein diets based on harvesting of other species to create food, which has its own impacts on natural ecosystems. Aquaculture can create local pollution issues and transmit disease, although these can be controlled in well-managed operations. As a solution to fishing down the food web in natural ecosystems, aquaculture can be a good option to provide high-quality protein to a large global population of humans.

Several forms of aquaculture occur, such as open pens in natural waters, farm ponds, rice paddies, and land-based (recirculating) systems. Each has challenges with the potential release of water high in nutrients, antibiotics, and other waste. There is considerable research on food use and nutritional balance in cultured species, which is intended to increase the efficiency of the conversion of food into growth and nutritional value of marketable fish or crustaceans. Greater efficiency of food use also reduces input costs into aquaculture operations and reduces waste. Some fish and prawns can be grown in rice paddies, which have the co-benefit of reducing the densities of aquatic invertebrates that damage rice, and thereby reduce the need for pesticides.

Aquaculture can also be a solution for the protection of threatened and endangered species from exploitation. In Chapter 12 we will see an example of sturgeon species, where aquaculture production has filled some of the demand for caviar (sturgeon eggs). Sturgeon are also grown in aquaculture for their meat and can reduce pressure on wild populations.

Whiteleg shrimp (*Litopenaeus vannamei*), which is really a marine or brackish water species, is the aquaculture species that contributes to more than half the global crustacean production from freshwaters (FAO 2020). These are grown in net pens and ponds are dug out to farm this species. Similarly, the giant river prawn (*Macrobrachium rosenbergii*) is well-known to aqua-culturists. Crayfish aquaculture is now common in Asia, Europe, and North America (Huner 1994) (see Box 10.2). Two main crayfish species grown in aquaculture are the redclaw crayfish (*Cherax quadricarinatus*), originally from Australia, and the red swamp crayfish (*Procambarus clarkii*), originally from the United States. These species are now being grown in aquaculture in many parts of the world, especially Asia (see Figure 10.4). Most of the crayfish (also known as crawfish or crawdads) production in the United States comes from Louisiana, and the majority is red swamp crayfish that is grown in aquaculture.

Aquaculture sites can be managed in a way that can be supportive of other freshwater bio-diversity. For instance, many aquaculture ponds can also provide habitat to a range of biologi-cal diversity and can be managed in a way to support other species. Using freshwaters that are primarily for aquaculture can provide co-benefits if their water quality and structure can allow for uncultivated species, and this can also include rice paddies, farm ponds, and even agricultural ditches.

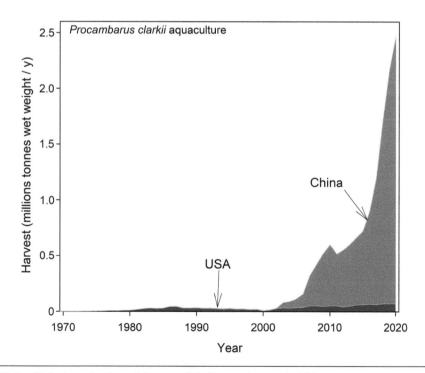

Figure 10.4 Global production of the crayfish *Procambarus clarkii* in aquaculture based on statistics from the UN's FAO. The rapid increase in aquaculture production since 2000 is largely attributed to the expansion of crayfish aquaculture in Asia, primarily China.

Rice Production

As a primary dietary item, rice is grown on large amounts of land, and depends on storing water to provide the appropriate growth conditions. In most places, rice is grown during the rainy (monsoon) season, so much of the water used is rainwater, rather than being drawn from surface waters. However, it is important to watch for potential contaminants since industrial rice production can use large quantities of pesticides. Better controls on the pesticides being applied and the concentrations of those have reduced contaminant loads in many parts of the world. Extensive research is producing new varieties, and some can be planted to produce up to three crops per year where irrigation allows.

Methane production in water-logged soils becomes an issue in rice production, and midseason drainage of paddies has been shown to be effective at reducing overall methane and N_2O emissions by more than a third, but there is still work to be done to determine the best timing for drainage. Much of rice cultivation uses N fertilizers to increase production, which causes pollution problems, and the N contributes to higher N_2O emission rates (Yao et al. 2017). Studies to limit the amounts of fertilizer used show that N application can be reduced to some extent with little or no reduction in yield. Draining water from paddies earlier than usual can reduce water use (including losses by evaporation) and may even increase yields. It may also reduce damage from crop pests, such as insects and blight. One reason for full submergence of rice is to reduce weed problems, but the depth of water is not a requirement for rice growth (Surendran et al. 2021). Hence, another technique used is to cover the soil surface with a plastic liner or plastic mulching (both usually polyethylene) to keep the soil moist, but not inundated (Yao et al. 2017). Replacing polyethylene with sheets of a biodegradable polymer is being tried to reduce plastic use and pollution from such plastics. Trials with different wetting schedules of paddies—that is, wetting and drying—have shown that there can be substantial water savings compared to complete flooding for the growing season. During the fallow period when rice is not being grown, the same field may be used to grow wheat in some regions (Surendran et al. 2021).

A great deal of research has also gone into selecting varieties that yield higher production, are more resistant to pests, and are more nutritious. Finally, some rice paddy operations use the flooded period to raise fish or shrimp, augmenting the productivity of these areas. Balancing rice production, water use, pollution, and greenhouse gas emissions requires ongoing research and innovation.

PERSPECTIVES

Fisheries of any sort are vulnerable to overexploitation or to shifts due to other kinds of management (stocking, etc.). They are also very difficult to monitor and rely on CPUE measures, which may not provide sufficient warning to managers to implement actions. Aquaculture has filled an important role in providing high-quality protein, but still needs to avoid issues with pollution, contaminant bioaccumulation, and depleting resources for native species, including harvesting of food items to feed the aquaculture populations. Capture fisheries and aquaculture do have some certification programs that can help consumers be aware of the sustainability of particular species and practices.

A number of countries are beginning to consider crayfish (and lobsters) as sentient beings. Cephalopods, such as octopus and squid, have long been considered sentient, and research with these animals requires animal care permits as with vertebrates. Work with crayfish may soon require animal care permits as well.

Box 10.1   Lake Erie and Its Commercial Fishery

There are many commercial-scale fisheries for wild populations of fish in lakes and rivers around the world. One large, inland fishery in North America is in Lake Erie, the smallest (area and volume) of the five Laurentian Great Lakes. Some of the key fish species harvested there are yellow perch (*Perca flavescens*), walleye (*Sander vitreus*), white perch (*Morone americana*), white bass (*Morone chrysops*), common carp (*Cyprinus carpio*), and rainbow smelt (*Osmerus mordax*), but there are several more. Cisco (*Coregonus artedi*) was once a major, commercially exploited species in Lake Erie, but it has not been part of the commercial catch since the mid-1960s and was already in serious trouble in the 1930s. Cisco, which was once a key predator, is now considered to be extirpated from the lake. The top predators remaining in the lake now include lake trout (*Salvelinus namaycus*), walleye, and burbot (*Lota lota*).

Estimates of Lake Erie's inland commercial fishery are 1,910,454 kg/y (range 1,703,555 to 2,182,930 kg/y; years 2011 to 2021) on the United States side (Ohio) alone (Ohio Department of Natural Resources). There are estimates of more than three million fish per year caught recreationally, again in Ohio waters alone. Lake Erie is also popular for ice fishing, which for readers from warmer regions might seem odd, but recreational fishers will set up huts on the ice, drill a hole, and angle for fish that are active below the ice.

One of the reasons for the high productivity of fish populations in Lake Erie is the high nutrient loading from the surrounding, mostly agricultural, landscape (Almeida et al. 2022). The highly productive algae support high production of zooplankton, and hence food to small fish that then are eaten by larger (harvestable) fish. However, eutrophication can also lead to anoxic zones, potentially toxic blooms of cyanobacteria, elevated turbidity, and elevated concentrations of contaminants (such as pesticides and pharmaceuticals). The lake is also plagued with invasive species, such as round goby and quagga mussels (see Chapter 11).

With the high nutrient loading in the 1960s and 1970s—in particular, phosphorus (P)—Lake Erie was overly enriched. Eutrophication led to reduced oxygen concentrations and diminished habitat for coldwater species of invertebrates and fish (deeper water often had lower oxygen), which resulted in large changes to the Lake Erie ecosystem. Pollution controls beginning in 1972 allowed the lake to rebound quickly (Scavia et al. 2014); however, by the beginning of the 2000s, the lake was again in poor condition due to eutrophication, both from nutrient runoff from agricultural lands and urban centers upstream, and from internal P loading from anoxic conditions in the lake bottom due to high organic matter loading. In particular, the western basin of the lake regularly experiences blooms of algae affecting the lake's food web. Management agencies are working to reduce P loadings in the lake in hopes of allowing the lake ecosystem to recover.

Box 10.2   Crayfish Aquaculture

Crayfish (order Decapoda) are consumed by many cultures, and as a good-sized crustacean it provides a food similar to some popular marine crustaceans. Commercial aquaculture for crayfish began over a century ago, and has developed into a relatively efficient food-production system. The predominant species grown in North America are the red swamp (*Procambarus clarkii*) and white river crayfish (*P. zonangulus*). Over 500 km² in the United States are given over to crayfish aquaculture. In the 1990s, the United States—primarily Louisiana—were responsible for about 90% of *P. clarkii* production globally, but by 2018 this was only about 4% with most of the remainder produced in China (see Figure 10.4). Global production of red swamp crayfish has expanded dramatically in the past couple of decades, and enormous increases in production in China have contributed to that. Estimates from the UN's FAO indicate that China is now producing over 1.6 million tons of crayfish per year. Many other species of crayfish and other crustaceans are also cultivated in aquaculture, such as redclaw crayfish (*Cherax quadricarinatus*), Chinese mitten crab (*Eriocheir sinensis*), whiteleg shrimp (*Litopenaeus vannamei*), and the giant river prawn (*Macrobrachium rosenbergii*).

One way to improve the efficiency of crustacean aquaculture is to take advantage of coproduction. For instance, farmers can take advantage of rice paddies. Some species of crayfish reproduce in spring, but then burrow into sediments, and can tolerate the emptying of rice paddies for summer before reflooding in the autumn. Some crustaceans can also be grown if rice fields are fallow or in crop production for the summer where grain crops can be grown, and the residue remaining after harvest and reflooding can form part of the food supply to the crayfish.

In the general management of such operations, a select number of individuals are allowed to breed and their young are reared in nursery ponds; then at a certain body size are moved to *grow-out* ponds. The physical structures of ponds for adults and juveniles are slightly different, and separating larvae from juveniles avoids some cannibalism. The density of juveniles placed into the grow-out ponds needs to be adjusted to avoid large numbers of stunted individuals. Ultimately, fully grown (commercial size) animals are harvested using baited traps.

Crustacean species grown in aquaculture are largely species that are large as adults, robust, have a simple life cycle, and are economical to produce. For instance, the redclaw crayfish from Australia takes about nine months to reach commercial size and tolerates high densities. In most cases the animals are fed with commercially sourced pellets with around 25% protein, often based on grain.

The western North American species, the signal crayfish (*Pacifastacus leniusculus*), was introduced to Sweden as an aquaculture species to replace native species that had succumbed to a widespread disease (see more in Chapter 11). Some of the signal crayfish escaped from aquaculture and the species is now considered invasive. The noble crayfish (*Astacus astacus*) of Europe is now an at-risk species, but efforts to grow it in aquaculture are being attempted. Any species introduced for aquaculture can also become an invasive species—either through escape from aquaculture facilities or through importation of live individuals for the pet trade or food markets, for instance, *P. clarkii* and *P. leniusculus*. As in any aquaculture system, there can be diseases that are easily transmitted through dense populations of monospecific crops, which can also be spread through escapees to wild populations.

ACTIVITIES

1. What are the fish and nonfish species exploited in your state or province?
2. Do any of the freshwater species (fish, crayfish, clams, etc.) you see in restaurants and stores come from aquaculture?
3. Is there a way to know?

11

INVASIVE SPECIES

INTRODUCTION

Invasive species are a global threat to native biodiversity and human economies (agriculture, forestry, and others), and usually once established, are nearly impossible to eradicate (Strayer 2010). Generally, invasive species are those whose introduction causes, or may cause, economic or environmental harm, or perhaps cause risks to human health. One estimate of the annual cost of invasive species in the United States (not solely aquatic species) is greater than $120 billion (Pimentel et al. 2005), and almost a third of the 1,300 species in the United States that are protected by the Endangered Species Act are thought to be at risk, in part due to invasive species (U.S. Fish and Wildlife Service). Thus, there is an economic and ecological cost from invasive species (Ricciardi et al. 2013).

Nonnative, nonindigenous, introduced, alien, pests, or exotic are obvious terms for species that were not originally part of a regional species pool, that is, the species has appeared as a result of human activity, and all six of these adjectives are found in the literature. One term used in the literature is *the homogenization of food webs* as some of the same nonnative and invasive species get inserted into food webs around the world (Baiser et al. 2012; Olden et al. 2018). The appearance of these invasive species in many parts of the world will make food webs there more similar to other places, or more homogenous, as those introduced species displace native species. The introduction of nonnative species can sometimes lead to *invasional meltdown* where invasive species facilitate each other's colonization of a local food web. Many species are carried beyond their natural range by accident, but as we'll see, a large number of species were purposefully transported to new places.

Invasive species can fit into almost any position within a food web, from top predator to microbes involved in decomposition. Some of these species are known widely beyond the scientific community, such as zebra mussel (*Dreissena* spp.), silver carp (*Hypopthalmichthys molitrix*) (also known in the United States as Asian carp), New Zealand mud snail (*Potamopyrgus antipodarum*), Asian clam (*Corbicula fluminea*), purple loosestrife (*Lythrum salicaria*), milfoil (*Myriophyllum spicatum*), water hyacinth (*Pontederia crassipes*), and others (see Figure 11.1). Once established, nonnative invasive species may have enormous impacts on freshwater ecosystems and be nearly impossible to eradicate. The impacts on indigenous species may result in local extirpation, and even extinction. There is a list of the world's top 100 worst invasive species based on their extent and impact, of which about 15 are freshwater species, including rainbow trout, bullfrog (*Lithobates catesbeianus*), common carp (*Cyprinus carpio*), zebra

Figure 11.1 Examples of nonnative, invasive species that cause changes to freshwater ecosystems. (A) Round goby (*Neogobius melanostomus*). Photo credit: Eric Engbretson, U.S. Fish and Wildlife. (B) Spiny water flea (*Bythotrephes longimanus*). Photo credit: NOAA Great Lakes Environmental Research Laboratory. (C) Eurasian milfoil (*Myriophyllum spicatum*) in a lake in Washington state. Photo credit: Roger Tabor, U.S. Fish & Wildlife. (D) A sea lamprey suctioned onto a rock in the Ocqueoc River. Photo credit: Dr. Andrea Miehls, Great Lakes Fisheries Commission.

mussel, and water hyacinth (see: www.iucngisd.org/gisd/100_worst.php). Many countries also track introduced species that are considered *injurious*, and in the United States, the Lacey Act (18 U.S.C. 42) facilitates keeping track of those species.

The term *invasive species* usually has the connotation that these are undesirable species. Not all nonnative species are invasive, and some have been unsuccessful at increasing in numbers and expanding their range beyond where they were initially deposited. There are several criteria for assessing whether a nonnative species is considered invasive (see Table 11.1). Some species may be introduced many times before becoming established and expanding in numbers and range. A *propagule* for an invasive species might be a seed, a cutting, an egg, larva, a reproductive individual, or any other stage reaching a new habitat with the potential of growing there. The *propagule pressure* is a term used to denote the number of individuals and the number of times a species has been introduced, and often, the more propagules and times introduced, the better the chances of them becoming established. The term *established* has a particular meaning in the context of nonnative species, but at the very least, includes breeding successfully.

Table 11.1   A set of criteria distinguishing which species are invasive, according to Ricciardi (2007).

1. The species appeared suddenly and had not previously been recorded in the watershed.
2. The species subsequently spreads within the watershed.
3. The distribution of the species in the watershed is restricted compared with native species.
4. The global distribution of the species is anomalously disjunct (i.e., contains widely scattered and isolated populations).
5. The global distribution of the species is associated with human vectors of dispersal.
6. The basin is isolated from regions possessing the most genetically and morphologically similar species.

Many alien, invasive species were introduced to new freshwater environments as a result of the aquarium industry and as ornamental species for gardens (Padilla and Williams 2004). Many of these species are raised for these uses particularly because they are hardy, which predisposes them to doing well in the wild if they escape. For instance, Eurasian milfoil is a plant that does well in many conditions, including aquaria, and is easily grown from fragments as propagules, and as such has spread widely in North America. The aquarium industry generates billions of dollars per year, and enormous numbers of fish, invertebrates, and plants (and probably pathogens) each year are shipped internationally and within nations. Unfortunately, these species are sometimes disposed of in ways that get into natural waters.

There are also many species that escape from aquaculture, including fish, crayfish, and others. As an example, the crayfish fauna of Europe and the United Kingdom includes relatively few species, but crayfish, including the most common European species, *Astacus astacus*, were a popular food source. In the late 1800s, a fungal disease (*Aphanomyces astaci*), referred to as the crayfish plague, began to decimate native European populations, either introduced through nonnative crayfish or ballast water from North America. To compensate for the loss of the native species, since crayfish are a delicacy, a North American species, the signal crayfish (*Pacifastacus leniusculus*) was introduced to European waters in the 1950s to provide an aquaculture opportunity. Signal crayfish were known to be resistant to the plague. However, signal crayfish also turned out to be a vector of the fungal disease, and they escaped and spread through much of Europe and the United Kingdom, further imperiling the few remaining populations of native crayfish.

Some freshwater fish are purposefully spread by people, including some of the 100 worst invasive species in the world, such as rainbow trout, largemouth bass, and brown trout. In order to create fishing opportunities, fishers often move fish species without permission from authorities, and those species are often detrimental to fish in the receiving waters (Dextrase and Mandrak 2006). Likewise, many farm ponds exist as small impoundments on streams and are often stocked with species from other regions, providing another avenue for invasive species to expand (Pfaff et al. 2023). California has many nonnative species, with almost half of the fish species in the state introduced purposefully by the California Fish Commission in the late 1800s (Dill and Cordone 1997). However, introductions are not always considered bad by all people—which many fishers will attest to when new trout and salmon fishing opportunities are created in parts of the world that originally had few or no salmonids (see Box 11.2). For example, rainbow trout, brown trout, brook char, Chinook salmon, and others have been introduced purposefully to many parts of the world outside of their native range—some for aquaculture and some to create world-class fishing opportunities.

Some species have been introduced initially because someone thought they would be good for biocontrol of another problem, such as the growth of algae or macrophytes. Bighead carp (*Hypopthalmichthys nobilis*), silver carp, and grass carp (*Ctenopharyngodon idella*), collectively referred to as Asian carp in North America, were imported and introduced on purpose to aid in reducing algal and macrophyte growth in aquaculture ponds in the United States (Norman and Whitledge 2015). Of course, all three species escaped and have established extensively across the United States (http://nyis.info/invasive_species/asian-carp/). A fourth Asian carp, black carp (*Mylopharyngodon piceus*), was introduced to the United States to control snail and clam problems that were associated with aquaculture and other areas where mollusks were problems (Poulton et al. 2019). All four of these carp species have become established in the wild, can grow to a large size and live a long time, and hence, create a range of problems for natural communities, especially outcompeting native species of fish or preying upon endangered mollusks.

Common carp were likewise introduced as a food fish to North America and other parts of the world. Large numbers of carp in a system can create turbidity issues since they forage for invertebrates on the bottoms of lakes and wetlands. With their large size and few predators capable of eating them, they have few natural enemies, thus allowing them to increase dramatically in population sizes, which can change entire communities and threaten recreationally and commercially important fish stocks.

Many species of fish, invertebrates, and plants (and probably microbes) have been spread incidentally in the ballast water of ships and other transfers of water, and probably unknowingly by being attached to boots, fishing tackle, and sampling gear. Some of these include invertebrates such as zebra mussel, the scud *Dikerogammerus*, and planktonic crustaceans such as *Bythotrephes*, Asian clams, New Zealand mud snail, etc. The round goby (*Neogobius melanostomus*), a benthic fish from the Black Sea in central Eurasia, is now firmly established in the Laurentian Great Lakes and is another species that is causing problems for freshwater ecosystems in that region (see Box 11.1). In recent years, most countries have adopted rules such that the exchange of ballast water in ships takes place at sea so that freshwater from one country does not get flushed into freshwater in another country that is an ocean (or less) away.

Introductions can take place through other means, either on purpose or by accident. Plants such as *Myriophyllum*, *Elodea*, *Eichornia*, and others have been moved widely, sometimes as materials for the aquarium trade or for gardens, and other times attached to boats and vehicles. Often there is a lack of restrictions on the travel of those commercial vessels that help spread nonindigenous species within states and provinces. Fishers will sometimes throw out their remaining live bait, even if it is not native to the waters in which they are fishing. There is even live trade involving species that subsequently escape, including species that are intended as food—and rules on such trade are generally weak.

IMPACTS

There are many potential impacts from invasive, nonindigenous species, including displacing and reducing native species, fouling water systems (such as increasing turbidity), carrying disease, hybridizing with native species, preying on native species, changing food webs and ecosystem functions, occupying space, and altering the stability of freshwater communities. Often nonindigenous species have no predators in the new environment and are able to increase in abundance—known as the *enemy release hypothesis*—changing food webs dramatically (Ricciardi et al. 2013). Invasive species may even support the increases in top predators that can also consume native species.

Predation, Competition, and the Disruption of Food Webs

Invasive species can alter food webs through predation, competition, and parasitism. The sea lamprey (*Petromyzon marinus*) was first documented in Lake Ontario in the 1800s, likely as a result of the Erie Canal. When the Welland Canal was upgraded in 1919, sea lamprey distribution expanded rapidly into all five Laurentian Great Lakes by the 1940s and 1950s. Sea lamprey feed on a few of the native species, including lake trout. Then within 20 years, lake trout were functionally extinct in the lower Great Lakes and the commercial fishery in the upper Great Lakes was reduced to less than three percent of its historical harvest (Pagnucco et al. 2015). The demise of lake trout as a top predator is considered to be the reason that numbers of the fish alewife (*Alosa pseudoharengus*) expanded dramatically as predation on it diminished, leading to changes in the plankton community of these lakes (Ricciardi 2001). Nonnative, predatory water fleas (Cladocera), such as *Bythotrephes longimanus* (an introduction to the Laurentian Great Lakes in ship ballast water), have long spines that render them safe from larger predators and at the same time have reduced other zooplankton through predation and competition by reducing food supplies of other planktivores (Barbiero and Tuchman 2004). Introduced species may also be predators and deplete species that have declined or been driven close to extinction. For example, see the effects of intentional Nile perch introductions to Lake Victoria in Africa in 1954, which appears to have led to the extinction of dozens of endemic species (see Chapter 12).

Nonnative species may outcompete native species. However, the success of a particular species might depend on location, such as the perennial plant the purple loosestrife, which does well in some places that it has expanded to, but not in others (Denoth and Myers 2005). For instance, several nonnative plant species have invaded New Zealand lakes, including the invasive species *Utricularia gibba*, which is competitively displacing the native bladderwort *U. australis* (Compton et al. 2012) and are causing damage to local food webs. Rainbow trout that were introduced into eastern North America occupy similar parts of streams and interfere with the feeding of the native Atlantic salmon (*Salmo salar*), which is a species of concern in that part of the world (Blanchet et al. 2007).

Increases in numbers and abundances of nonnative, invasive species can alter community biodiversity, often leading to simpler communities. This, in turn, can reduce the resilience of food webs by reducing diversity and increasing biotic homogenization (Olden and Rooney 2006). These introductions can disrupt food webs by altering existing species-to-species interactions; for instance, the introduction of opossum shrimp (*Mysis diluviana*) to western North American lakes where it competes strongly for food with young stage salmonids and other species (see the *Physical Removal* section on page 190). Another example is the invasion in 2009 of Lake Mendota, Wisconsin, by *Bythotrephes longimanus* (Walsh et al. 2016). As an effective predator of other zooplankton, this species' invasion resulted in a trophic cascade by reducing algae-eating zooplankton and degrading water quality, particularly clarity and greater algal growth. Estimates are that it would cost at least $140 million USD to reverse the water quality problem, but would still not remove the invasive *B. longimanus*.

Waterways around the world have been linked with canal systems to facilitate transportation, which in turn has enabled species to invade. The Rhine and Danube Rivers of Europe provide one example where species have expanded into new watersheds across the divide through canal systems. For instance, *Dikerogammarus villosus*, is a predatory Ponto-Caspian amphipod crustacean spreading from the Danube River watershed westward through Europe's canal systems and causing problems as a competitor and predator on other invertebrate species (see other examples in Pagnucco et al. 2015). Several other amphipods, fish, and other freshwater species have moved westward into western Europe along the canal systems—disrupting food webs and also expanding the range of parasites to new hosts via these invasive species (Emde et al. 2012).

Damage to Infrastructure

Nonnative, invasive species can cause direct damage to infrastructure, such as the zebra (*Dreissena polymorpha*) and quagga mussels (*Dreissena bugensis*). Zebra and quagga mussels originate from the Ponto-Caspian region (western Asia) and have spread via larvae carried in the ballast water of ships and barges, and through connecting canals. These two encrusting species are problems for the fouling of pipes and other infrastructure (see Figure 11.2), the overgrowing of other species (such as other mussels, many of which are endangered), and overtaking substrates that could be used by native species (Ricciardi et al. 1998). These mussels do not require a larval host, as is the case for native mussels, and can disperse rapidly by releasing their young directly into the water. Studies in the Hudson River, New York (from Troy to New York City), showed that by about 1992 this invasive species had reached densities of nearly 1,000/m^2 and their filtering of the phytoplankton from the passing water reduced suspended algae by 85% (Caraco et al. 1997). At this filtration rate, there was little food for native species in the river, and while the water was clearer (lower turbidity) there were higher concentrations of phosphorus due to lower uptake by plankton. These two invasive species attach to any hard

Figure 11.2   Zebra mussels, an invasive species from eastern Europe can end up encrusting any hard surface, which can include other species such as freshwater mussels. In this case, they are covering a suspended receiver unit for acoustic tags in Lake Winnipeg, Manitoba. Photo courtesy of Doug Leroux (Fisheries and Oceans Canada).

surface—this includes native mussels (many are species-at-risk; see Chapter 12) that can be suffocated by the load of *Dreissena* mussels growing on them. These invasive mussels clog intake and outflow pipes, cover boat hulls, docks, rocks, and other hard surfaces, incurring large maintenance costs and reducing some places' recreational use.

Environmental Changes that Alter Behavior

Introductions of nonnative species can endanger native species, as previously mentioned and as you will learn more about in Chapter 12. This can happen by invasive species reducing available resources, preying upon native species, and even interbreeding—thereby changing the environment they are in. One example of where changes in the environment are detrimental can be seen in the behavior of mating by two co-occurring forms of threespine sticklebacks (*Gasterosteus aculeatus*) in the same lake, a phenomenon caused by sympatric speciation (more in Chapter 12). This pair of forms of stickleback were reproductively isolated, even in the same lake, and one of the ways they remained separate was a difference in color—males of the pelagic form are bright red when ready to mate, and males of the benthic form are very dark—and the depth at which they nest. When a native crayfish colonized the lake and increased in density, the crayfish caused a large increase in turbidity, and the color of males was no longer available as an isolating factor, and subsequently, the two forms hybridized to the point that neither of the original two forms remains (Taylor et al. 2006). Another example comes from the intentional introduction of salmonids in the western United States, particularly rainbow trout. One of the subspecies of cutthroat trout, the westslope cutthroat trout (*Oncorhynchus clarkii lewisi*), has hybridized with introduced rainbow trout to the point that pure westslope cutthroat trout are now a subspecies at risk.

Spreading of Disease

Some water-borne pathogens of humans have been distributed through human movements. For instance, *Giardia* (also known as *beaver fever*) is a protist that has become widely distributed in surface waters of the world. Other pathogens, such as cholera, *Cryptosporidium*, and *E. coli*, have also been spread through surface water contamination. There are also pathogens associated with freshwater insects (see Chapter 14 on climate change) that have expanded their geographic range as their host mosquitos or other vectors have increased their range.

Diseases of fish and other freshwater organisms have also been introduced by nonnative species and have caused difficulties for populations of native species. Diseases and parasites might be nonindigenous and expand with new and naïve hosts. Some examples include muskie pox (*Piscirickettsia* cf. *salmonis*), viral hemorrhagic septicemia virus (VHSV), and spring viraemia of carp virus (SVCV) (Pagnucco et al. 2015). Some diseases affect multiple species of fish, such as the SVCV originating from Europe that has impacted several native species in North America since carp have been moved around the continent. VHSV originated from salmonids in aquaculture in Europe and in North America and affects about 50 species of wild fish, sometimes causing mass mortality (Pagnucco et al. 2015). Muskie pox was first detected in the Great Lakes in 2002 and potentially threatens the fisheries for muskellunge (*Esox masquinongy*), a large freshwater predatory fish that is native to North America. As previously mentioned, the introduction of the western North American signal crayfish to Europe also contributed to the increased spread of the crayfish plague (itself introduced to Europe a century before) to more populations of European crayfish species, which has contributed to the decimation of populations of many of Europe's crayfish species.

There are debates about how much we should worry about nonnative species. In some cases, nonnative species are relatively benign and do not cause large impacts—for instance, the freshwater jellyfish (*Craspedacusta sowerbyi*) introduced to North American waters from China (Ricciardi et al. 2013). However, most of these species known to come from freshwaters are moved to places that lack their native predators and pathogens, and therefore, are often able to reach high densities (discussed more in the following section). Another concern is biotic homogenization globally, that is, a community with many invasive species that leads to having similar sets of species globally, and the loss of native biodiversity and ecosystem resilience (Olden et al. 2018). Most of the impacts of nonnative invasive species are relatively small; big impacts are somewhat less common and may take time before their populations are sufficient to be considered big impacts (Ricciardi et al. 2013).

INTERACTIONS WITH OTHER STRESSORS

Impacts of nonnative invasive species may depend on productivity and other existing stressors on a community; thus, impacts can be variable across environments. Nonnative invasive species can cause changes to productivity and communities. In some cases, they grow to such huge numbers or biomass that they can deplete food resources that are needed by native species. As an example, after their invasion in the early 1990s, the dreissenid mussels (zebra mussels and quagga mussels) were so effective at filtering out organic matter particles in the Hudson River in New York, that they cleared the water and improved water quality, but they also reduced food supplies to native filter feeders (Strayer et al. 2020). As filter feeders, such mussels can also become biomagnifiers of contaminants, which could create problems if native consumers could eat them. In some cases, these invasive species can also become a *trophic dead-end*, wherein native species cannot consume them, so energy cannot continue to be shared through the food web. This is particularly common with invasive mollusks that few predators can cope with.

Nonnative species may colonize more easily when native communities are stressed, and there is less competition or predation as a result. Climate change will make some habitats more easily colonized by invading species or it may reduce biotic resistance of communities to invasion (Rahel and Olden 2008; Ricciardi et al. 2013). Nonnative, invasive species may establish easily in the absence of their native predators or pathogens, may be strong competitors, may occupy an *underutilized* niche, or all of these. For instance, the signal crayfish (*Pacifastacus leniusculus*) was introduced to replace endangered crayfish species in Europe, but ultimately made the native species of crayfish even more endangered and moved across much of Europe (see Preau et al. 2020).

The water hyacinth, *Eichornia crassipes*, is effective at removing cadmium, chromium, and other toxicants from freshwater, and has been introduced in many places for such restoration purposes. Likewise, other plant species have been introduced for particular purposes, and removing them may create other problems. Not all nonnative, invasive species are negative for our ecosystems.

Sometimes nonnative invasive species become a new food source for native species. For instance, the formerly endangered Lake Erie watersnake (*Nerodia sipedon insularum*) preys heavily on the highly invasive round goby and may even owe part of its recovery to the extreme abundance of an easily captured food source (King et al. 2018). Likewise, the lake sturgeon

(*Acipenser fulvescens*)—considered threatened in the Great Lakes—appear to be adding the invasive zebra mussel and quagga mussel to their diet, which might help the recruitment and growth of young sturgeon (Jackson et al. 2002).

SOLUTIONS

Solutions include reducing the numbers of invasions and propagule pressure, removals and eradication, limiting the rates of spread, policy changes, and public education. Many activities are aimed at reducing the spread of species, and myriad attempts at reducing population sizes of exotic species have been tried. Early detection and control are important elements of the response of many government agencies; and surveys are a key component of these programs, including using environmental DNA (eDNA—fragments of DNA shed from skin or feces, etc. into water and identified through genome sequencing of water samples; see Chapter 16). For instance, survey programs for the round goby, which have rapidly increased and spread in the Laurentian Great Lakes, have shown that eDNA can be an effective and semiquantitative tool to detect spread and population growth (Nevers et al. 2018). Environmental DNA is also being used to detect the spread of the carp species *Hypopthalmichthys* spp. that was mentioned before (Nevers et al. 2018)—but DNA can move beyond where the species occurs and lead to false positives.

Policy and Public Awareness

One means of invasive species reaching additional bodies of water is through human transport of propagules on boats, boots, nets, etc. Government agencies and private companies often inspect boats and boots to ensure snails, mussels, algae, milfoil, and other species do not get moved from waterway to waterway, although this is rarely thorough. Often agencies place signs at marinas and other areas where fishers and other people using water for recreation can learn about the risks of species movement. Education and awareness of people can be a significant tool in reducing rates of spread. It makes sense to avoid transporting nonnative species by any method. For instance, better rules for importation of live organisms set by states and provinces in coordination might help (Pagnucco et al. 2015).

A ship's ballast is one of the primary ways that exotic species have colonized new places, and in the Great Lakes, about half of the invasive species were a result of discharging ballast water. Policies may cover whole mechanisms for transporting invasives, such as rules around the flushing of ship ballast water. Rules for the exchange of ballast water at sea—flushing tanks with seawater before entering foreign freshwaters—have been applied in the Laurentian Great Lakes since 1993 and were made even more rigorous in 2006 and 2008 (Pagnucco et al. 2015). The Nonindigenous Aquatic Nuisance Prevention and Control Act (a United States federal law) has required saltwater flushing of residual ballast sediments of transoceanic vessels since 2006 (Sturtevant et al. 2019). This is coupled with inspections for compliance. These rules reduced the propagule load from freshwaters overseas that were being dumped into freshwaters in North America. The United States Geological Survey (USGS) and the National Oceanic and Atmospheric Administration (NOAA) maintain a database of invasive species in the Great Lakes.

Many species are intentionally introduced, such as the stocking of fish for recreational fisheries. This action, usually by government agencies, has been recognized as a serious problem. Attempts to reduce risk have included creating triploid fish (three sets of chromosomes instead of two), which is intended to render such fish sterile (also used in some aquaculture). However, triploid fish also can have abnormalities (skeletal, eyes, etc.) which may reduce the recreational values of such fish. There are also technologies that kill female embryos, so that only male fish can be introduced. While the use of triploids and other methods reduces one problem of fish stocking, it does not remedy the competitive and predatory impacts on local resident species from such introductions.

Public education to draw attention to the threats of invasive species may be worthwhile (Strecker et al. 2011). This can include notices at public docks, fishing areas, aquarium clubs, and in pet shops. Another option would be to better regulate distribution of species that are strong threats to invade natural waters or provide some kind of option to return unwanted organisms to the pet shop (Strecker et al. 2011). In some places it is illegal to transport live fish in any circumstance, except by permit, in case they escape.

Many states and provinces have programs to remind people to check and clean boats, waders, and nets when they leave a body of water. Many species of invasive organisms, such as *Myriophyllum*, mud snails, zebra mussels, and others, can easily attach or get lodged in crevices in boats and their trailers. Washing boats (and associated equipment) and trailers thoroughly can help stop spread. The eruptive and damaging alga, *Didymosphenia geminatum*, is thought to be transported from place to place on the boots, waders, and fishing gear of anglers (and probably other users), although some jurisdictions consider this to be native to many freshwaters. Nevertheless, it is an excellent practice to clean and dry nets, boots, and even scientific sampling gear in between locations to prevent the possible spread of nonnative species. Various suggestions for cleaning include washing in moderately warm water, freezing, drying for several days, or washing in bleach (4% to 50% solution depending on target species and equipment).

Physical Removal

Eradication of almost any invasive species is unlikely because it is difficult to get every last individual from every local habitat, and those individuals that do avoid capture or detection are likely to have higher reproductive rates due to relaxation of density dependence, higher *per capita* resource supplies, and perhaps higher rates of dispersal. The eradication of small, local populations might be feasible if sufficient efforts are expended to capture every single individual of all life stages and exclude immigrants. In one example, the invasive rusty crayfish (*Faxonius rusticus*) was intensively removed from a small lake over an eight-year period. The goal was to see if it was possible to aid the population growth of the scarcer, native virile crayfish (*Faxonius virilis*) over the removal period and for 11 years afterward (Perales et al. 2021). In that study they were able to reduce the rusty crayfish to less than 5 percent of its initial numbers, and their numbers remained relatively low over the following 11 years, while the virile crayfish numbers increased almost 20-fold (although still fewer than the rusty crayfish), and other components of the lake rebounded, such as increased numbers of sunfish and an abundance of macrophytes (see Figure 11.3). It is possible to reduce the numbers of invasive species, usually with enormous efforts, but almost impossible to eradicate and difficult to sustain the reduced numbers of a nonnative species. The actions taken depend on the extent of the species, rate of spread, and methods for control.

The opossum shrimp, *Mysis diluviana* (formerly *M. relicta*), is native to eastern North American lakes, and was first introduced to a lake in British Columbia in 1949, thousands of kilometers outside its native range, by a misguided government attempt to augment production

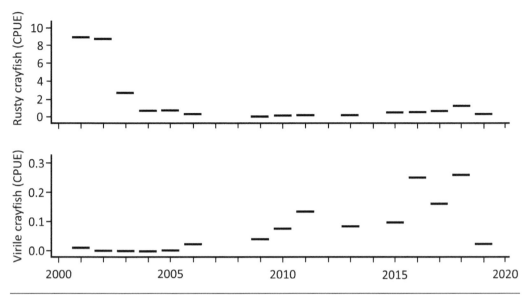

Figure 11.3 Example of a persistent effort at removal of rusty crayfish (invasive species) from a lake in Wisconsin, in order to improve the populations of the virile crayfish (native species). Numbers are catch-per-unit-effort (CPUE) and not densities. Note that even after removals, rusty crayfish are much more common than the native species. The reduction in the number of invasive species provides an opportunity for the native species to increase. Note that medians are shown for each year, but that there was considerable variation around the central estimates. Data redrawn from Perales et al. (2021).

of kokanee as prey for a trophy fishery of rainbow trout (Schindler et al. 2012). Kokanee (*Oncorhynchus nerka*) is the land-locked version of sockeye salmon, and its adults are considerably smaller than sockeye. Kokanee thus provides good prey for large rainbow trout. It was imagined that *M. diluviana* would provide an appropriately sized prey for the growing kokanee. However, *M. diluviana* migrates vertically downward during the day, and thus is not generally available to young kokanee that also migrate vertically, but in the opposite direction. Moreover, *M. diluviana* competes for similar prey as the young kokanee (Levy 1991; Spencer et al. 1999; Schindler et al. 2012), and in some of the lakes where the opossum shrimp invaded, they have lost native salmonid populations (Devlin et al. 2017). The opossum shrimp species has since been introduced to other large, western lakes, with the intent to fill in a size *gap* in food webs leading to sport fish, and with similar detrimental outcomes. As a result of this unfortunate mistake, the government created a removal program, along with other interventions, such as lake fertilization and stocking of fish. However, in one of these lakes, the British Columbia government supported a harvest of opossum shrimp starting in 2000 to try to reduce their numbers—and hence, their competition with species of interest—and also to generate a fishery. The annual harvest over the first 20 years was approximately 39.2 tons dry mass—far short of the suggested target of 200 T/y, or the approximately 350 T/y predicted to make a difference in the impacts of *Mysis* on the rest of the food web (Kay 2002). This is typical of such removal programs, that the intensity of catch is rarely sufficient or sustained long enough to deplete numbers, and the relaxation of density dependence allows the populations to replace themselves even more rapidly.

A similar example is the removal of bullfrogs (*Lithobates catesbianus*), a species of eastern North America and one of the world's 100 worst invasive species. The bullfrog was introduced to the Pacific region of the continent, and other parts of the world, ostensibly as a food production enterprise for humans. Attempts at reductions have included control of hydroperiods (autumn or winter drying) to dry out overwintering tadpoles, passive trapping of adults and tadpoles (for instance, using Fyke nets; Louette et al. 2013), biomanipulation by the addition of larger predators (such as northern pike, *Esox lucius*), and targeted adult killing—all leading to euthanasia of the captures. As with other species, removals have to be intense and sustained to reduce populations, and eradication is unlikely (Govindarajulu et al. 2005).

Fish can be depleted, and in some cases eradicated, by catching them with different kinds of gear. Of course, success depends on the lake or river size. In small lakes or streams, it is possible to use gill nets (lethal), seine or Fyke nets, or their variants (are nonlethal and require another means of euthanization). In lakes of the high Sierra Nevada (see Box 11.2), small lakes with introduced rainbow trout and brook trout had fish removed successfully using gill nets, and given there were no other fish, there was no concern over nontarget species (Pope et al. 2009). With any removal method, it is important to recognize that individuals that are not captured are likely to grow faster and have more offspring as density dependence is relaxed. This suggests that the removal rate must be very high to exceed the replacement rate. For instance, Holbrook et al. (2016) studied trapping results for sea lamprey near Michigan and determined that the current rates of capture were insufficient to reduce the numbers overall.

Models exploring the potential to decrease (or eradicate) common carp in a shallow, North American lake suggested that all methods together would be insufficient to reduce carp numbers sufficiently at the target numbers that were hoped for (Pearson et al. 2019). The methods included increasing commercial harvest rates as food, trapping, electroshocking embryos, and additional mortality factors (Pearson et al. 2019).

Removal of plant biomass cannot generally be accomplished with herbicides, which would affect nontarget species and cannot easily be controlled, yet pesticides are still sometimes applied. Some nonnative, freshwater plants can be removed using mechanical harvesters. Some species are controlled to an extent by mechanical roto-tilling, such as for Eurasian milfoil (*Myriophyllum spicatum*) in some North American lakes. Some government agencies have restricted the use of mechanical harvesting because it causes too much damage to the rest of the freshwater ecosystem. On a small scale, such as a recreational property on a lake, hand (or rake) removal is possible for some nonnative plants (depending on local laws), although it is important that the plants being removed are dried or bagged in a way that they (fragments or seeds) cannot reenter the water. There is evidence that it can be effective for divers to remove the milfoil and other plant species by hand. Biocontrol of aquatic plants through herbivorous invertebrates has been tried, but there can be problems if the herbivores are not as strict in their diet as hoped for and they also consume indigenous species. However, many aquatic plants can reroot from fragments, so these actions sometimes make the problem worse and aid in spread through water networks.

The Asian clam has invaded many waterways and is often a food web *dead end*; that is, there are few to no predators. An extreme method to reduce its numbers is by waiting until low flows when some individuals are exposed near the surface and burning them with what is effectively a flame thrower (Coughlan et al. 2019). This method was able to kill *Corbicula* when water was shallow, or they were at the surface but was less effective in mud or underwater. Regardless, this is unlikely to become an effective method to deal with the large geographic extent of this invader.

Biological Control and Poison

Introducing other species, often exotic, as biocontrol agents is common; but be careful because these nonnative species often escape and frequently prey on a wider range of food organisms than they were brought in to control (Bajer et al. 2019). For instance, black carp (*Mylopharyngodon piceus*) were initially introduced as biocontrol to help control mollusk problems, but since being established, they have expanded their range, are not as effective at biocontrol as expected, and may even consume some threatened mollusks (Lydeard et al. 2004). Consideration of other control programs such as targeted pathogens has been put forward, but the prospects of unanticipated impacts make this a dangerous option, as in any introduction of a nonnative species.

Efforts at biological control, such as the introduction of a novel predator or parasite, has been proposed for some species. However, as you have already learned, introductions rarely go well. There are few to no examples of successful biological control for aquatic invasive species. It is useful to remember that purposeful introductions are often the source of later problems. However, it depends on one's perspective; after all, the introduction of mosquito fish around the world (Cucherousset and Olden 2011) has probably reduced the mosquito population and the human diseases they spread, while also displacing native organisms—and it is difficult to gauge the balance of the value of those two outcomes.

An extreme method that has been used in the past is the application of poisons. Poisoning of lakes and other water bodies with antimycin or rotenone for nonnative fish has had success (Rytwinski et al. 2019), but has collateral impacts on native species. There are other chemicals used for more targeted interventions, such as lampricides specifically to control the invasive sea lamprey in the waters of the Laurentian Great Lakes (Young et al. 2021). These lampricides have been developed and tested to ensure low toxicity to desired species of fish, and applications to their nursery tributaries are generally of short duration to minimize exposure of other species. However, lampricide use has severely impacted populations of native lamprey species (Neave et al. 2021).

Barriers

After the invasion of the Great Lakes by sea lamprey in the 1920s, responses included poisons (lampricides; see the previous paragraph) and the building of low-head weirs (see Figure 11.4A) to limit their ability to expand their spawning areas to protect habitats of native species (McLaughlin et al. 2007). These barriers were modestly successful, but also blocked access for native species to upstream environments. In addition, these low-head barriers also posed a safety risk to people because of the hydraulic jump created, similar to an undertow, immediately downstream of the structure. Electric fences have also been used to create a barrier for this species (see Figure 11.4B) and others. Newer barriers are coupled with traps that will allow for the removal of the sea lamprey and provide for access to upstream sites by native species, particularly by adding fishways (McLaughlin et al. 2007).

Canals and other water diversions can result in movement of nonindigenous species between watersheds, as has been observed in Europe's canal systems, such as the Rhine-Danube, and in the United States. In Illinois, the Chicago Sanitary and Ship Canal connects Lake Michigan (and the rest of the Laurentian Great Lakes) to the Mississippi River system. Establishment of several species of carp (Cyprinidae), mentioned in the introduction of this chapter, into the Mississippi River would present a hazard if they were to move through the canal and into Lake

Figure 11.4   Barriers to dispersal of sea lamprey. (A) Sea lamprey barrier and trap on the Carp Lake Outlet in Michigan, used to block migrating sea lamprey from spawning grounds. Photo credit: Dr. Andrea Miehls, USGS. Public Domain. (B) An electrical barrier used to block sea lamprey passage in the Black Mallard River in Michigan. Photo credit: Dr. Andrea Miehls, USGS. Public Domain.

Michigan. A large electric *fence* was built by the U.S. Corps of Army Engineers to impede or stop the potential invasion of the Great Lakes by these three species of carp coming from the Mississippi River. This method may work, and in the absence of other practical solutions, the U.S. agencies have increased the power of the fence. However, recent studies have shown that the passing of barges weakens the fence and allows fish to move across the barrier (Davis et al. 2016). A solution that is not being considered because of the economic damage it would cause to shipping is to close the canal and sever the direct link. More effort to test technologies such as electric fences, bubble curtains, and sound (Feely and Sorensen 2023) will be needed.

Box 11.1   Case Study: Laurentian Great Lakes

The Laurentian Great Lakes ecosystem has been severely damaged by more than 188 invasive and nonnative species, mostly from Eurasia (Pagnucco et al. 2015) (see Figure 11.5). Many people consider these lakes to be the most invaded aquatic systems in the world. The Great Lakes have experienced invasions through all the potential pathways expected. Not every species that has arrived from elsewhere has expanded, but many of these species have been given multiple opportunities. The methods of arrival include in the ballast water of ships, purposeful introductions, releases of unwanted pets (red-eared slider turtles, goldfish, etc.), release of baitfish, and removal of barriers to dispersal for shipping by the creation of surface connections between water bodies (for instance, the canals and locks of the St. Lawrence Seaway to the upper lakes). Animals, algae, vascular plants, microbes, and perhaps other taxa, have invaded, and more species are likely to join them (Avlijaš et al. 2018).

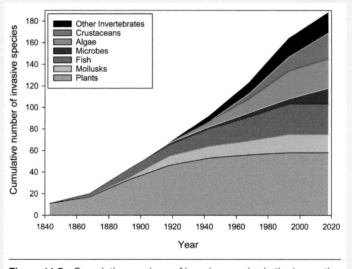

Figure 11.5   Cumulative numbers of invasive species in the Laurentian Great Lakes (data from Sturtevant et al. 2019).

   Some species have become commonly known to the public, such as the zebra and quagga mussels, the round goby, and the sea lamprey (see Figures 11.1 and 11.2). These species have spread widely and, at times and in some places, have reached extremely high densities, sometimes due to the lack of predators in the new ecosystem, for instance, zebra mussels or the spiny water flea, *Bythotrephes* sp. These species degrade habitat quality, and may outcompete native species, which can have knock-on effects on whole communities and ecosystems. The same is true of plants, such as Eurasian watermilfoil, which easily spreads vegetatively creating solid beds of plants that

continued

displace native species and provide little by way of food resources to herbivorous species. These changes also have dramatic economic effects and can even be a threat to the health of humans. Encrusting species such as zebra mussels can interfere with infrastructure, particularly pipes carrying water. Fisheries in the Great Lakes have been impacted by species such as sea lamprey and others, which predate upon or displace native species of fish (https://www.regions.noaa.gov/great-lakes/index.php/great_lakes-restoration-initiative/invasive-species/).

There are at least 28 nonindigenous fish species that have established themselves in the Great Lakes, some of which were first recorded in the 1800s, such as lamprey, alewife, common carp, and several Pacific salmon (Sturtevant et al. 2019). Four species of Pacific salmon (*Oncorhynchus* spp.), along with rainbow trout, have provided recreational opportunities, and there are even fishing derbies for some of these nonnative salmonids. Many other nonnative fish species, such as common carp and tench, are destructive by stirring up sediments resulting in increased turbidity, excluding other species, aiding in the dispersal of vegetative parts of invasive plants, and perhaps even spreading diseases and parasites to native species (Avlijaš et al. 2018).

Box 11.2   Trout Introductions into Fishless Lakes

Rainbow trout and other salmonid species have been widely introduced into freshwaters outside their range, and even into waters within their native range where they did not previously occur. As mentioned, rainbow trout is among the 100 worst invasive species as a result. One example of salmonid introductions that started more than a century ago is the deliberate planting of rainbow trout and brook trout into the High Sierra Nevada lakes of California to create recreational fishing opportunities in previously fishless lakes (Knapp et al. 2001). Some lakes were regularly stocked with fish in the absence of reproduction, and others included reproduction of the introduced fish and stocking.

A suite of changes to the original community of these lakes followed introductions. The tadpoles of the mountain yellow-legged frog (*Rana muscosa*) and Cascades frog (*R. cascadae*) were mostly eliminated, leading to large reductions of the overall populations (Knapp et al. 2001). Moreover, the larger benthic and planktonic invertebrates were seriously depleted by these salmonids (Knapp et al. 2001). In one set of such lakes in California, 16% of invertebrate taxa were eliminated due to trout introductions (Knapp et al. 2005). The specific effects of trout introductions on the community of any given lake were modified by other environmental variables, such as elevation, surrounding forest cover, and lake depth (Piovia-Scott et al. 2016).

Recovery of these lake ecosystems after fish died out (because of no further stocking and no reproduction) or fish were actively removed showed that these ecosystems were, in general, quite resilient. California has stopped stocking some lakes, and in others, they have instituted removal programs—partly motivated by the conservation of threatened amphibians and other species. In many lakes with salmonid removal, the biological community became similar to that of the reference lakes (never stocked with fish). This occurred within a few years in some cases, and in other lakes, it took up to 20 years; but still, eventually, most recovered after fish removal (Knapp et al 2001). Recovery depended in part on nearby sources of propagules (for example, adult aquatic insects) or frog populations, and in some cases, those sources were regionally diminished, especially for the yellow-legged frog (Knapp et al. 2001). Nevertheless, it is encouraging that some ecosystems have the resilience to recover from these kinds of disturbances.

ACTIVITIES

1. Where do you find information on nonnative invasive species for your state or province?
2. Which government agency is responsible for the control of invasive species and what are the government's responses to invasive species?
3. How can those agencies responsibly reduce future invasions?
4. Is there any example you can find of extirpation of an invasive species? Anywhere?

12

ENDANGERED SPECIES

INTRODUCTION

There is a huge number of species around the world that are at risk—that is, endangered (potentially at risk of extinction if not protected) or threatened (at risk of becoming endangered)—and freshwater organisms are no exception. In fact, some estimates suggest that freshwater organisms are becoming endangered at six times the rate of species from other ecosystems (Dudgeon et al. 2006; He et al. 2017; WWF 2022). For freshwater species, it is estimated that around 30% of species are threatened or endangered (Böhm et al. 2020). Another estimate suggests the extinction rate of North American freshwater species is five times that of species from tropical forests (Ricciardi and Rasmussen 1999). More than 42% of amphibian species globally are in decline and most of those are considered endangered. Given the range of threats we have explored in this book, it is probably not surprising that many freshwater species are at risk. The World Wildlife Fund (WWF) indicates that of monitored populations of freshwater vertebrates (for 3,358 populations of 880 different species), the declines in numbers have averaged 4% per year since 1970 (WWF 2018). The danger of extinction of freshwater taxa affects large and small species, plants and animals, and probably protists and microbes; and this is occurring in most parts of the world (Strayer and Dudgeon 2010).

There is an important distinction between risk factors that make a species susceptible to being threatened and the threats that a species is exposed to (anthropogenic or natural disturbances). One aspect of risk is being rare in some form, which Rabinowitz (1981) defined as combinations of range extent (geographic area), habitat specialization, and population abundance (see Figure 12.1). An important aspect of freshwater ecosystems that makes our aquatic species more vulnerable to threats is that freshwater populations have limited means of moving between suitable habitats. For example, it is not simple for a lake species to move between lakes if it has to move through a river system, past the gauntlet of predators it is unfamiliar with, in an environment it may not tolerate, and where it is not able to find its normal food, until it finds another suitable lake. There are many things that humans have done to alter the water landscape, sometimes called the *waterscape*, such as dams and other changes that further restrict movements between habitats.

The abundance of a species may begin to decline for a variety of reasons. Declines may be due to habitat loss or fragmentation, exploitation, invasive species, changes in their food webs, disease outbreaks (for instance, chytridiomycosis in amphibians), or other causes. There may also be cumulative effects of combinations of the aforementioned causes (see Chapter 13), or

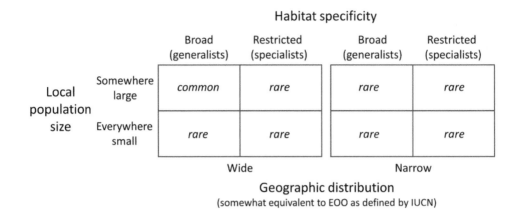

Figure 12.1   Rabinowitz' seven forms of rarity. Combinations of geographic extent, population sizes, and ecological specialization intersect to create different possibilities that might define a species as rare. Ecological specialization is a risk factor that can predispose populations to be susceptible to threats. Note the dichotomous nature of the categories, but there is no absolute distinction, and these binary classes are more useful in a comparative sense than specific criteria. Redrawn after Rabinowitz (1981).

persistent and intensifying impacts of those causes. The reasons for declines are sometimes not obvious, especially when a species uses multiple habitats during its life cycle since any one of those habitats might be the limiting factor, even if only used for a short period of time. When a single population goes extinct—for instance, if a fish species disappears from a lake—we refer to that as extirpation (or local extinction, that is, the local population goes extinct); not to be confused with global extinction where the species is gone from the wild (some may continue to exist in zoos or aquaria). As the numbers of a local population decline, the population can enter what is called the *extinction vortex*—using the analogy of water going down the drain in faster and faster spirals. This vortex happens as populations get smaller for several reasons, which can include higher *per capita* population pressure (such as predation), genetics (reduced genetic diversity, declines in fitness), demographic stochasticity (random variations of numbers in age and sex classes), environmental stochasticity (random variations in weather or other conditions), and even the Allee effect (Gilpin and Soulé 1986; Caughley 1994).

Most of the world's countries (with one notable exception) have agreed to an accord to conserve biodiversity in all its forms (genetics, species, communities, and ecosystems), known as the Convention on Biological Diversity or CBD (1992). As signatories to that agreement, those countries have a process for deciding which species are most at risk, either threatened or endangered. International guidance for these classifications has been established by the International Union for the Conservation of Nature (IUCN), which has good explanations of the criteria-based classification system, along with online lists of the global status of many species. There are several categories, including *critically endangered* (some jurisdictions use the term *imperiled*), of *least concern*, and others (see Figure 12.2 and Table 12.1). The IUCN Red List (https://www.iucnredlist.org/) is an excellent resource for information on species at risk that are being assessed at the global level.

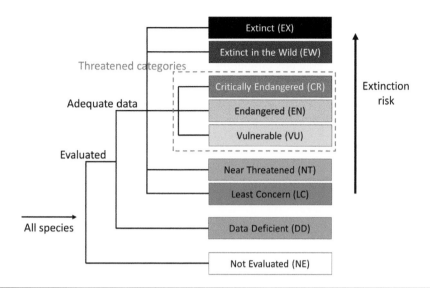

Figure 12.2 Categories used by the International Union for the Conservation of Nature (IUCN) and in most countries' assessment and listing processes. Redrawn after IUCN 2012. IUCN Red List categories and criteria, version 3.1, second edition. Gland, Switzerland, ISBN: 978-2-8317-1435-6.

Table 12.1 A simplified list of criteria used by the International Union for Conservation of Nature (IUCN) for determining species' status. Extent of Occurrence (EOO) is based on the area within a polygon including all occurrences of a species (or a designatable unit—usually below the species level, such as a subspecies). The Index of Area of Occupancy (IAO) is based on the summation of 2 km by 2 km grid cells positioned over all known occurrences. For freshwater species, this can be based only on the size of the freshwater habitat (freshwater species do not occur on the surrounding land), although that is still sometimes debated. See IUCN for greater detail. See also Fig. 12.2 on other listing categories.

	Critically Endangered	Endangered	Vulnerable
A. Reduction in breeding population size over the last 10 years or 3 generations	>90%	>70%	>50%
B. Geographic range: EOO—extent of occurrence, IAO index of area of occurrence, and number of occurrences	EOO <100 km² *or* IAO <10 km² *and* only one occurrence *or* severely fragmented *or* continuing decline *or* extreme fluctuations	EOO <5,000 km² *or* IAO <500 km² *and* only 5 or fewer occurrences, *or* severely . . .	EOO <20,000 km² *or* IAO <2,000 km² *and* only 10 or fewer occurrences, *or* severely . . .
C. Population size (breeding individuals)	<250 individuals *and* declining or extreme fluctuations in numbers	<2,500 individuals *and* declining . . .	<10,000 individuals *and* declining . . .
D. Population size (breeding individuals)	<50 individuals	<250 individuals	<1,000 individuals
E. Quantitative analysis—probability of extinction in the wild	>50% in 10 years or 3 generations (whichever is longer)	>20% in 20 years or 5 generations	>10% in 100 years

In general, there are five criteria for how a species is assessed for their risk of endangerment. These are:

A. Rate of decline in overall breeding numbers
B. Small range of distribution and few occurrences
C. Decline in breeding range of distribution
D. Increasing threat with no relent
E. Demographic modeling indicating sensitivity to threats

See Table 12.1 for the quantification of these rates and numbers. These are considered standard criteria that are applied globally. Results from these criteria determine the assessment of the risk level proposed for a species, and the next steps depend on the outcome of this assessment. The other critical element of the assessment is a threats analysis, which in combination with risk, determines the proposed status.

There are different terms used for the number of places a species is found, such as *occurrences*, *subpopulations*, and *locations*, which can be confusing (these are not exactly equivalent terms, but the subtle differences are often overlooked). Even if there are many individuals, the number of occurrences and the extent of the range of distribution are important. For instance, even with many individuals in a single occurrence (or location), there is a great risk to that species if that occurrence is disturbed and there are no other occurrences.

Each country does its own assessments, on the assumption that it cannot rely on its neighbors. Most countries use the IUCN categories of risk and their process for assessments, but there can still be differences within a nation or between nations. See IUCN-standardized criteria for ranking the degree of imperilment. Once a species (or subspecies) has been assessed, the listing of the species-at-risk requires legal authorization under legislation. Most countries have legislation that identifies how the status of a specific species is assessed and enables protection of the species that is at risk once they are listed as such.

If a species' populations are separate and unique, they may be assessed as separate units called *designatable units* (DUs) or *evolutionarily significant units* (ESUs). This means that a population within a species that has a unique genetic structure, behaviors, or is disjunct from other populations, may be assessed separately as DUs under various endangered species assessments. For example, breeding Chinook salmon show strong fidelity to their natal streams and thus, populations show strong evidence of unique and important genetic structures (see Beacham et al. 2021). In Canada alone, there are 28 DUs of Chinook salmon. As a result, separate DUs of Chinook salmon are considered individually for the purposes of assessment and listing—and subsequent management and recovery efforts. This is also important to remember when captive breeding is used so that DUs, and even populations within DUs, are maintained independently, and genetic structure is not contaminated by hybridization.

Freshwater species are also more likely to become differentiated into ESUs because of isolation; that is, they are restricted in dispersal by being limited to water. Three key processes resulting from isolation lead to multiple ESUs (DUs). First, some amount of isolation usually means populations are relatively small and vulnerable as a result of somewhat low numbers in particular places. Second, small populations are more subject to genetic drift, the random accumulation of genetic differences based on nonrandom frequencies of who mates. Third, specific habitats can result in the natural selection of suites of adaptive traits that lead to populations adjusting to specific local habitats, which can eventually be reinforced by mating isolation. An example of this would be the threespine sticklebacks (*Gasterosteus aculeatus*), which have differentiated across many habitat types. However, sticklebacks have even formed discrete pairs of DUs sympatrically (without complete isolation) within specific lakes, and the two DUs have little to no interbreeding and maintain distinct lines, despite the ability to hybridize in the laboratory (Schluter and McPhail 1992). This sympatric radiation of forms within lakes has occurred often, such as the African Great Lakes and their cichlids (Cichlidae) (Meier et al. 2017), the amphipods in Europe (Jardim de Queiroz et al. 2022), or multiple forms of char or whitefish in many lakes and streams (Doenz et al. 2019).

IMPACTS

The loss or degradation of specific habitat types (such as breeding sites, overwintering sites, etc.), overexploitation, loss of prey, fragmentation of habitats by dams and other barriers, pollution, and competition with other species, including nonnative invasive species (see Chapter 11) can all greatly impact the populations of freshwater species. A particularly dramatic example of the impacts of an invasive species on native populations is the intentional introduction of the Nile perch (*Lates niloticus*) to Lake Victoria in Kenya, in an attempt to produce a larger, harvestable fish (Kaufman 1992; Marshall 2018). That introduction resulted in the extinction of hundreds of unique fish species (mostly small, endemic cichlids) as a result of competition and predation from the Nile perch. This loss of biodiversity is an ethical and practical dilemma. A species can become endangered due to one or several impacts affecting a population. Consequently, a chapter on endangered species could be very long, so here I simply give some examples to help illustrate what is happening to various species in freshwater systems.

We tend to be most aware of large-bodied species disappearing, such as river dolphins, sturgeons (*Acipenser* spp.) (see Box 12.1), crocodilians, turtles, and others (He et al. 2017). Larger-bodied and later-maturing species are also more vulnerable because their populations are generally smaller; therefore, they need larger habitats per individual (or population) and are frequently subject to exploitation. However, even very small species may go locally or globally extinct with little attention; for instance, the small flatworm (*Crenobia alpina*) in Wales that has declined in population and has been lost from some watersheds, most likely associated with the warming of waters (Durance and Ormerod 2010). A small plant known as the Louisiana quillwort (*Isoetes louisianensis*) is known to grow in only five locations, and its occurrences are under threat—both reasons to list this species as endangered.

One example of freshwater organisms that are particularly at risk is the unionid mussels around the world (Ricciardi et al. 1998; Vaughn 2010) (see Box 12.2). Pearl mussels (*Margaritifera margaritifera*) in Europe and eastern North America were important for pearl fisheries in freshwater, they occurred in high densities that had significant ecological roles, but they are now endangered. Numbers of this species had declined by more than 90% by the 1990s (Geist 2010). No single cause for its demise is known beyond general habitat degradation, fragmentation, flow regulation, and serious reductions in salmonid fish, which are intermediate hosts for the pearl mussel larvae. In North America, estimates are that more than half of the approximately 300 species of freshwater mussels are imperiled; they are closely followed by freshwater gastropods, crayfish, amphibians, and fish (Warren and Burr 1994; Ricciardi and Rasmussen 1999).

Many more species may be at risk, but for many, we have very limited information, particularly smaller species. Often a species that seems to be at risk may be considered to be *data deficient*; that is, someone could ask whether the species has been suitably searched for in most places it could occur, and perhaps the search effort by knowledgeable people has not been sufficient. For example, the stonefly *Lednia tumana* (Nemouridae) was petitioned to be listed as endangered and was only known from a few high elevation streams in Glacier National Park in Montana. Further collecting confirmed not only that it was rare and restricted but resulted in the discovery of two new species of *Lednia* from other western United States mountain ranges, both from only a few locations each (Baumann and Kondratieff 2010; Muhlfeld et al. 2020). *Lednia tumana* was eventually listed in the United States as threatened. It is difficult to list every species that appears at risk since that takes considerable time and may incur legal protections, which may be costly. However, the larger risk is losing species that are legitimately in peril.

INTERACTIONS WITH OTHER STRESSORS

Loss of native species reduces the complexity of communities and may weaken their resilience and resistance to disturbances. A great deal of literature indicates that simplified food webs are less productive and less stable. Depending on the trophic level of a species at risk—that is, whether they are a top predator or prey—changes in population sizes may have broader effects on communities. As mentioned before, there are many reasons that a species' population may begin to decline in abundance. As a result, they may be outcompeted by invasive species, or the reduction in their numbers might enable introduced species (nonnative or other native species) to increase. Once invasive species are established, it might be difficult for an endangered species to increase again (Vaughn 2010). Depending on trophic position, the loss of a species may cause large-scale alteration of communities and lead to other species in their community becoming at risk; for example, if the initial species that is threatened is a key prey species. As a species' numbers decrease, they may become more vulnerable to additional environmental stressors (referred to as the extinction vortex). Stressed communities may become even more susceptible to invasions and other changes.

Changes to communities can lead to shifts in rates of various ecosystem functions, which can have other consequences for connected communities or downstream communities. These changes also impact the values that humans derive from nature (ecosystem services). Within communities, a range of species traits affects the roles of each species, and there are many dependencies—for instance, resources (prey), complementarity, facilitation, mutualism, etc.—within interaction webs of communities (Vaughn 2010). The decline or loss of a particular species can have a range of possible outcomes for the community and ecosystem, depending on its role. Species are not generally substitutable, so the loss of a particular species leads to changes in the rates of ecosystem functions and other properties of communities (Vaughn 2010). Moreover, individual species are considered to have a right to exist and should be protected for their inherent value.

SOLUTIONS

Laws, Rules, and Agreements

There are several international, national, and subnational (state or province) laws that contribute to the protection of species. Most of the United Nations member countries that ratified the international agreement on conservation of biodiversity in 1992, the CBD, created legislation that went into force in 1993. Another international agreement called the Convention on

International Trade in Endangered Species of Wild Fauna and Flora (CITES) restricts which species can be imported across international borders. This is helpful in reducing the movements of endangered turtles, fish, salamanders, and others (http://www.cites.org/), thereby curtailing the use of endangered species as souvenirs and pets, as well as body parts being used for medicines. The aim is to protect rare species that become more attractive to collectors as they get even more rare. The convention currently protects about 37,000 species worldwide. There are strong penalties for violations, but it depends on effective enforcement at border crossings, which may not be present or be less focused on wild species. There are many other laws intended to protect nature.

One of the first stages for protecting threatened and endangered species is to have them officially listed. There are several stages in the process. First, species that are of conservation concern are tracked by a state or provincial data center or heritage center, and those data go toward assessments of a species' risk. Those government agencies record occurrences of each species in an attempt to determine the number of occurrences and any changes in those numbers. Data for North America are also tracked by NatureServe (a nonprofit organization based in the United States). Evidence that a species is rare or declining can lead to a petition for listing from the government or the public. Second, a status report is prepared to summarize all available information, and specialist committees evaluate those reports to recommend a status. For rare species, it is often difficult to get adequate data due to that rarity, which may leave them data deficient. Finally, the recommendation is considered by government agencies for legal listing. In the United States, the *Endangered Species Act* (ESA) requires that someone or an agency petition the government to list a candidate species, and the ensuing process can be lengthy. The equivalent to the ESA in Canada is the *Species at Risk Act*. In Europe, they have the Habitats Directive, and there are similar entities in other countries. Once a species is officially *listed* as at risk, most jurisdictions require recovery plans to be developed and implemented to aid the species, hoping to reduce their level of imperilment.

Number of Occurrences

One of the criteria for listing species as endangered or imperiled is the number of occurrences, which could also be considered as a number of independent populations (sometimes called subpopulations). One idea from conservation biology is to encourage multiple populations in order to spread the risk of threats and to provide the possibility of metapopulations, where the rescue effect may be possible. This might also be considered related to the debate about whether it is better to have a single large area or several small protected areas. This is mostly in the context of allocation of land for protection and has been debated for years within conservation biology. The implication for at-risk species is how the allocation of space might affect the persistence of local population sizes versus the ability to disperse across a broader landscape.

A relatively recent method of detecting occurrences of rare species (or invasive species) in freshwater is through the use of environmental DNA (eDNA). This method is less disruptive to the natural habitat and the individual species and may detect occurrences more easily than expensive field sampling since it is based simply on the amplification of DNA from water samples for which there are primers available. For instance, Davy et al. (2015) surveyed nine freshwater turtles and estimated the cost as between 10 to 50% of normal field surveys. Likewise, sampling of eDNA for assemblages of species, such as a group of stream fish, showed that eDNA metabarcoding was more sensitive and efficient at detecting the presence of specific species than electrofishing (McColl-Gausden et al. 2021). However, one must note that there is always a possibility for false positives and false negatives.

Habitat Protection and Connectivity

Of course, the primary solution to all endangered species issues is to remove the threats. However, this is more easily said than done, given that loss and fragmentation of habitat, overexploitation, dams and barriers, pollution, and other threats continue. Protection of habitat before it is disturbed is preferable to having to restore habitats after the damage is already done (see Chapter 15). Habitat can be protected as long as there is habitat remaining that can be managed, restored, or reconnected to other existing occurrences. However, there are also provisions in most countries to respect private property rights, and if a landowner is unwilling to protect habitat, there are few options. Many countries build a component of habitat protection into plans for species' recovery. One term that has a legal meaning is *critical habitat*, and that refers to an aspect of habitat that is absolutely essential to the completion of a species' life cycle, even if it is only used briefly during their life—for instance, mating sites (Richardson et al. 2010). The definition of critical habitat is sometimes controversial due to the legal ramifications of when it is identified on private land.

Re-establishing a natural flow regime is important to many species (see Chapter 8), along with removing barriers where possible and in whatever way possible; for example, building fish ladders, removing perched culverts, and reducing water withdrawals (Januchowski-Hartley et al. 2013). One often imagines this in the context of fish habitats, but it can also apply to many other species, for instance, freshwater mussels (see Box 12.2). Texas wildrice (*Zizania texana*), a critically endangered grass, is found only in the San Marcos River in Texas, and only about 150 individuals remain. The decline in numbers is attributed to low water levels that are caused by the Spring Lake Dam (Terrell et al. 1978). Environmental flows are one solution to aiding some of the species that are endangered due to flow regulation.

Among the many barriers that may cause problems with connectivity, we can include roads where large numbers of amphibians, turtles, and crayfish can be killed as they move between habitats. In many places the solutions include dispersal tunnels under busy roads, fitted with drift fences to guide the animals toward the tunnel. These may be effective where there is a concentrated dispersal corridor between aquatic and terrestrial habitats; however, several studies show that they are not yet as good as one might hope. Risks from roads may even extend to species like dragonflies flying along roads and being killed by cars, but proposals to reduce road speeds near habitats of endangered dragonflies have rarely been successful.

Captive Breeding, Headstarting, and Reintroductions

A common response to deal with endangered species is captive rearing or captive breeding, which is costly. This can include breeding in hatchery conditions, rearing through critical life stages known as *headstarting* (for instance, rearing tadpoles through to metamorphosis), or even maintaining broodstock in laboratory conditions for future reproductive attempts. These broodstock programs are sometimes referred to as an *ark* or captive survival-assurance populations. Examples include fish hatcheries, which are used often for species of recreational and commercial importance, such as trout, salmon, sturgeon, and bass. However, such programs are also used for other fish, unionid mussels, amphibians, turtles, and others. *Ex situ* (not in the natural habitat) breeding and culturing are important methods that are used for species recovery (Preston et al. 2007), and attention needs to be paid to population genetics, which can show a lot of local variation (Geist 2010). This requires the use of multiple individuals as genetic sources, different individuals from year to year, and even maintaining heterogeneity in culture conditions so as not to select for particular genotypes (Geist 2010). It is important to recognize that the effectiveness of these kinds of programs are not often evaluated, and when they are, they do not appear to be the panacea that one might imagine (Rytwinski et al. 2021). IUCN and the American Fisheries Society have developed guidance documents for these kinds of species-recovery programs.

The European and eastern North American freshwater pearl mussel (*Margaritifera margaritifera*) is a classic example of a long-lived and previously common bivalve that has become endangered (see the beginning of this chapter). Pearl mussels, like other species, can be raised in laboratory conditions through reproduction and provision of hosts for the larval stage (glochidia). Once they drop off of the host, they can be released into the wild. Sometimes, adding in host fish is enough to kickstart a population of freshwater mussels, as long as the fish survive and it is the right fish host. However, the habitat may no longer be suitable for reintroduction.

There are various programs and methods of captive breeding for reintroduction. For instance, the Amphibian Conservation Action Plan records over 213 species of amphibians in such programs for the sake of conservation, but an increasing fraction of such programs are for survival-assurance breeding populations (Harding et al. 2016). Some of these programs take years to develop methods that are successful. For instance, captive breeding and rearing of the Devil's Hole pupfish, which occurs as a single population in Nevada, has taken more than a decade to get the right combinations of environments to be successful (Hausner et al. 2016). Endangered turtles, crayfish, fish, mussels, plants, and others are often bred and reared in captivity for conservation reasons. However, all of these programs are expensive in terms of personnel, costs of space, environmental control, food, medications, and specialized instruments. One also needs to remember that the reproductive value of a juvenile in a population is tiny, relative to that of a breeding adult (basic population ecology), and therefore, ensuring the survival of breeding individuals is more important than simply providing lots of propagules.

When organisms are taken into captive breeding and captive rearing programs, it is important to avoid loss of genetic variation. In some cases, it is sufficiently important to perform genotyping of individuals in order to ensure outcrossing of matings—that is, to avoid close relatives being mated. If one has a good genetic characterization, it can be possible to maximize genetic variation using pedigrees and outcrossing of broodstock. In some hatchery situations, a few individuals for breeding may be all that is available, but where possible, locally adapted strains and genetic variation should be maintained.

The release of individuals from captive breeding (and rearing) where the species is still present in its native habitat—also known as population reinforcement or augmentation—brings its own set of concerns. In some cases, agencies may use translocation of individuals from one wild population to another location. As previously mentioned, it is important to be aware of the potential for loss of genetic variation, and this might mean not overwhelming remnant wild populations with large numbers of individuals that are likely to be closely related (such as families of full siblings). There is also the danger of introducing disease into wild populations, so precautions against that can be vital—and testing for disease, antibiotic applications, and other actions can be taken. In some cases, organisms from hatchery conditions have not learned how to find food on their own or avoid predators. These *naïve* individuals may be more vulnerable to predators, and there are suggestions that they may attract predators that feed on wild individuals. It is also worth avoiding competition with wild individuals by not introducing more individuals than the currently available habitat might sustain. Limits of this captive breeding and rearing approach for at-risk species is that suitable habitats may have disappeared, which may be why the species is threatened in the first place. When possible, introductions from hatcheries should be to unoccupied, yet similar habitats in order to avoid any potential impacts on remnant populations.

As we learned in Chapter 11, invasive, nonnative species may threaten many species. We have already covered problems from invasive species, along with solutions that may be helpful to recovering some of our endangered species in the wild (*in situ*). Even native species may be a problem since species that prey on endangered species may then be considered a threat. There are many means of predator *control*, which can include culling predators in some situations or somehow protecting endangered species directly from being vulnerable to predators, such as predator-proof cages or netting. Dealing with pollution, sediment (which we have discussed in previous chapters), and habitat homogenization (simplification of habitats) will also help, but these are all major challenges and we need to realize that some endangered species are likely not recoverable.

Box 12.1   Sturgeons

Sturgeons are among the most endangered group of fish (Order Acipenseriformes: sturgeons and paddlefish); there are only 27 species globally—all in the northern hemisphere. There are only two living species of paddlefish (family Polyodontidae), although one of them, the Chinese paddlefish, has recently been declared extinct. These are relatively primitive fish, and along with being long-lived, they can grow to enormous sizes; for instance, the white sturgeon (*Acipenser transmontanus*) can get to six meters long and weigh nearly 1,000 kilograms (see Figure 12.3). As big fish, they typically occupy large habitats such as rivers and lakes, and are primarily freshwater for most of their lives, particularly their early life stages. Out of the 27 species in the order, all are listed by CITES,

Figure 12.3   A 240-centimeter-long white sturgeon. This white sturgeon was tagged with a passive integrated transponder, measured (fork length and girth), and released back into the Fraser River as part of the Lower Fraser River White Sturgeon Monitoring and Assessment Program. Photo courtesy of Troy C. Nelson, Director of Science, Fraser River Sturgeon Conservation Society, Vancouver, British Columbia.

continued

25 are on the IUCN Red List of at-risk species, and at least 19 are listed as endangered or critically endangered (Pikitch et al. 2005; Haxton and Cano 2016).

Fisheries for sturgeon were important at the end of the 19th century and the start of the 20th century, but overfishing and damming of rivers prevented access to spawning habitat and drove many populations close to local extinction (see Figure 12.4). One of the other commercial values of sturgeon is for caviar, their unfertilized eggs, which are considered a delicacy in many parts of the world. The most important fishery for sturgeon is the Caspian Sea, the home of six species of sturgeon, and even that fishery is now at only a small fraction of its historical levels (Pikitch et al. 2005). The demise of many populations of sturgeon has resulted in a lucrative black market for caviar, and as a result, illegal harvesting (poaching) is extremely high in some parts of the world. Now, aquaculture provides for half or more of the legal supply of sturgeon meat and caviar (Pikitch et al. 2005).

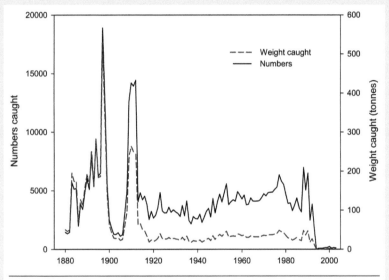

Figure 12.4 Catch statistics for white sturgeon in the Fraser River, British Columbia, Canada, as an example of a collapse in a commercial fishery for white sturgeon, similar to what has happened to many other sturgeon species (see Pikitch et al. [2005] for other examples). Note the large numbers and weight of white sturgeon harvested from the beginning of catch records in about 1880 and the declines in the mid-1910s. After about 1915, there were still many individuals caught, but those were mostly smaller individuals, accounting for the much lower total weight of the catch. After 1994 fishing for white sturgeon was restricted to a catch-and-release sport fishery. Redrawn with data from Echols (1995).

Sturgeons are also severely affected by habitat changes from flow regulation, armoring of channels, and alteration of sediment fluxes. Flow regulation by dams has alienated sturgeon from some of their habitat, changed thermal conditions, and affected their food supplies. As an example, dams on the Volga River in Russia have fragmented previously contiguous habitat and blocked sturgeon from 30 to 90% of their historic spawning habitats (Pikitch et al. 2005). For sturgeon, typical fish ladders are not a viable option to enable movements past dams, as most fishways are designed more for salmonids than the larger sturgeons. Fishways can be designed to accommodate

continued

the movements of sturgeons, even if the rate of those movements is still constrained by dams. Moreover, the lack of natural flows due to flow regulation has resulted in apparent changes to the open cobble, gravel, and boulder substrates that are needed for spawning—the sturgeons deposit sticky eggs that adhere to bottom sediments, and they may not stick to *fine* sediment. Some species are primarily supported by captive breeding in hatcheries. There are many things unknown about how some habitat changes affect these species, given the scale of the habitats they occupy.

As long-lived species living in large rivers and lakes, they are exposed to pollution in all its forms. The range of chemicals, plastics, and other inorganic particles they live in was described in Chapter 4. Their habitats are some of the most altered and polluted on earth. As bottom feeders, their food may be another pathway of contaminant exposure, and the productivity of their food supplies could also be compromised. Restrictions on legal harvest, international agreements (CITES), management of flow regimes, and other conservation efforts are the main hope for the sturgeons and paddlefish of the world.

Box 12.2   Freshwater Mussels

As mentioned before, freshwater mussels (see Figure 12.5) are among the most threatened groups of freshwater organisms. There are many reasons that these species are at risk, from channelization, loss of host species, pollution, flow modifications, burial by sedimentation, dredging, damming,

Figure 12.5 Western pearl shell (*Margaritifera faclcata*)—an example of a species of freshwater mussel in a bed of a stream in the state of Washington. Mantle expanded and inhalant siphon for taking in suspended particles shown clearly. Photo: Roger Tambor, U.S. Fish & Wildlife Service.

continued

invasive species, and others. In particular, the Mississippi River basin is a global hotspot for diversity of Unionida mussels with around 300 species of the world's approximately 900 species. That region of the United States is also a densely populated area.

Mussels are often patchy in their distribution, and those patches are sometimes referred to as *mussel beds*. This also means that detailed distribution maps are valuable in identifying target areas for conservation or restoration (for example, Ries et al. 2016). Adults are capable of small movements of one to a few meters (estimated annual maximum of 100 meters) and thus have limited capacity to respond to changes in their habitats (Vaughn 2010). Mussels are typically long-lived; a general assumption is that they take approximately five years to reach maturity (Ries et al. 2016), and it is possible to find populations of adults with few or no juveniles, which can indicate recruitment failure. Some populations have been described as *living fossils*; that is, as the adults die with no youngsters in their populations to replace them, the population will disappear locally.

The availability of hosts is critical for many species because in most species the glochidia (larvae) released from adult mussels must spend some time developing as gill parasites of host fish, sometimes having a very limited range of host species. One species, the salamander mussel (*Simpsonaias ambigua*) uses a salamander, the mudpuppy, as a host. Some mussels have curious ways of attracting fish, such as pieces of flesh (their mantle flaps) that they can move like a worm or other invertebrate or a small fish, and then the mussel expels its larvae at the fish. Some species (combshells, *Epioblasma* spp.) attract fish and then trap them between their valves to pump their glochidia onto the fish's gills. The particular host fish species is important to the success of the mussel larvae and provides a dispersal mechanism. When glochidia have grown to the point that they are ready to find their final home, they drop from the host. Unfortunately, they are vulnerable to getting washed along by the currents and may end up in unsuitable habitat, and while there is some small amount of movement possible to select habitats, it is on the scale of perhaps meters and not much more (Amyot and Downing 1997). Thus, there can be considerable mortality of larvae if they do not land on suitable habitat, or if their habitat is altered after they have settled.

Mussels are filter feeders and can filter large volumes of water, thereby providing a large capacity to remove particles and nutrients from water flowing by, as long as the mussels are abundant (Vaughn 2010). One estimate of this ecosystem service provided by mussels is that the native mussels in the Mississippi River 724 kilometers upstream of St. Paul, Minnesota, filter over 86 times the amount of water processed by a wastewater treatment plant in that city (the latter estimated at about 700,760 m³/d) (*Source*: United States Geological Survey). This filtering capacity is part of why freshwater mussels are considered ecosystem engineers, as not only do they provide physical structure that allows other organisms to find suitable habitats, they also convert large amounts of particles from transport into nutrient recycling and feces (and pseudo-feces) that become organic matter that is available to other consumers. However, this also means that mussels can ingest large amounts of contaminants and are well-known for bioaccumulating toxins (sometimes also used as biomonitors—see Chapter 16). It is important to recall that not all mussels have the same dimensions of filtering cilia, the same spatial occurrence and abundance, or the same thermal performance, therefore, species are not simply redundant, and one species cannot substitute for another (Vaughn 2010).

ACTIVITIES

1. Search for which freshwater species are considered to be *at risk* in your state or region. This may be on a natural heritage or NatureServe website maintained by your region.
2. Find out how your state or province deals with endangered species. For instance:
    a. Do they have stand-alone legislation regarding endangered species?
    b. Do they develop recovery plans or strategies?
    c. Are there captive rearing programs?

13

MULTIPLE STRESSORS AND CUMULATIVE EFFECTS

INTRODUCTION

Many of the impacts we have reviewed in this book are generically referred to as stressors. This chapter is intended to expand on how we address the cumulative effects and potential nonadditive interactions of multiple stressors. Most of these stressors do not occur in isolation and hence, we must consider the impacts of multiple stressors acting on individuals, and consequently, on species and ecosystems (Crain et al. 2008; Seitz et al. 2011; Piggott et al. 2015a). The term *multiple stressors* includes consideration of the effects of mixtures of chemicals found in surface waters (Schäfer et al. 2023). Stressors affect the growth, reproduction, and general metabolism of individuals, and if certain thresholds are exceeded, death may follow. It is important to note that anthropogenic stressors, many of which are discussed in this book, also interact with more natural stressors, such as floods, drought, and temperature. In addition, stressors can modify ecological interactions between populations, such as in predator-prey relationships and competition within freshwater ecosystems. Impacts of stressors can be dependent on ambient conditions, such that the intensity of response might be greater when temperatures are higher than when it is cooler (or vice versa). Consequently, cumulative effects can result from impacts of more of the same kind of stressor or from multiple stressors adding their impacts. Cumulative effects may increase stress across space, for example, accumulating across a watershed. They can also accumulate over a period of time with additional sources of stress being added. Cumulative effects might be detected at a single location, or across space when decreases in regional biodiversity or productivity are observed (for example, Streib et al. 2022). Moreover, the impacts of multiple stressors can differ from additive effects, that is, they can interact with each other, and the effect might be bigger (synergistic) or smaller (antagonistic) than the simple addition of individual effects (see Figure 13.1). Synergistic effects occur when an individual or population is negatively impacted by one stressor, such as a contaminant, then the second stressor (for instance, heat) is added, and the combined effect is more serious than the addition of the individual effects. For an example of an antagonistic effect, a pesticide might have a negative effect on a freshwater species, but slight nutrient enrichment may result in increased growth, slightly offsetting the negative impact. Multiple stressors and cumulative effects are current themes in applied biology (see Sabater et al. 2018).

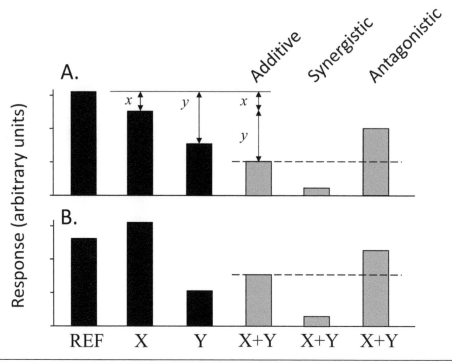

Figure 13.1   A schematic of the possible range of effects between individual stressors X and Y and a reference state (REF). REF represents an undisturbed condition to which these impacts would be compared. These impacts can be measures such as changes in growth rates, population densities, species richness, ecosystem rates, or others. The impacts usually decrease in one or more of these types of measures. (A) Additive effects are simple additions of two or more impacts. In the top panel, stressor X results in x reduction from the reference condition, and stressor Y results in y reduction, so the additive (combined) impact of stressors X+Y is simply the REF level minus x and minus y, shown as the distance from the top of the reference to the dashed horizontal line. The synergistic case results from the combined effects of stressors being greater than the additive situation (more negative than the additive case), and the antagonistic case results in a lesser effect than in the additive case (less of a negative impact of the two stressors combined). (B) Other situations can occur when one stressor, such as nutrients or loss of a predator, might increase the growth or survival over the reference state, while the second stressor reduces growth (or other response measure). Figure redrawn from Crain et al. (2008).

A meta-analysis of the effects of pairs of stressors from 88 studies indicated that there were more antagonistic outcomes (41%) than synergistic responses (28%), in part because stresses like nutrient additions may produce higher productivity offsetting the negative effects of other stressors (Jackson et al. 2016). However, a review of multiple stressor impact studies found that the majority of tests for interactions were not synergistic, and mostly were additive (Côté et al. 2016). However, predicting outcomes rather than describing impacts remains a big challenge (Suleiman et al. 2022). These results provide a glimpse into why multiple stressor effects are so tricky to predict.

There is a myriad of possible consequences of the combinations of multiple stressors on individuals, populations, and communities, and this is also a source of uncertainty for managers.

Piggott et al. (2015a) pointed out that beyond possible synergistic and antagonistic impacts, these interactions can result in reversals and other complicating differences in effect sizes beyond the additive case. One also must be clear on the appropriate null, statistical model since different models make different assumptions about the sensitivities of individuals to pairs of stressors (Schäfer and Piggott 2018; Spears et al. 2021). If species' sensitivities (tolerances) to two stressors are negatively correlated, then the action of each stressor may separately kill individuals and the magnitude of population consequences from the two stressors is likely additive (see Figure 13.2A). The multiplicative model assumes sensitivities are not correlated, and thus sensitivity to stressor A is independent of sensitivity to stressor B (see Figure 13.2B). On the other hand, the *dominance* model occurs when the worst, or dominant, stressor has the greatest effect, and there is no greater impact from adding the second stressor (see Figure 13.2C). This can also be seen in the concept of co-tolerance where, if a species is relatively tolerant to one stressor, they may also be tolerant to another, and so adding stressors may yield no greater impact (Vinebrooke et al. 2004). These individual and population-level effects propagate to the community level, accounting for changes in composition. For example, with co-tolerance, if the community is already represented by species that are tolerant to the first stressor, then an additional stressor might not have additional impacts. For example, some stressors may mask other effects, such as warmer water may have a strong effect or no additional effect depending on sensitivity to the first stressor (Morris et al. 2022). Moreover, the temporal order of exposures to multiple stressors may influence the outcomes (Birk et al. 2020; MacLennan and Vinebrooke 2021).

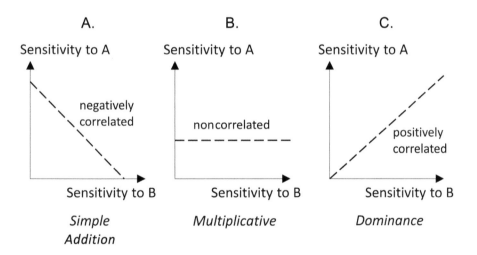

Figure 13.2   Different sensitivities of an individual (or by extension a population) to two stressors, A and B. The effects of these stressors can be described in three general models: simple addition, multiplicative, and dominance models. With the *simple addition* model, the sensitivities of a species to one stressor are unrelated to the mode of action of the second stressor, and so both stressors have impacts. The *multiplicative* model assumes that the mechanism of action of each stressor is different, and therefore the organismal sensitivity to one stressor is not correlated with sensitivity to the other stressor. The *dominance* model occurs when there is some degree of positive correlation between sensitivities to stressors; for instance, once a population is reduced due to one stressor, the second stressor has a smaller effect than it would if the population had been exposed to the second stressor alone. That is, one stressor is sufficient to do all the *damage* that will occur. The sensitivities are often specific to that particular stressor. Redrawn after Schäfer and Piggott (2018).

There is an important distinction between additive versus interactive (multiplicative or dominance) models of cumulative effects (see Figure 13.2), and deviations from each prediction that would be evidence of synergism or antagonism (see Piggott et al. 2015a; Côté et al. 2016) (see Figure13.1). For example, additive models would simply mean that if one stressor reduced a population by 10% and another stressor reduced that population by 10%, then we would see a 20% decrease. However, imagine that one stressor reduces a population of some sensitive organism by 60% and another stressor in isolation has a similar effect of reducing that species by 60%. We cannot have −20% of a real number as a prediction. This is where the multiplicative model becomes useful as a guide to impacts on populations, communities, and ecosystem processes (see Figure 13.2). However, the emphasis on deviations from the additive model has not always been coupled with a mechanistic understanding of the processes and appropriate null models of interactions, which will be key to developing solutions.

Many kinds of legislation refer to cumulative effects and the need to avoid impacts accumulating in space or through time. For instance, cumulative effects are mentioned explicitly in the U.S. National Environmental Policy Act and the Canadian Environmental Assessment Act. However, how to predict such impacts and monitor for cumulative effects remains vague (Spears et al. 2021). Part of the challenge with prediction is that the mechanisms of interaction of many kinds of stressors are poorly known, and even the combined effects of several contaminants may not be known (Spears et al. 2021). Ecotoxicology has studied the effects of combinations of contaminants for decades, but there is still much to learn. Moreover, most of what we know about multiple stressors and cumulative effects comes from experimental studies, therefore, knowing what to monitor in the field remains challenging.

IMPACTS

In earlier chapters we have seen the impacts of individual stressors and a little of how they interact to affect freshwater systems. Most experimental studies have shown that the combined impacts of multiple stressors can be additive or even more than additive, that is, synergistic (interactions) (Beermann et al. 2018; Chará-Serna et al. 2019). Such laboratory or field-based mesocosm experiments are highly instructive in how stressors can interact, although they do not often identify the mechanistic basis of impacts. Comparisons can be made between the individual and combined effects of fine sediment, salinity, insecticides, nutrients, and others in replicated, factorial designs. On the other hand, these kinds of experiments cannot possibly test interactions between all the possible stressors or the levels of those stressors under a range of environmental conditions. Moreover, most of these experiments test only two levels of each stressor, and usually it is only ambient versus some elevated level—this is done to allow for replication to test for different models of interactions between stressors but is not useful for predicting effects across a range of stress levels. It is important to note that the effect sizes of interactions can change, depending on stressor intensities such as concentration (Turschwell et al. 2022).

Most of our freshwater ecosystems are subject to more than a single stressor, and often these are in the forms of multiple kinds of chemical contaminants. In field surveys, more than 95% of river sample sites in Germany (Schäfer et al. 2016) and 68% of streams (of 434 sites) in the United States (Waite et al. 2021) had two or more stressors that exceeded established risk thresholds. In a survey of 101 streams in Germany for 75 pesticides, 33 pesticide metabolites, and a further 257 substances, Liess et al. (2021) found that 83% of streams exceeded the combined pesticide concentration targets for the protection of aquatic life. However, it is difficult to compare the expected effects of the combinations of so many kinds of stressors. To determine the potential effect on stream invertebrates, they used a summation of toxic units (TU_{sum}) based on the addition of the LC_{50} values for each pesticide for invertebrates. They also considered the maximum toxicity of the highest concentrations of single contaminants (TU_{max}) and showed a clear negative relationship with the invertebrate communities. In this case, the single most toxic compound best explained the observations of degradation of stream communities, while the concentrations of the second most toxic compound were often an order of magnitude lower (Liess et al. 2021). This illustrates the challenges of evaluating multiple stressors, many of which are not regularly monitored (see Figure 13.3).

Lake Victoria in Africa is a good example of a lake that is showing cumulative effects. Lake Victoria had stressors added in the form of nutrients, nonnative fish, increasing exploitation of fish populations, increased sediments, and of course, climate change (Hecky et al. 2010). Introduction of Nile perch and Nile tilapia to the lake, where they had not occurred previously, coupled with large increases in nutrients, resulted in a great expansion of the fish harvest from the lake. However, all the changes also resulted in dramatic alterations to the ecosystem, and large losses of endemic species of fish (and perhaps other species), as we saw in Chapter 12.

Oil and natural gas developments can exert multiple stressors on stream fish. These industries can impact sediment loads, nutrient fluxes, and contaminants, while decreasing riparian areas, all of which can interact with flow variations, especially at low flows (Walker et al. 2020). Evaluation of the effects of low flows and land use on stream fish abundances showed a more-than-additive, negative effect of the two sets of stressors combined, resulting in a synergistic effect (Walker et al. 2020). Moreover, the magnitude of these interactive effects differed

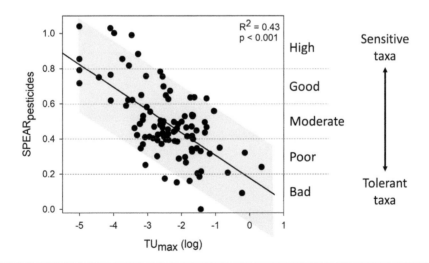

Figure 13.3 Cause and effect relationship between toxic units (TU_{max}) of the highest concentration pesticide from each of 101 streams in Germany. $SPEAR_{pesticides}$ is a measure of the condition of the invertebrate community, showing a clear reduction in condition as the maximum toxic unit increases. On the right are classes associated with the condition of the benthic community. The blue band represents the 90% prediction interval. From such relationships one can estimate the numbers of streams not meeting acceptable conditions. Redrawn with data from Liess et al. (2021).

depending on which fish species was assessed, and whether population abundance or colonization rates were considered.

Another challenge is that there are several pathways through which a particular stressor can create impacts. Just considering contaminants, such as heavy metals, some substances may cross membranes in dissolved form, but a large burden may come from direct consumption of food (Poteat and Buchwalter 2014). Some organisms may be exposed to heavy metals and other contaminants that interfere with osmoregulatory pathways, and chemicals may be taken up that way. However, the pathways of uptake of particular contaminants do not work the same in all taxa or in all conditions of water quality, since water hardness or pH can affect rates.

The impacts of multiple stressors may be through changes to one part of a food web, such as reductions in the number of basal resources or a key prey species, and that alteration might be transmitted through the food web even if there is no apparent stressor acting on other species. This could also happen if an influential predator is reduced by some set of stressors. Stressors can also shift size distributions of members of a food web (often to smaller bodied, fast life-cycle species), so that appropriately sized prey may be unavailable to some consumers (Sprules and Barth 2015; Pomeranz et al. 2019). Measures of shifts in food web composition and size-frequency distributions can be useful indicators to assess impacts of single and multiple stressors.

INTERACTIONS WITH OTHER STRESSORS

This chapter is mostly about interactions with other stressors, including climate change (see Chapter 14). However, there are still many aspects that are not considered in the literature and provide opportunities for future research. Most of the published studies have considered existing communities in one or two places. However, environmental gradients, such as latitude, productivity, or community composition, can also affect the magnitude of impacts of multiple stressors. There are additional components that are needed before we can comfortably extrapolate the results from one or a few studies to other places and ecosystems. The response of communities may also differ depending on whether the community has lost species or gained a nonnative species. For instance, an experiment by Greig et al. (2012) found that the presence or absence of the top predator led to very different outcomes of community structure in response to experimental manipulations of temperature and nutrients (see Figure 14.1). Atwood et al. (2013) showed carbon dioxide (CO_2) emissions from water were much higher when the top predator was removed from each of three different types of freshwater ecosystems. In Lake Erie, increases in nutrient concentrations increase hypoxia, but also increase productivity of prey species, which may create opposing responses depending on a fish's ability to escape zones of hypoxia (Almeida et al. 2022). When a stream community was subjected to both elevated fine sediment levels and the presence of a predator, the mortality rate of prey was lower than predicted from simple addition of the individual effects, an example of the multiplicative and dominance models that were previously discussed (Louhi et al. 2017). As a final example, stream periphyton was exposed to triclosan (an antimicrobial chemical), grazers, or both triclosan and grazers. In that experiment, exposure to triclosan reduced the ability of the periphyton to rebound from grazing pressure (combined effect), whereas grazing alone enhanced the productivity rate of the periphyton (Guasch et al. 2016). These examples indicate that changes to community structure will also modify the mechanism by which other stressors impact a particular ecosystem. However, scaling up from small-scale experiments to ecosystem-level effects is a challenge that requires broader landscape analyses (Lemm et al. 2021).

Stressor impacts within freshwater ecosystems can be exported to terrestrial systems or the marine environment. For example, adults of aquatic insects carrying a body burden of contaminants from their larval aquatic habitat are consumed by spiders, birds, bats, or other predators, leading to detectable concentrations and effects in terrestrial species (Walters et al. 2008; Graf et al. 2019). There are also contaminants that might be associated with marine-derived subsidies, such as salmon, eels, or lamprey returning from the sea and carrying contaminants that were acquired through biomagnification in marine food webs and then carried back to freshwaters during migration (Christensen et al. 2005).

It is important to remember that even short-term impacts might leave long-term legacies. For instance, a hurricane or landslide might occur in a very short time window and leave effects that are detectable a decade or more later, and perhaps even centuries. This could also be a result of a spill, dam failure, or other event. Thus, we need to consider variation (especially the extremes) in these kinds of disturbances, as well as the average. There is also considerable evidence that the sublethal, diffuse pollution of many of our surface waters may have already resulted in shifts to communities of more tolerant species (Reiber et al. 2021), which we might consider reference sites in the modern world. In addition, recall that ecosystems are nonstationary, as mentioned in the introduction, which also alters what we think of as reference sites. These changes through time result in what we call a *shifting baseline*. What we perceive as reference conditions now may actually be seriously impaired from a century or more of impacts and, therefore, not representative of preindustrial and undisturbed ecosystems (see also Chapter 16).

A related outcome of multiple stressors and cumulative effects is that a population, community, or ecosystem may pass a threshold beyond which recovery becomes less likely. These thresholds are also known as regime shifts or tipping points. One management response to this threat is to look for signs that a system is approaching a threshold so that action can be taken. However, a synthesis by Hillebrand et al. (2020) of over 4,600 studies suggests that thresholds will rarely be detectable due to the large amount of variation within natural systems. This is further evidence that we cannot wait until the last moment to act in order to protect the environment because we cannot yet reasonably predict at what point irreversible damage will occur. However, there is a general consensus that multiple stressors diminish the resilience of communities and ecosystems to further change. As pointed out by Côté et al. (2016), application of these ideas by managers will mostly be important if different interaction types require different management responses, or if tipping points to be avoided are identified.

SOLUTIONS

Solutions to treating multiple stressors and cumulative effects start with recognizing the range of possible impacts and with having mechanisms to incorporate that recognition into policy and monitoring. For instance, as currently devised, the European Union's Water Framework Directive (WFD) largely deals with contaminants as if they were independent; although progress is being made (Spears et al. 2021). As discussed before, cumulative effects are mentioned explicitly in some legislation, and multiple stressors are increasingly entering policy discussions. However, agencies have difficulties determining how to implement these concepts as enforceable guidelines. As Orr et al. (2020) point out, solutions must acknowledge ecological complexity, better integrate realistic scales, and work toward better predictive capacity. New methods are being refined to use extensive monitoring data, such as in the European Union, to detect the effects of multiple stressors (Spears et al. 2021). Having good detection tools will make it increasingly feasible to determine how managers should evaluate and prioritize measures to protect and restore our freshwaters.

Management of single stressors may have little value if other stressors are not considered, and as a result, most environmental regulatory agencies will consider cumulative effects. For instance, most watersheds in developed regions have some pollution, channelization, habitat loss, managed flows, higher peak discharges due to altered hydrology, and other problems (Craig et al. 2017). These all combine with each other in some ways that modify species' responses. Further, we need to consider that ecosystems are nonstationary; that is, environments are changing through time, such as by climate change, atmospheric deposition of contaminants (nitrogen, pesticides, acidification, etc.), and increased concentrations of CO_2 in the atmosphere.

Better Statistical Models and Experimental Tests

Detection of synergistic or antagonistic effects of multiple stressors depends on a good definition of the statistical assumptions needed. There are different null models for such tests, which make different assumptions about how stressors interact and require a more explicit explanation of the mechanisms by which a stressor affects an individual (Schäfer and Piggott 2018). The stressor-effect relationships can take several forms, such as linear or sigmoidal, which need to be incorporated into the choice of the appropriate statistical model. Individuals (for instance, of different age classes or sexes) may be differentially sensitive to stressors. The choice of a statistical model for examining the effects of multiple stressors could be additive models, which is the simplest version, but the choice of models should consider the mode of action and organismal sensitivity (see Figure 13.2). There are other models, such as the multiplicative model and dominance model, that must be considered in light of the mechanisms of effect (Schäfer and Piggott 2018). The multiplicative model uses proportional (or percentage) responses and assumes that there is no correlation in sensitivity to the two stressors. The dominance model assumes that the main impact will be from the worst stressor and that additional stressors will have lesser impacts than they would have had independently.

Experimental work can provide insights into interactions between multiple stressors, but usually only a couple of stressors and only at a couple of levels (magnitudes), limiting the generality we can derive from this approach (see Box 13.2). Many laboratories around the world are testing interactions of stressors on freshwaters, generating a greater understanding of the possible range of outcomes (additivity, synergies, antagonisms, etc.). The contribution of many of these studies is not the specific outcomes of a single experiment (with the number of stressors and stressor levels limited by replication), but the generality of outcomes that can be used to predict other classes of interactions. For instance, Bracewell et al. (2019) used a meta-analysis of many multiple-stressor studies to define typical outcomes for groups of stressors. One further aspect of this kind of experimental work is to extend studies to sufficient duration and spatial scales to assess the indirect effects (through food webs) of stressors on species' interaction (see Gessner and Tlili 2016; Bracewell et al. 2019).

There can be a lot of temporal variation in the intensity of stressors, such as temperature peaks, sediment pulses, and contaminant concentrations. There remains uncertainty about actual exposures to stressors, largely because much of the monitoring that is done is based on discrete sampling, which may miss extremes. For instance, high-flow periods may expose organisms to contaminants that were washed into water that could be missed, or low-flow periods may concentrate the contaminants. The development of integrated sampling tools, as mentioned in Chapter 4, helps to address this. For instance, Schreiner et al. (2020) showed that by using passive samplers for pesticides, many peak exposures that were associated with floods were missed by monitoring schedules for discrete samples and that even the duration of sampling with continuous samplers can affect estimations. More attention to how stressors are monitored (frequency and timing relative to other field conditions) will provide better data for exposures. Some agencies and nongovernmental organizations have developed tools to aid in studying multiple stressor effects, such as material on the Freshwater Information System website (http://fis.freshwatertools.eu/).

Advances in Ecotoxicology for Multiple Stressors

Many new chemicals are tested for their toxicity in a standardized laboratory setting, but they are typically tested individually with tolerant, laboratory organisms (see Chapter 4). More attention to the combined effects of chemicals has been promoted in the past decade (Gessner and Tlili 2016; Schäfer and Piggott 2018). Ecotoxicology provides a better understanding of the potential interactions of stressors and modes of action (see Boxes 13.1 and 13.2). For instance, contaminants may bind to particles (organic or inorganic) and find their way into food webs more easily than if those same contaminants were dissolved in ionic form (for example see Bundschuh and McKie 2016).

The awareness of potential incremental effects of multiple contaminants has led to a few different ways of addressing these potentially additive or interactive impacts. One somewhat traditional approach to predicting the combined effects of stressors includes the concentration addition and effect addition methods—simply adding the effects of individual stressors. These methods assume that individual toxicants have separate mechanisms of impact, and their effects can be added as a first approximation. There are a variety of terms and calculations, such as toxic equivalents, concentration addition, TU_{max}, TU_{sum}, cumulative criterion units, or the hazard index (Clements et al. 2010; Schäfer and Piggott 2018; Posthuma et al. 2020). However, looking at the community-level impacts of toxicants in laboratory studies, the synergistic impacts of stressors on toxicant sensitivity may be 10 to 100 times higher than predicted by government safety levels based on those traditional methods (Liess et al. 2016). This indicates that sublethal effects on individuals and populations may be greatly amplified by other stressors.

Liess et al. (2016) surveyed aquatic studies of multiple stressors in a meta-analysis and found that when combined with toxicants, other environmental stressors increased sensitivity by up to 100-fold. They developed a *stress addition model* to help predict the combined, additive effects of multiple stressors as a tool to guide management. Nevertheless, many chemicals in the marketplace that can end up in freshwaters are not monitored. In the European Union, only about 24 priority substances of the estimated 144,000 compounds (approximately 0.2%) that might enter surface water are used to determine *chemical status* (Posthuma et al. 2020). In addition to the priority substances, each member state in the European Union identifies a number of river-basin specific pollutants, often more than 100, to measure for water quality assessments. As noted, many water quality standards still use individual assessments for each chemical monitored, rather than their combined effects.

Increasingly there are tools for evaluating the chronic effects of multiple contaminants (mixtures), such as changes in physiology and reproductive rates. This was described in Chapter 4 under *The Field of Ecotoxicology* to address methods beyond the acute toxicity measures that come from 96-hour exposures and LC_{50} values for single contaminants. In some experimental trials or even field exposures, measures of contaminants may not reflect actual exposures since it may take weeks for an uptake of contaminants to reach a steady state with the exterior environment (Poteat and Buchwalter 2014).

It is helpful to understand that contaminants and other stressors can enter freshwater food webs from many pathways. Some of these can be via cross-ecosystem resource subsidies, such as terrestrial invertebrates or leaf litter that have taken up chemicals in the terrestrial environment. In areas with airborne or groundwater pollutants, these chemicals could end up on or in the leaves, or in the bodies of consumers, such as insects. As previously mentioned, we also need to keep in mind that these contaminants can also be carried back to terrestrial ecosystems via adult aquatic insects or by the consumption of salmon and other anadromous fish by terrestrial consumers.

Bioaccumulation and Biomagnification

The term *cumulative effects* shows up in many government policy documents, especially in environmental protection policy. However, while there is an intuitive sense that multiple stressors or more of the same stressor will cause greater impacts, this is difficult to make operational. As previously discussed, the interactions of stressors could offset each other, they might be additive, or strongly synergistic. It is difficult to predict the outcomes, and therefore, cumulative effects work is generally based on the monitoring of outcomes. However, we would be able to manage environmental damage more competently if we could somehow forecast possible outcomes from multiple stressors.

Although managing to protect freshwater systems from multiple stressors is complex, it is not an excuse for ignoring the problem. Many groups have provided frameworks for how to advance the management tools, including better identification of appropriate statistical models and better use of empirical data (Craig et al. 2017; Spears et al. 2021; Schäfer et al. 2023). This is one environmental problem that we recognize as crucial, and one for which progress on the management aspects is essential.

PERSPECTIVES

There is a great interest in cumulative effects by policymakers. The types of stressors are broad, and we have few general principles at the moment to predict interactions; thus, this is a fertile field for new research. However, Côté et al. (2016) point out that management is often about reducing or removing stressors (eliminating a stressor is less likely), and we actually have relatively few examples of how disturbed ecosystems or populations respond to completely removing a current impact, which could also be a useful research direction to pursue, as noted in Vos et al. (2023). The designs of removal experiments are likely to need a temporal component since most stressors cannot be removed instantly, nor will natural systems recover quickly. Stressor levels in experiments should also be tied more closely to field monitoring in order to confirm the realism of experiments (Côté et al. 2016).

Box 13.1   Experimental Mesocosms Testing for Multiple Stressor Effects

Experimental work to identify the additive and interactive effects of stressors on freshwater ecosystems has been underway for decades. These replicated, mesocosm-scale studies can help identify the mechanisms by which stressors have their impacts. Such experimental studies are necessarily small scale, such as aquaria, small flumes, cattle tanks, and other containers that are large enough to support functioning ecosystems, but small enough to be replicated and managed (see Figure 13.4).

Figure 13.4   A mesocosm array for experimentally testing complex interactions of stressors for stream organisms (University of Otago, New Zealand). Each circular container has water inflow to generate a current and is colonized by organisms (invertebrates and microbes) from a nearby stream. The full array includes 128 containers (left) and each one is a circular container with its own water (and contaminant) supply (right).

Matthaei, Piggott, and colleagues in New Zealand have carried out many experiments testing the interactions of multiple stressors—including temperatures, pesticides, flow velocities, sediments, and nutrients (Piggott et al. 2015a; Macauley et al. 2021) (see Figure 13.5). Their experimental system is a highly replicated (n = 128), but small dimension (3.5 L, external diameter 25 cm), flow-through mesocosm system. Examination of Figure 13.5 (also see Figure 14.1) shows that each of the factors interacts with the others in complex ways, and the combinations are not simply additive. Such experiments provide important evidence for interactive effects, and multiple stressors or multiple levels of some stressors can be manipulated. However, this approach cannot address all the possible interactions, and a global challenge is how to predict other complex interactions from what is learned in such studies.

continued

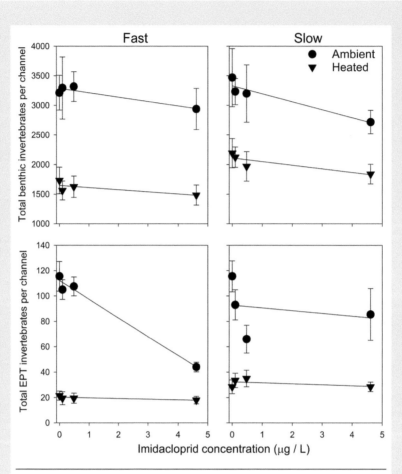

Figure 13.5 Results of an experimental manipulation of temperature, flow velocity, and concentration of the insecticide Imidacloprid. Two panels on the left are for fast flows and the two on the right are for slow current velocities. Two different response measures are shown, the total numbers of invertebrates (top two panels) and only the Ephemeroptera, Plecoptera, and Trichoptera (EPT), where the EPT are considered to be generally more sensitive than many other freshwater invertebrates. Redrawn with data courtesy of Macauley et al. (2021).

The value of such experiments is both in identifying interactions of particular stressors and their interactive effects, and also in building our ability to generalize to broad classes of stressor interactions. It is not possible to do experiments with all stressors, magnitudes of stressors, ecosystem types, and other conditions (such as temperature, solar radiation, ambient nutrient levels, and others). We need to be able to look for general patterns to inform management actions because enhancing our predictive capacity is a primary goal of science.

Box 13.2 Predicting Multiple Stressor Impacts for Natural Freshwater Ecosystems

As noted in Box 13.1, it is not possible to investigate experimentally all of the possible interactions between types of stressors, concentrations, exposure durations, etc. However, for agencies that are managing the protection of freshwaters, it is necessary to have tools to consider the status of real ecosystems across national and continental scales. Using data from over 100,000 streams that are monitored in Europe as part of the WFD, Lemm et al. (2021) considered the combined and interactive effects of a range of stressors. This large-scale, WFD monitoring program uses fish species richness, abundance of sensitive macroinvertebrate taxa, and diatom biomass as indicators. There is also a wealth of data on land use, nutrient status, alteration of flow characteristics, and toxic mixture pressure. Lemm et al. (2021) differentiated 12 stream types and then used the various indicator and seven predictor variables to assess the individual and multiple stressors as explanatory measures of stream condition. They found that the interactions among the seven variables accounted for over half the deviance (a measure of variation) in their data, so interactions among stressors were clearly an important explanation of the patterns observed for European streams.

It is difficult to monitor all of the possible contaminants in surface waters, and it is also problematic to capture data on the variation in concentrations through time and space. Weisner et al. (2021) took 830 water samples to test for 464 pesticides in Germany and found the risk from the mixtures was on average 3.2 times higher than estimated from exposure to the pesticides individually. Field studies of the taxonomic richness of stream invertebrates and rates of decomposition found that there was significant impairment even at pesticide (toxic equivalents) concentrations 100 times lower than levels that agencies, such as the European Union, considered protective (Schäfer et al. 2012; Beketov et al. 2013). Combinations of contaminants (and other stressors) will often be more toxic than the safety levels if only a single contaminant was present.

Two significant challenges in the multiple stressor world are (1) how to predict ecosystem-level outcomes and (2) how to prioritize which stressors to start controlling first. There is a lot of information on the effects of single stressors (such as nutrients and pesticides), as we have learned in earlier chapters. Although there is still a deficiency of data on interactions among the myriad of interactions among stressors, we cannot wait until we have complete knowledge before agencies respond to the issues. One example was developed for agricultural ditches in the Netherlands, where knowledge about community-level responses to single stressors was used to build predictions about the impacts of multiple stressors and food-web consequences (Bracewell et al. 2019). This exercise assigned relative sensitivity scores to taxa in order to generate a sensitivity-response matrix. In predicting a food web's response, it is important to incorporate indirect effects of changes in composition due to stressors. Putting the pieces together in a model framework will not be exact, but guidance for agencies to set priorities regarding how to address multiple stressors is useful—and we cannot afford to wait for more complete knowledge.

ACTIVITIES

1. Pick two or more stressors that you know of from your region (for instance, nutrient pollution and water withdrawal) and try to predict how changes in each of them would affect your favorite freshwater ecosystem.
2. Consider how climate change might modify your predictions.

14

CLIMATE CHANGE AND CLIMATE PROOFING

INTRODUCTION

The headlines about global change are most often about the world heating overall, glaciers melting, more fires due to drought, and an increased frequency of intense storms. These are all important, but there are additional aspects as well, and in this chapter, we will work through each of these. Climate change is one component of global change that was referred to earlier. There are many reviews of the data and projections for climate effects, most of which refer primarily to average global temperatures and the measured effects of the burning of fossil fuels in the last century on greenhouse gas concentrations in the atmosphere. Some of the climate change effects that will affect freshwater ecosystems include increases in average temperature, other changes in temperature regimes (timing, range, geographic variation), shifts in precipitation patterns, extreme weather, alterations of wind patterns, humidity, and the overall increase in atmospheric carbon dioxide (CO_2) (Schindler 2001; Woodward et al. 2010).

Predicting climate change effects on freshwater ecosystems is complicated because there are many dimensions in flux. A lot of the global circulation models for climate are at relatively large scales, so local effects are difficult to predict. We anticipate changes to precipitation patterns (timing, magnitude, and duration), temperatures (higher on average), hydrological flowpaths, and the relative contributions of snow and ice to the timing and magnitude of flow generation, along with altered evapotranspiration patterns and perhaps reduced light intensity (due to more vapor). Changes to periods of ice and snow cover on lakes might affect timing and duration of lake stratification. There is abundant evidence for northern hemisphere lakes of later freezing (about 11 days per century) and earlier ice melt (6.8 days per century) in many places (Sharma et al. 2021). Data on extreme precipitation events globally (1964 to 2013) showed large increases in frequency and intensity, with a 7% increase in the last decade of the study alone (Papalexiou and Montanari 2019). Predictions of the specific and net effect of these changes are complex in general, and especially for particular sites (see review in Stephens et al. 2021). Some regions are predicted to get more precipitation and others less—some will have more seasonal distribution of precipitation and others less, etc.

The definition of extreme weather is difficult but generally includes events well beyond the normal range of conditions, in whatever way the normal range of variation gets defined. For example, extreme might mean outside of the 95% confidence intervals or more than two standard deviations beyond the mean, etc. (Papalexiou and Montanari 2019). This includes extreme winters, intense precipitation events, hot and dry events, hurricanes, tornados, and others (for example, see Estrada et al. 2021). Even without climate change, our climate systems produce a lot of year-to-year variation (see Box 14.1) that human infrastructure struggles to cope with. However, climate change means bigger floods, more intense droughts, greater fluctuation in the water levels of lakes and wetlands, and other events. Warmer temperatures and droughts can also mean more wildfires and reduced vegetation cover. These all have impacts on fresh waters. One thing that will not change is the intensity of sunlight, even though many people seem to think this might be a part of climate change. If anything, it is possible that solar radiation might diminish slightly if there is a higher moisture content (vapor) in the atmosphere. These changes in weather are complex and most projections are made at coarse spatial scales, requiring downscaling to local predictions, which are still relatively general. The increasingly variable inputs of water (bigger storms, more intense drought) are one of the most difficult things to manage. Humans are very good at dealing with averages, but increasing extremes require new thinking about management, and many countries lack the economic means for engineering-type solutions (Vörösmarty et al. 2010).

Many temperate parts of the world depend on water that is generated by the slow melting of winter snowpacks. However, with global warming, less precipitation will be stored as snow (warmer on average in winter) and what is stored will likely melt away faster. Glaciers cover 10% of the Earth's terrestrial surface, but in many places, glaciers are rapidly shrinking, and glacial meltwater is diminishing. However, in very high elevations, glaciers might still be increasing in volume as long as most of the glacier is above the point where melting takes away the annual increment, or more, of ice (Clarke et al. 2015).

Greenhouse gases, including CO_2, methane, and nitrous oxide, are all naturally produced by functioning ecosystems. However, the use of fossil fuels and other human activities has increased the rates of production of all these gases, which now contribute to the trapping of heat and global increases in temperatures. One effect of elevated CO_2 is that leaf tissue may be tougher and harder to break down, contributing less to freshwater food webs (Tuchman et al. 2002; Kominoski et al. 2007). Another direct effect of increased concentrations of atmospheric CO_2 is that it will affect hydrology through water-use efficiency of plants since stomata can open for shorter periods to get all the CO_2 they need, yet lose less water. This reduction in transpiration rates can have a significant positive effect on water yield from watersheds, but the specific amounts depend on plant type, plant cover, temperatures, and other variables (Knauer et al. 2017). This is yet another of the complications in predicting net effects of all the components of climate change.

Increasingly, the values of services or benefits provided by nature are enumerated to indicate what the costs of the loss of freshwaters might mean—that is, the value of nature to man, also known as ecosystem services. There are many kinds of ecosystem services, and they are generally lumped into four categories: provisioning, regulating, supporting, and cultural services (see www.cices.eu/). This provides a mechanism to put a value on what humans get from nature and to find an economic balance between the benefits that some individuals or organizations might get from the consumption of those resources that reduce the value of those services that are available to other users. Shifts in climate are likely to alter the value of these services in particular locations in ways that will require economic investments to compensate for the loss.

IMPACTS

There are many impacts of climate change on fresh waters. At one extreme, there is less water, less habitat, less biodiversity, more reservoirs, and more diversions. We might anticipate warmer water in summer, greater or lesser stratification of lakes, potentially smaller and warmer hypolimnia in lakes, and perhaps no winter stratification (see Chapter 2). At the other extreme, we have more intense storms with greater flood damage from streams, and there are fluctuations in lake and wetland water levels (and area). Even in Lake Superior, the world's largest lake by surface area (third by volume), sampling by cores to examine 9,000 years of the lake's past showed a very large increase in primary production in the last century that was attributed to climate warming of surface water, along with longer periods of stratification and ice-free duration (O'Beirne et al. 2017).

Most of the public's concern over climate change has to do with temperatures, but this change manifests in many ways. Warmer summer temperatures may exceed the tolerances of some species, or at least alter their interactions with prey, predators, and competitors, all of which can be temperature dependent. Warmer winter temperatures can mean a lack of ice cover on temperate lakes, which could affect whether there is winter stratification and lead to overall warmer hypolimnetic temperatures. Warmer water in winter may also speed up the development of vulnerable egg and larval stages. For instance, sockeye salmon fry may leave the gravels of their streams too early compared to the productivity of the lakes to which they move to rear, which would result in slower growth or even starvation. Lack of snow storage or earlier melting can reduce meltwater contributions to flows or lake levels, and advance the timing of spring freshets, which may not match with the phenology of some organisms.

Extreme Weather Events

Climate change is causing more frequent heavy rainfall and flooding events, storm surges, hurricanes (cyclones), hotter temperatures, more fires, and droughts, all of which affect freshwaters. Extreme weather events can disrupt normal seasonal patterns—for instance, delivering enough energy to severely deepen the thermocline in a stratified lake in summer and to change the temperature, oxygen, and chemical gradients. Such an event in a lake in Germany resulted in the mixing of a cyanobacteria bloom from deeper water into surface waters, leading to intense productivity, changes in surface-water chemistry (especially elevated pH), altered temperature profiles, and a large increase in turbidity (Kasprzak et al 2017). Large and rapid inputs of precipitation can affect lake levels and stream flows, with many consequences for ecosystems from flooding. At the other end of the precipitation spectrum, long periods of low precipitation and higher evaporation rates due to warmer temperatures lead to more intermittency of flows in streams, which can restrict the distribution of many organisms (Datry et al. 2017).

A heat wave across Europe in 2003 illustrates the expected consequences of warmer summer temperatures. In two lakes in Switzerland, mean summer (June, July, and August) air temperatures were 4.4°C above the long-term average, and summer epilimnetic (water surface) temperatures were the hottest ever recorded, at more than 2.5 standard deviations above the long-term mean (1.5 to 2.0°C warmer) for summer maxima (Jankowski et al. 2006). Such high temperatures reinforced the stability of the stratification of the epilimnion and extended the period of stratification, resulting in more pronounced depletion of oxygen in the hypolimnia (lower water layer) of the two lakes. Further reduction of oxygen in the hypolimnia might also be compounded with oxygen demand as a result of eutrophication, which combined to create anoxia in the bottom of these lakes. Over time, such heating can also lead to warmer water in the hypolimnion as circulation takes place and integrates very warm surface waters.

One example of an extreme event in Italy had approximately 340 millimeters of rain in about six hours, which led to massive landslides, erosion, and flooding, with large impacts on streams and riparian areas (Segadelli et al. 2020). In that region, 30 millimeters per hour is considered extreme. The magnitude of impact also depends on what was happening to that freshwater system beforehand since it may already be disturbed from prior events, or it may have even adapted to extremes. Much of the literature on the consequences of extreme events is focused on running waters, largely because of the forces applied against the landscape. However, the materials that are mobilized typically end up in lakes and wetlands causing large, and sometimes persistent, changes. For instance, in that same event in Italy, a lake downstream was turned extremely turbid, with a variety of impacts on the entire ecosystem that persisted for many months.

Quantity and Quality of Water

Shrinking glaciers have received considerable attention around the world. As glaciers retreat, it will mean less water supply in streams during the summer and also sea level rise, which will impact shorelines and estuaries. In streams with some glacial contribution to their flows, the summer flows are sustained and cooler, thereby protecting downstream environments and cold-water species such as salmonids from other changes, but as glacier volume decreases, the amount of water supplied diminishes, and that component of streamflow will eventually disappear (Milner et al. 2017). Moreover, many agricultural areas, such as the North American Great Plains or areas south of the Himalayas, depend on glacial melt to provide water for the irrigation of crops and livestock. Such water supplies are also critical for drinking water supplies for many people. Proglacial lakes at the toe of glaciers are increasing in number and volume. The geological features that retain the water in these lakes can sometimes fail catastrophically and may entrain unconsolidated sediments that have been exposed by the retreating glaciers, which could negatively affect downstream environments. Glacier shrinkage will also change fluxes of dissolved organic carbon, nutrients, and even contaminants that have accumulated via airborne pathways, with consequences for the receiving environments (Milner et al. 2017). Biological communities of glacial-influenced streams are uniquely adapted to those conditions and may become locally extinct as glacier cover disappears from watersheds (Milner et al. 2017; Cauvy-Fraunié and Dangles 2019).

Greater use, diversion, and storage of water systems due to climate change will reduce the availability of surface waters. Flow intermittency in streams will diminish their values as habitats and reduce the rates of some ecosystem processes (see Datry et al. 2018). As noted in other chapters, storage typically alters habitats and flow regimes, creates barriers for organisms, changes water quality, results in more evaporation, etc. It is reasonable to expect more storage of water for drinking water supplies, for irrigation, and for hydro-electric power generation (especially as we look to reduce use of hydrocarbons) in response to climate change and increasing human populations.

Thermal Stress on Organisms

Fish kills can occur in lakes that reach higher temperatures or those that have long periods of ice cover in extreme winters. Evidence from Wisconsin and Minnesota shows that massive fish kills in lakes occur primarily under conditions of warmer-than-average surface waters. The direct effect of thermal stress might be combined with lower oxygen levels (perhaps due to high productivity and/or high temperatures) and toxic algal or bacterial blooms (Phelps et al. 2019; Till et al. 2019).

Fish distributions have been shown to be changing, apparently in response to increases in average water temperatures; two good examples from France demonstrate these patterns. In one study, freshwater fish have clearly expanded their ranges to higher elevations within drainage basins, while almost no species have expanded to lower elevations. Many of these species have moved from their lower limit of distribution to higher elevations as well (Comte and Grenouillet 2013). Similarly, the distribution of fish in France has apparently moved predominantly northward over 35 years (from 1980), with species that were primarily characteristic of the south now in more northern watersheds (Maire et al. 2019).

There are many additional concerns about how climate change will alter the composition of biological communities as the ranges of some species change due to alterations in their local environments. One aspect that is of direct concern to humans is that climate change may also lead to an increased risk of certain diseases. There are many diseases associated with water—some are waterborne (such as cholera, *Giardia*, *E. coli*) and some use arthropods or mollusks that live in water as vectors (mosquitos, black flies, Ceratopogonidae, snails, etc.). These diseases include West Nile Virus, river blindness, malaria, schistosomiasis, and many others. Changes to water supplies and warmer conditions may allow some of these diseases and their vectors to spread to regions beyond their current range.

Potential Return of Acid Rain

Acidification of the oceans due to increased CO_2 concentrations is reasonably well known. The effects of higher atmospheric CO_2 and, hence, the elevated acidity of precipitation on freshwaters have not been a major concern in recent decades. However, in the 1970s and 1980s, the northern hemisphere experienced a long period of acid rain, generated by industrial emissions. While atmospheric concentrations of CO_2 are unlikely to create rainfall with pH as low as it was during that period, it can serve as a case study as to what could happen again, even if it happens more slowly (Schindler et al. 1985; Likens et al. 1996). One consequence of the acid rain era was that some of the acid-neutralizing capacity of soils and water was lost due to buffering and leaching of calcium ions (Ca^{2+}) from past and ongoing acid inputs (Jeziorski et al. 2008). This has led to the depletion of Ca^{2+} from some lakes and is having an impact on a number of aquatic species since most organisms require calcium for a variety of biological functions (see Jeziorski and Smol 2017; Weyhenmeyer et al. 2019).

INTERACTIONS WITH OTHER STRESSORS

Changes in temperatures, hydrology, and loss of glacier melt (and snowmelt) result in large changes to water supplies that are available for human and ecosystem support. These changes will also affect the amount of water that is available for hydroelectric power production; just as the world seems to be leaning toward hydroelectricity as a greener alternative to fossil fuels. As you may recall, we discussed the negative impacts of dams and reservoirs in Chapter 7. Balancing the availability of water to support human needs and ecosystem needs will be challenging, and the large variability in actual weather in any year or season compounds that task.

Climate change (and past forest management) has resulted in greater fire risk; and fire affects hydrology, geomorphology, and biology in many ways—not all of which can be addressed here. Fire removes vegetation, reducing transpiration and interception, and allows for more of the incoming precipitation to run off into streams and lakes, increasing water yield and flooding (Niemeyer et al. 2020). The reduction of vegetation also leads to bank instability with greater erosion risk and the moving of sediments downstream, frequently leading to debris torrents and enormous destruction downstream. Rapid run-off can also result in lower low flows during dry periods. Fire can also mobilize contaminants, such as mercury and arsenic, which can get into waters. Burned soils also become hydrophobic—less water infiltrates the soils, resulting in more overland flow and soil erosion, which exacerbates the effects of flows into streams and leads to an increase in stream power and erosion.

Some of the consequences of climate change are difficult to predict and some concerns cannot be assigned directly to climate. Another of these impacts is the brownification of waters, which has been observed predominantly in lakes and wetlands in northern latitudes. The specific mechanism is not entirely clear, but the organic molecules in soil are being broken down and mobilized as dissolved organic carbon (DOC), leading to higher concentrations of colored DOC. The mechanisms have been attributed to land-use changes (less forest cover), nitrogen deposition from the atmosphere, climate change, and the reduction in the extreme acidification of rain that was a serious environmental issue of the 1970s and 1980s. All of these contributing processes have resulted in lakes that are more colored, which changes productivity, temperature regimes, light attenuation, and even oxygen concentrations (Solomon et al. 2015; Meyer-Jacob et al. 2019).

With greater evaporation rates and more groundwater withdrawals, we can expect salinity increases in surface waters of arid areas. This is happening globally, such as the in the Great Plains of North America (Elmarsafy et al. 2021), and Australia has encountered salinity increases for decades in areas such as the Murray-Darling River basin (for example, see Hart et al. 2020). This will become a more widespread and profound issue affecting agricultural lands and eventually, our freshwaters. In general, changes in the timing and supply of water flows will mean less water and thereby, less dilution of nutrients and contaminants, furthering eutrophication and toxicity.

The complex interactions of climate-change-induced temperature patterns with other factors will need greater study. In one mesocosm experiment, manipulations of temperature (+3°C increase versus control), augmented nutrients (ambient, added), and the presence or absence of a small predatory fish (threespined stickleback) resulted in complex results and also varied seasonally (Greig et al. 2012). In spring, heating reduced production, whereas in summer, heating increased production; the latter being in contrast to what one might predict (see Figure 14.1) (see also Box 13.1). In this experiment, adding a moderate amount of nutrients usually increased production, as one might predict, but not in summer without heating. Within an ecosystem, the diversity of such interaction pathways makes predictions difficult, so we need

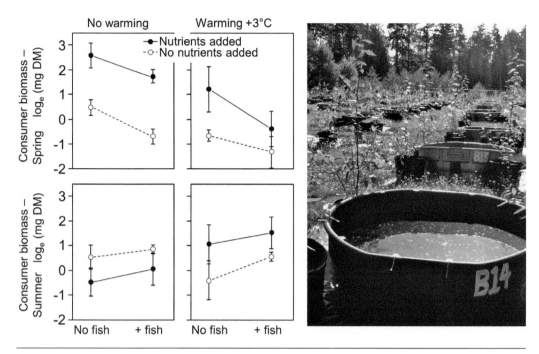

Figure 14.1   Experimental manipulations of three factors, including a 3°C increase of temperature over ambient temperatures, nutrients, and presence/absence of a predator fish (threespine stickleback). The results show the complex interactions between these three variables. Such results give some sense of the difficulty with predicting the net effects of climate change versus other interacting global changes. Modified from Greig et al. (2012).

to take advantage of observations from *natural experiments* that are based on extreme weather events, as well as controlled experiments where they are feasible.

SOLUTIONS

An optimistic view is that humans will sort out our carbon problem, and the amount of warming and concentrations of CO_2 and other greenhouse gases will be reduced. Even so, it possibly will take a century or more for the effects to be reversed. There are some extreme but unlikely solutions. For instance, in Switzerland (and other alpine countries) for a couple of decades, geotextiles have been placed over some glaciers in summer to reduce glacier melt rates—but thus far, only around 0.18 km² are actually treated this way in that country, and estimates are that this is not a scalable solution (Huss et al. 2021). Some people have considered geoengineering of the atmosphere by *sprinkling* fine particles into the atmosphere to reduce solar radiative heating, but this represents a global experiment that could also have unpredictable outcomes. One of the main solutions in the foreseeable future is *adaptation* to a changed reality. One simple adaptation is water conservation in the home, in agriculture, and elsewhere. Reducing the watering of lawns alone could be a large savings in water use (see Box 6.2).

As noted before, less water results in less dilution of nutrients, contaminants, salts, and other materials conveyed through our water systems. One solution to this is better control of nutrient inputs to freshwaters, and some major advances have been made toward this, but more could be done. One example is how southern Australia dealt with the increasing salinity of the Murray-Darling River basin (an area of nearly 1 million km^2, but mostly arid), where natural salinity was combining with anthropogenic salinity increases from various land-use practices (Hart et al. 2020). The first step was to agree to specific targets for salinity concentrations that management agencies would work to meet. Among the solutions were managing river regulations to raise groundwater levels (dilution of saline groundwater), increasing the efficiency of irrigation with less dependence on groundwater, and restoring native vegetation where feasible (Hart et al. 2020). The example shows that water-quality issues from reduced flows due to climate change are solvable but require innovative practices.

Water Storage Strategies

Greater water storage in reservoirs for consumption, irrigation, power production, and other uses is a very likely response to climate changes, with all the attendant impacts on freshwater systems (see Chapters 6 and 7). Of course, that is only in countries with the economic ability to invest in such reservoirs and their management. In addition to the impacts on connectivity and downstream flow regimes, more dams will impact a greater amount of the world's surface waters. Creating more reservoirs is also likely to heat water intended for downstream ecosystems causing water quality problems and increased evaporation rates.

Water is becoming increasingly limited in a heavily populated and warming world. Issues of water security for people and nature show that areas that are very vulnerable are insecure for both (Vörösmarty et al. 2010). One solution is more storage as previously noted, but water losses from reservoirs and conveying canals due to evaporation may be quite large. Some estimates from reservoirs in the southwest United States and even Lake Superior (a cold lake) indicate evaporative losses of 40 to 60% of annual water yield, and numbers are similar for other reservoirs around the world (Friedrich et al. 2018). Estimates of evaporation from two of the largest reservoirs in the arid southwest United States (Lake Mead and Lake Powell) reach 1.4 trillion liters per year combined, which could easily supply three million households at current use rates (Friedrich et al. 2018). Other estimates for reservoirs in warm, arid parts of the world similarly suggest that up to 50% of the water might be lost to evaporation from small reservoirs, and substantial volumes even from large ones (Aminzadeh et al. 2018). Increasing reservoir storage capacity alone will not solve water issues if such large amounts are evaporated away. In Los Angeles, California, the addition of black, high-density polyethylene (HDPE) balls to the surface of reservoirs of treated water had success in reducing evaporation (see Box 14.2). However, rules for the protection of water quality for human consumption do not allow plastic in contact with treated water. This required rethinking to provide a cover that does not touch the water, at least on their smaller reservoirs. Methods for choosing sites that will have lower-than-average evaporation based on wind speeds, vapor pressure, and air temperatures are being developed by engineers. One approach that has also been attempted is the diversion of surface waters to deeper aquifers where available, thus avoiding surface evaporation (Friedrich et al. 2018).

Limiting Increases in Water Temperatures and Evaporation Rates

One of the means to reduce peak water temperatures is shading with forest canopies (buffer zones) along streams, sometimes referred to as *climate proofing* (Thomas et al. 2016), while reducing erosion and enhancing nutrient uptake at the same time. This can be accomplished with deciduous or coniferous trees that provide shade during summer when solar radiation and resulting temperatures peak. The evidence for lower thermal inputs to streams by providing for vegetative stream cover comes from extensive study, primarily related to forestry and the protection of riparian zones (Moore et al. 2005; Broadmeadow et al. 2011; Leach et al. 2012). Some people have suggested using forest harvesting to increase water yields since after a forest harvest, there is often a several-year increase in water yield, however, it is becoming clear that this is not so simple (see Box 8.1). In general, we should expect less forest clearing alongside water to help with several aspects of climate change.

In many parts of the world, reservoirs and canals store and transmit water, but can also be a very large source of loss of water through evaporation. In particular, these areas provide large areas of unshaded surface water that are subject to heating. *Solar canals* are emerging as an interesting solution in parts of the world, for instance in India, Spain, Brazil, and Egypt (Kumar and Kumar 2019; el Baradei and al Sadeq 2020). A lot of photo-voltaic energy generation (solar power) comes at the expense of farmland being occupied and thereby not available for agriculture. Solar canals are water canals that are covered with and thereby shaded by solar panels. These canals are usually highly modified and are often lined with concrete that will not support natural ecosystems (see Figure 14.2). Kumar and Kumar (2019) estimated a 29% reduction in evaporation loss from covered tanks, and other studies have shown significant reductions in evaporation rates. There are also floating versions of these solar energy systems that were designed for reservoirs. There is evidence that the humidity associated with being near water may reduce the efficiency of power production, but it still produces substantial renewable energy. This win-win of producing renewable energy and reducing the loss of water is a promising solution. The World Bank estimates that more than 400,000 km^2 of canals and reservoirs might be available for such installations globally (Kumar and Kumar 2019).

Figure 14.2 Water canals in many places are open to the sky and lead to heating, algal production, and high rates of evaporation. In India, large projects of covering canals with electrovoltaic systems (*solar panels*) use space that cannot be used for anything else, with the added benefits of electricity production and large savings of water due to reduced rates of evaporation. Photos with permission of Elsevier, courtesy of Dr. Manish Kumar (see also Kumar et al. 2018).

Defenses against Extreme Weather

Flood defenses against larger peak flows will be needed to guard human infrastructure. For protection of ecosystems, this will require enabling mechanisms for delays in runoff generation, such as more infiltration to groundwater (as opposed to surface flows across pavement and through pipes) and more storage in wetlands or floodplains. Conservation of floodplain areas to function as water storage areas can be a valuable strategy for protecting ecosystems for humans and other life. Restoring river floodplains to more natural, hydrogeomorphic designs is emerging in Europe and North America (Feld et al. 2011). One design that has been used are two levels of dikes, with one set closer to rivers designed for average flood stages, and another level set further away from rivers to protect from high peak flows. The area between is available as agricultural or recreational land but would be risky for other infrastructure. Such a design has been used in the Netherlands (see Chapter 7, Figure 7.2B).

Novel Ideas

One of the most important solutions will be new ways of thinking about water, particularly about the extremes of supplies (intense floods, profound droughts) and how to manage for variation, not averages. As mentioned before, engineering solutions, such as bigger reservoirs and diversions, will serve wealthier nations with the economies to build such works. Those all come with their own environmental impacts on biodiversity, altered flow regimes, evaporation losses, and other impacts that we have learned about in earlier chapters. Novel solutions will depend on new ideas from you and your colleagues.

PERSPECTIVES

All the stressors we have considered in previous chapters potentially interact, and their connections with climate change are not easily predicted (reviews in Davis et al. 2013; Larsen et al. 2016). Predicting the future is difficult, but being prepared for possibilities through adaptation is more powerful (and efficient) than responding after the change has occurred.

Box 14.1   Laurentian Great Lakes Water Levels

The Great Lakes are the largest lake system in the world and contain about 20% of the world's surface freshwater. The balance between inputs and losses can be quite dramatic, and in the 2010s, lake levels rose significantly by 0.5 to 1.0 meter to the highest levels in the recorded data series, causing a number of issues. An estimated 34 million people live in the basin of these five lakes, and thus many people depend on the stability of water supplies, shorelines, infrastructure, and even commercial shipping. In contrast to general expectations of reduced water balance due to higher temperatures, there is evidence that climate change may lead to more water in eastern North America, associated with changes in atmospheric patterns (Gronewold et al. 2021). In addition, observations indicate lower rates of over-lake evaporation and higher rates of over-lake precipitation in the past decade (Gronewold et al. 2021). However, water levels in the lakes have varied as

continued

much in the past century as in the last two decades (also see Chapter 16, Figure 16.3), so one needs to be cautious of the time periods one is comparing (https://tidesandcurrents.noaa.gov/).

High water levels lead to the loss of beaches, damage to shoreline infrastructure, and possibly interference with water intake (or wastewater) systems. This occurs not just by inundation, but because wave action in these large basins can be extremely powerful and easily erode shorelines (Kramer 2020). Stormwater drainage systems in some municipalities may lose the height difference for water to drain into the lakes. High water levels and erosion can also damage coastal wetlands because most freshwater wetland species are very sensitive to soil moisture and inundation (Smith et al. 2021). Evidence along Lake Ontario after several years of high-water levels demonstrates the impacts on wetlands, including the loss of some species. In particular, the cattail *Typha* was reduced in coverage since they drown when their roots are submerged, while other species increased in coverage (Smith et al. 2021). Such shifts in plant community composition can result in related changes in wetland species—aquatic and terrestrial.

Chicago, Illinois, like many cities built near lakes and rivers, is built in a relatively flat area that is not much higher in elevation than Lake Michigan, which it is beside. Variation in lake levels causes a number of problems for such cities. Given the marginal difference in elevation between the city and the lake, heavy precipitation events or snowmelt do not drain quickly, particularly in years or seasons when the lake is elevated above its long-term average (see Chapter 16, Figure 16.3). Moreover, when the lake is at higher levels, water can back up into the Chicago River, causing flooding of the harbor and city. Chicago has responded with the Chicago Harbor Lock, a dam or barrage intended to keep the lake out of the harbor and river when lake levels are above the normal river level. This is one example of adaptation to protect cities against changes in the environment. Other cities and towns face similar problems and will have to adapt infrastructure to adjust to the natural variation that is not likely to be controlled through any kind of engineering.

Whether or not the general water balance is toward greater water yield in the Laurentian Great Lakes, variability in hydrological balance at such a large scale is unlikely to be managed by people. Adaptation to variations in lake levels and variability, in general, is an important element of preparation for human activities and the protection of nature.

Box 14.2   Covering Reservoirs

Evaporation of water from reservoirs and canals can be a major loss in warm regions, especially since water is already in short supply in many of those areas. In some cases, reducing water loss from human infrastructure could result in more water remaining for natural ecosystems, for instance, covering water canals in India (see Figure 14.2) and similar proposals for canals in the United States. In Los Angeles, California, and elsewhere, floating HDPE balls were introduced to cover the surface of treated water in a reservoir to reduce potentially harmful algal growth, including cyanobacteria blooms, and other impacts on water quality (see Figure 14.3). It turns out that they also reduce evaporation rates. These balls are also known as *bird balls* because one of their original uses was to keep ducks and other birds out of reservoirs. Since water in these reservoirs is treated with chlorine, reducing the amount of light diminishes the input of ultraviolet (UV) radiation that can cause harmful chlorination byproducts, such as bromate. Some other reservoirs in the area that held untreated water were provided with floating covers.

continued

Figure 14.3   Bird balls on a reservoir surface help to reduce evaporation, as well as limit algal production. These black HDPE balls float on the surface of some reservoirs—in this case, in Virginia, USA. Photo: Walter Gills, Virginia Department of Environmental Quality.

The floating balls that were introduced in the Los Angeles area reservoirs were estimated to save about 1.15 million m³/y in water losses due to evaporation (Haghighi et al. 2018). However, the manufacturing of the 96 million 10-centimeter balls was estimated to require 0.25 to 2.9 million m³ of water depending on the thickness of the wall (up to 5 mm). Thus, water is saved in one way (reservoirs), but used in another aspect, so we need to consider costs and benefits. Thinner-walled balls sit higher on the water and allow more water to be exposed to the air, and also do not stay in place as well, so five millimeters is a preferred thickness. The balls may also be partly filled with water to limit their ability to be blown about.

One also needs to consider the full life-cycle analysis of products. For instance, water is used to produce petroleum for the manufacturing of the balls and for other steps in distribution. Balls could be water efficient in a net accounting if the balls last more than 2.5 years (Haghighi et al. 2018). Fortunately, the balls are expected to have a 10 to 20-year life span and are manufactured with a compound to avoid UV degradation of the plastic. In some cases, the expectation is to recycle the balls after 15 years. Other environmental costs that were incurred included their carbon footprint associated with petroleum extraction, HDPE production, and transportation. In some places, these kinds of balls may also provide a large surface area for biofilm development and other changes to the reservoir ecosystem. So, although there are benefits from these actions, there are also costs in terms of water, and one needs to evaluate the overall balance, which is still net positive.

Such shading and reducing of evaporation may not be helpful during nondrought periods, depending on the politics of water allocations, but most parts of the world are so overdrawn on their water supplies that any conservation measures will help. Water supply districts monitor a number of water chemistry parameters in order to ensure that no toxic chemicals leach from the polyethylene balls. Another solution to evaporation losses and algal growth is floating covers, which have been installed on other reservoirs. United States federal law now requires treated water (already treated to drinking water standards) to be covered entirely.

ACTIVITIES

1. Is your state/province doing anything different to safeguard water supplies against climate change?
2. What are they doing and do you think their actions are likely to succeed?
3. What else can you imagine trying that might reduce the impacts on freshwater ecosystems from climate change?
4. Some people have suggested that we can reduce vegetation cover in forests to increase water yield. Given the other consequences of removing forest cover, do you think this is a good idea on the whole?

15

RESTORATION

INTRODUCTION

It is very obvious that humans have done a lot to degrade freshwater environments through our activities. In many cases, it is tempting to think we can patch up nature and put it back on its way to recovery—bringing back the health and vitality of its natural processes and structures. However, this attitude may be simplistic since we tend to overestimate our ability to fix the problems and we allow damage to occur that we probably cannot fix (Moore and Moore 2013). It would certainly be simpler, easier, and cheaper in the long run to protect our ecosystems rather than to try to fix them after they've been damaged. Thus, the first priority should be to protect ecosystems and avoid damage that needs repairing—in other words, practice good stewardship of our natural environments. Most freshwater restoration projects are expensive (estimated at more than one billion dollars per year in the United States alone) and take a heavy engineering approach to habitat elements, while not often meeting restoration goals (Roni and Beechie 2013; Palmer et al. 2014; Johnson et al. 2020). There are also many *nature-based solutions* where natural processes toward recovery are facilitated, such as through the planting of riparian vegetation. Nevertheless, when an ecosystem has been seriously degraded, some form of restoration can assist the recovery process.

The recovery process should not always be considered as returning a freshwater ecosystem to some historical condition last seen centuries before industrialization. Restoration is broadly meant to include rehabilitation, remediation, enhancement, and other activities. Most of the time we are removing stressors and performing actions that put an ecosystem on a trajectory of recovery. Complete restoration may take many years or possibly centuries to achieve—if at all. A general aim for restoration is to re-establish dynamic, self-regulating ecosystems, and not solely to create static structural features, which is sometimes done with the placement of large wood and boulders (Beechie et al. 2010; Johnson et al. 2020). In some cases, we may try to replace ecosystem functions that have been lost, such as with fish hatcheries, where we add juvenile fish to impaired freshwater systems. As seen in Chapter 9, we *daylight* a buried stream by bringing it back to the surface instead of flowing through pipes. Sometimes we try to enhance the functioning of a system to better than a predisturbed state, although this is difficult. For instance, we may fertilize unproductive waters with inorganic nutrients or even salmon carcasses. All of these examples are considered components of restoration (see Figure 15.1).

One has to be very clear and specific when defining targets of restoration actions. Stating that you will "improve fish habitat" or promising similarly vague objectives does little for the

241

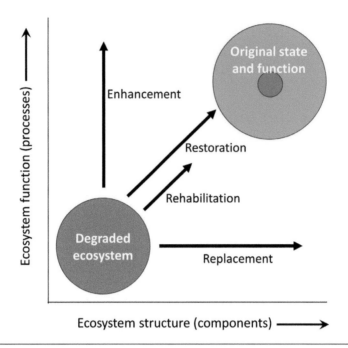

Figure 15.1   There are different possible endpoints for restoration activities and not all practices are intended to end up with the recreation of the original ecosystem. All of these actions are included within the scope of restoration. Based on Bradshaw (1984).

tax-paying public that imagines your actions will bring back the same number, species, and sizes of fish as centuries ago. For some projects, improving fish habitat is measured by the success of placements of physical structure alone (such as pieces of large wood in a section of stream) (Frissel and Nawa 1992; Pierce et al. 2013). When fish numbers or growth are actually measured, the results may be disappointing (Pierce et al. 2013). Having well-defined and measurable outcomes is important for judging the success or failure of any restorative action (Wagner et al. 2015). One can measure and monitor the target of establishing 100 fish of some species in a kilometer of stream reach over a five-year period, as opposed to just *fixing habitat*. This type of clear objective setting also makes it easier to learn from experience and to convince funders to continue providing resources. A criticism of restoration practices is the scarcity of follow-up monitoring (also known as *effectiveness*) or when there is only monitoring of the physical structures and not of the valued biological components (Palmer et al. 2014).

What are we trying to restore? A certain amount of diagnosis of the source of the problem for a stream, lake, or wetland is needed in order to determine the appropriate answer to this question (Roni and Beechie 2013; Carvalho et al 2019). For instance, for streams, if the problem is pollution or warm water from upstream or alterations to watershed hydrology, then actions at a reach scale, such as creating complex structural habitats, will not do much or anything to help with recovery since the underlying issues were caused at a much larger spatial scale (Booth et al. 2016). This leads to a *process-based approach* to restoration versus static structural approaches (Roni and Beechie 2013; Booth et al. 2016). Adding physical structures at the local scale, such as large wood and boulders, may do nothing positive for the ecosystem if that is not the limiting factor(s) at the landscape scale. In fact, applying the wrong intervention for the problem may make things worse (Roni and Beechie 2013).

Another consideration when answering the "what are we trying to restore" question is to ask whether the target is either to restore a functional and diverse freshwater ecosystem or to increase the numbers of a single species. In many cases, the objective may be driven by a single species, such as salmon, and may be only for a single life stage of that species. A target to increase a single species is also not conducive to a process-based approach. Many *single-benefit* projects have narrow aims and may provide poor habitat for nontarget, but native species (see examples under *Impacts*).

Restoration of freshwater ecosystems is big business, and there are many studies about which actions are effective and which are not. Several authors have reviewed large numbers of restoration projects with the general conclusion that of those that are properly monitored, success (defined as some improvement in ecosystem condition) is only slightly more frequent than failure (Bernhardt et al. 2011; Roni and Beechie 2013; Palmer et al. 2014). Note that this trend is only for projects that are monitored. Palmer et al. (2014) emphasize that the majority of projects are not monitored in any rigorous way. However, this highlights the fact that there is still a lot to learn in order to *fix* aquatic ecosystems reliably.

Nilsson et al. (2015) summarized seven possible reasons for the lack of ecological success of some restoration projects:

1. Objectives were poorly defined
2. Factors limiting populations were poorly understood
3. Monitoring was inadequate or nonexistent
4. Species chosen as targets or indicators were poor choices to represent the community overall
5. Colonists or propagules were not available or able to reach the project area
6. Recovery times may be longer than anticipated
7. Restoration may have occurred naturally, prior to actual interventions

All of these considerations need to be built into the planning, monitoring, and evaluation of restoration projects. It is sometimes argued that the primary response of fish as a target species is for aggregation around physical structures, which could be mistaken for increases in numbers when there has been no demographic response. Roni (2019) reviewed evidence for increased survival and reproduction of target species resulting from wood installations, and he concluded there are often positive demographic responses. One also needs to keep in mind that recovery of ecosystems may not be linear and that there may be thresholds that require longer time frames to overcome (Clements et al. 2010). Priority effects may even occur; for instance, tolerant species that establish during the disturbed state may take time to displace. Lastly, there may be dispersal limits or total barriers to colonists arriving at restored sites, as they may need to immigrate from distant locations.

Restoration actions may take a considerable time to show positive effects. Louhi et al. (2011) found that instream restoration activities in Finland actually had significant consequences by damaging the moss growth on rocks during operations, and mosses can take decades to grow to the size where they provide important habitat. In stream restoration projects for trout in Montana, some sites required more than a decade, along with additional actions, to show positive effects in terms of trout numbers (and some showed no benefit even after a decade) (Pierce et al. 2013).

The responses monitored are also worth considering. Many projects are simply judged by whether the structural features have remained in place, such as log and boulder placements (Frissel and Nawa 1992). Biological measures, when they are monitored, include fish, benthic invertebrates, and sometimes algae or mosses. For proper comparison in a before-after-control-impact (BACI) design (see Chapter 8), it is essential that there are appropriate reference sites monitored at the same times and for the same measures. This also requires that there is sufficient time after the actions to detect changes since there are many sources of interannual variation—such as floods, droughts, population size, etc.—that might obscure the detection of changes in the short term. Clear targets, appropriate actions, sufficient monitoring relative to reference sites, realistic expectations, and duration of surveying adequate to the rates of change are all key to evaluating the effectiveness and efficacy of restoration projects.

There is a social aspect to any kind of restoration action as well. People have to agree to changes within local environments because of the fact that freshwaters are used for many recreational and other activities, and restoration actions might compromise those uses. Also, many freshwaters impinge upon private lands, and not all landowners are willing to have modifications made to their properties. Specific actions may also be seen as undesirable. For instance, in many parts of North America, large and small wood accumulations in streams (and hence wood placements) are seen as natural, but some people—and particularly in some countries—might see wood in streams as dangerous and unsightly (Piégay et al. 2005). Consequently, approval by local residents will be needed for most projects.

IMPACTS

In general, we expect that restoration activities will improve ecosystem functions and the species they support. As previously discussed, not all actions are monitored in a way that allows us to objectively measure improvement, and some projects have outcomes that can cause more harm than good (Palmer et al. 2014). It is critical that we define what the outcomes could look like from the population, community, and ecosystem perspectives. We also need to define what we compare those outcomes against—essentially, the point of reference that would indicate that the condition of our target freshwater ecosystem was good.

When considering what our targets should be, we may consider whether we want to facilitate the recovery of ecosystem functions, some structural aspects, or particular components of the biota. Some single-benefit projects may benefit one species, or even one life stage of a species, but be detrimental to other species. Physical alteration of habitats for salmonids in California may reduce the numbers of the Sacramento Pikeminnow (*Ptychocheilus grandis*) by changing habitats in a way that is not conducive to them (Romanov et al. 2012). In one study in England, wood placements had a positive effect on the numbers of brown trout that were one year old or more, but had a negative effect on the numbers of young-of-the-year brown trout and other fish species (Langford et al. 2012).

Many projects have shown that placements of large wood and boulders can add to the heterogeneity of channel features and contribute to greater storage of organic matter (Negishi and Richardson 2003; Lepori et al. 2006). However, in some cases, these changes make little difference to the biological community in terms of composition, productivity, or abundance (Lepori et al. 2006; Stewart et al. 2009).

INTERACTIONS WITH OTHER STRESSORS

It is important to recognize that restoration activities are completed against a background of many other changes, such as climate change, pollution, and landscape modification of hydrology—all of which may work against successful outcomes of restoration work. Depending on how restoration is done, it may reduce the impacts of some stressors, but in other cases, may exacerbate them. For instance, restoration activities could result in the mobilization of contaminants that are stored in sediments, thus creating another problem for the ecosystem that is intended to be restored.

In the Laurentian Great Lakes, which account for 80% of North America's surface fresh water, large efforts to restore those ecosystems are underway, at costs in excess of a billion U.S. dollars. These lakes are subject to a large array of interacting stressors that we have covered in this book, including pollution, invasive species, endangered species, harvesting of resources, etc., and will require innovation beyond current methods (Allan et al. 2013).

A limitation to the recovery of freshwater ecosystems after restoration actions are performed is the possibility of priority effects. Several studies have indicated that degraded systems may be occupied by populations of highly tolerant species, including invasive species. After restoration, these tolerant species may persist and restrict colonization by the more desirable species for which restoration was intended to promote (Barrett et al. 2021). There may also be limited sources for colonization by target species (Sundermann et al. 2011). These limitations may require additional interventions to affect the biological assemblages.

SOLUTIONS

When considering restoration, a good starting point is to evaluate the processes that are adversely affecting ecosystem functions, as suggested by Roni and Beechie (2013). For instance, extensive engineering of structural features of a stream reach is unlikely to be successful if the problem is related to water quality or hydrological effects from land clearing upstream (Beechie et al. 2010). Taking a landscape view of how watersheds (catchments) affect a freshwater system through hydrology, water quality, and other aspects will help. Ultimately, the restoration objective should be to recreate a self-sustaining system that will not require continued input costs. One should also have clear, measurable, and quantitative targets against which one can evaluate success and efficacy. It is also essential to objectively determine what is considered *success*, such as increases in abundance and biodiversity of some groups of organisms (often fish or invertebrates), an improvement in water quality, or better ecosystem function. However, there is still a lack of monitoring of many projects beyond noting that the action (structural modification) still persists. Fewer than 50% of projects objectively evaluate outcomes, and yet the managers involved believed that they were successful, even when the objective evidence is contrary to that (see Figure 15.2) (Jähnig et al. 2011; Palmer et al. 2014). While aesthetic aspects are important, having data allows one to measure success and efficiency, and potentially leads to improved practices to yield better outcomes.

It is important to consider watersheds as units, but sometimes, in terms of conservation biology, forest biodiversity and freshwater biodiversity are often considered as two isolated entities. For instance, a recent book on forest biodiversity and forestry mentions water or riparian areas only a few times in 640 pages (Krumm et al. 2020). Better consideration of the joint benefits of conservation actions for forests *and* freshwaters might lead to better and broader outcomes.

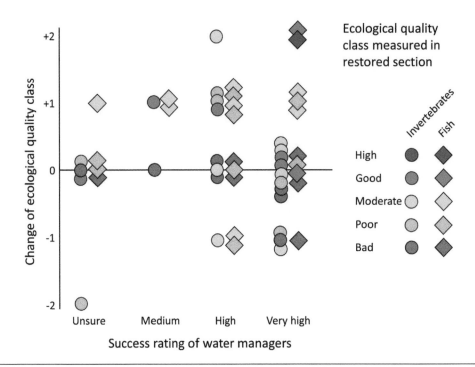

Figure 15.2   Comparison of perceived restoration project success versus objective assessments. The practitioners' response to a questionnaire about their perception of the outcomes of their projects (x-axis) as compared with empirical measures of actual outcomes (y-axis). There was no obvious relation between the perception of practitioners and real measures. Figure after Jähnig et al. (2011).

For example, practitioners who are interested in riparian forest regeneration may be tempted to consider the use of herbicides to control invasive plants, even though we know that herbicides have a number of negative effects on water quality and may be toxic to aquatic life (Weidlich et al. 2020; Florido et al. 2022). Better coordination and integration of restoration and protection efforts across ecosystems, taking a watershed approach (Roni and Beechie 2013), will likely advance the overlapping aims of practitioners. A lot of restoration activities for freshwaters include protecting and enhancing the condition of riparian areas surrounding the water body.

Stewardship

Practitioners have a hierarchy of the extent of environmental damage to address, from allowing it to occur—to avoiding it entirely, under authorization from government agencies. These categories can range from permitting degradation, to avoiding damage during development, through to conservation offsets. Although a project may be allowed to proceed and incur a loss of habitat or function, it may be possible to minimize the impact by changing the project design or location (mitigation). In some cases, projects will go ahead with plans to rehabilitate the site after the fact or will include an approach to augment the ecosystem (hatcheries). When projects that cause habitat harm proceed, there may be *offsetting* applied, referring to other locations that are protected, and sometimes augmented, to provide similar habitat elsewhere outside of the project area as compensation for harmful alterations (see Figure 15.3) (Theis et al. 2022). Each of these categories has variable successes at meeting their objectives. Unfortunately, in some cases, there is little or no monitoring of outcomes.

Sometimes recovery due to intervention takes considerable time, even decades, to occur (White et al. 2011; Pierce et al. 2013; Louhi et al. 2016). Lorenz (2020) studied the re-meandering of three small streams in a BACI design over 10 years and found negative effects in the first year, and then measurable improvements in community structure several years later (see Box 15.1 for another example of a remeandering project). In some instances, it appears that removing the stressor and doing no additional work was as successful as intensive engineering solutions (Louhi et al. 2011).

Protect whenever possible—in other words, practice stewardship and conservation. In particular, controls on what enters freshwaters as a result of human activities, including nutrients, contaminants, sediments, invasive species, etc., is a general first step. Diagnosis is needed to

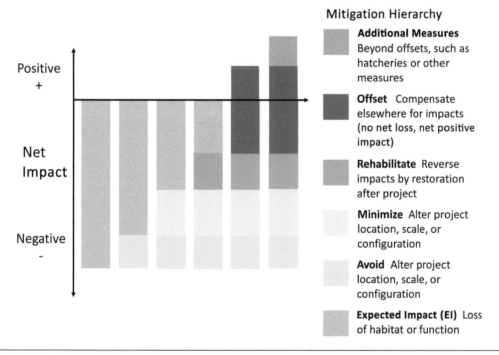

Figure 15.3   A practitioner's *hierarchy of considerations for actions* from left to right based on degree of avoidance, restoration, or offsets to achieve no net loss for biodiversity (represented by the horizontal arrow). The worst outcomes would be expected impact—that is, loss of habitat, but some projects are permitted by government agencies even when this is the result. The best outcomes are to the right of the figure—that is, offsets or additional measures, such as enhancements to balance any losses. In some cases, there is little that can be done to protect the environment, so plans will include rehabilitation after the fact. It may sometimes be possible to minimize or avoid impacts by changing project designs or locations. The worst outcome would be a loss of habitat or function. Redrawn after Theis et al. (2020).

determine the source and scales of the degradation because working on the wrong problem will not likely provide the expected benefits (Roni and Beechie 2013). Often restoration actions take place at a local scale, such as stream reaches, although the actual problems are at the watershed scale; no amount of investment at the local scale can compensate for problems with water quality, hydrology, and temperatures stemming from upstream impacts (Beechie et al. 2010; Booth et al. 2016). In such cases, a broader approach to restoration is needed.

There are many restoration practices that are available, depending on the source of the problem. Methods for many practices are outlined in a variety of books and online resources. One such resource online is from the U.S. Department of Agriculture titled *The Stream Corridor Restoration Handbook*. There is a wide array of solutions, ranging from mostly engineered habitats to actions that take advantage of natural processes, also known as nature-based solutions.

Proper functioning condition assessments of freshwater riparian corridors are the tools for understanding the state of physical and biological processes in these linked ecosystems (Dickard et al. 2015). This can indicate how resilient the riparian ecosystem is and how well its current state functions to provide the ecological benefits expected. This method is applicable to any riparian area, whether associated with a lotic or lentic freshwater ecosystem. The proper functioning condition method is used by many government agencies and is based on a field assessment of 17 attributes to judge whether riparian areas are properly functioning, and what can be done to remedy any shortcomings, which can include restoration actions.

In addition to a range of activities to restore freshwater ecosystems, there can also be the creation of replacement ecosystems, such as the digging of holes to create new wetlands or new stream channels, usually referred to as offsetting or compensation, which will be discussed further in the following sections. Creation for replacement is thus one action toward attempting to fill in for lost ecosystems. This may include fish hatcheries (or other husbandry practices such as raising crayfish or mussels) to replace or augment natural production of species that are at risk or at lower-than-expected densities. Another alternative is *habitat banking*, also called *biodiversity offsets* or *mitigation banking*, which allows for enhancing another freshwater ecosystem to compensate for the degradation of one that may have been allowed to be developed or with no potential fix. There are many discussions about the social dimensions and politics of allowing for such things, and you may be interested in exploring those.

One aspect of restoration that may not be immediately obvious is the colonization potential of sites that have been physically modified. It is possible that the *Field of Dreams* concept (if you build it, they will come) (Palmer et al. 2014) may not work if some species have been extirpated from watersheds. In such cases, one could consider active stocking of species that are missing from a watershed where they may reasonably have been expected previously. However, as we have seen in previous chapters, one needs to do so cautiously to ensure such actions do not introduce diseases and parasites or cause a limited range of genetic variation among the stocked species or the nonnative species (see Chapter 11).

Riparian Zone Restoration

In many situations leaving a strip of natural vegetation as a *buffer* or planting vegetation alongside water can have tremendous benefits for stabilizing banks, reducing erosion, taking up nutrients, providing shade and organic matter inputs, and contributing large wood. Earlier we also learned that vehicles should be kept away from shoreline areas where they can compact soils and damage streambanks. Planting vegetation or protecting the natural vegetation along water bodies can be an advantage to municipalities, landowners, and farmers by reducing erosion, protecting freshwater resources, and supplying habitat for pollinators and other species (Cole et al. 2015). If vegetation is sufficiently dense and tall, it can also provide shading and organic matter inputs. Note that there are jurisdictions where retaining natural vegetation along water may be against regulations, such as some agricultural areas. However, a strip of vegetation along the water's edge is unlikely to be sufficient if the watershed is still a source of elevated nutrient fluxes, contaminants, and altered hydrology (Le Gall et al. 2022).

In many livestock operations, planting along shorelines will not help if cattle and other livestock continue to have access to surface waters. Continued access to wetlands and streams is convenient for many livestock operations but is detrimental to freshwaters because the amount of erosion and sediment transport causes degradation. Fencing that is designed to keep livestock out of streams and wetlands can solve that problem (Riley et al. 2018), as long as the animals are given an alternate source of water. Sometimes that is as simple as periodically filling watering troughs. However, in some rangeland situations, it may be more effective to keep cattle or sheep from some pastures for four or more years, rather than using fencing of limited reaches (Herbst et al. 2012). Fencing can work when it is maintained properly to exclude cattle and if the fenced reach is long enough that the effects of upstream grazing do not propagate downstream.

Planting of shoreline vegetation along streams, lakes, and wetlands is often an effective restoration initiative. In general, one should use natural vegetation rather than heavily armored banks with boulders and rip-rap, which alienate freshwaters from their floodplains and riparian areas. Although armoring banks in this way may be the most immediate option, this activity moves sediments downstream instead of allowing them to settle, and there is nowhere for organisms to take refuge from floodwaters. A simple way of revegetating is the use of live stakes (or whips or stems)—preferably, the cuttings of poplars, willows, and red osier dogwood—that will root and grow quickly. These are usually cut fresh during the growing season but can be dormant, woody stems with intact lateral buds, cut to lengths of 40 centimeters or longer. The fast growth of stakes within a growing season, coupled with bundles of dead, stout stakes laid on the ground parallel to the stream (or lake), can increase bank stability, provide organic matter, and facilitate those species of plants that need some shading or other requirements for neighboring plants (see Figure 15.4). Plantings of other species, usually seedlings, can follow, depending on the intentions for the site. Such erosion control also typically contributes to the sequestration of nutrients by riparian vegetation, or the mineralization of nitrogen (N) and carbon.

An efficient approach to restoration involves providing room for streams to flood and have more natural dynamics (Staentzel et al. 2020; Cooke et al. 2022). In many places, streams have been diked, channelized, or both to protect shorelines and property. One solution is to move dikes further back from stream edges to allow for a floodplain area where water can spread out during a flood without the increased depth and tractive force that comes with narrowed channels (see Chapter 7). However, it can be expensive to set homes, farms, and other structures back from water once allowed to be established.

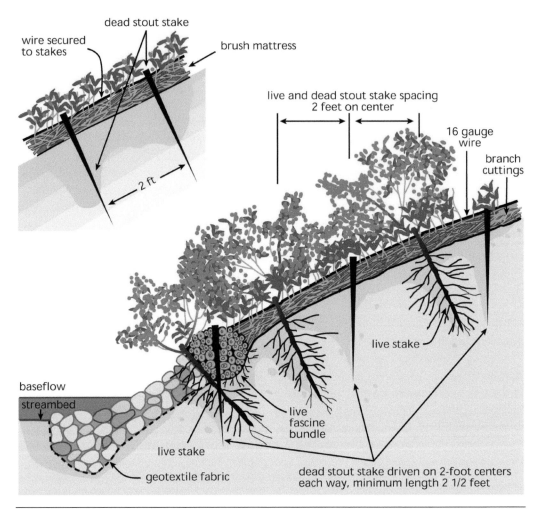

Figure 15.4 Streamside planting to provide bank stability, reduce erosion, and eventually provide shade and organic matter inputs. Live stakes can be cut from some trees, particularly willows and poplars. Source: United States Department of Agriculture-NRCS, Stream Corridor Restoration. https://www.nrcs.usda.gov/wps/portal/nrcs/main/national/water/.

Stream and River Restoration

Among the common approaches used in restoration of lotic waters are structural modifications (Roni and Beechie 2013). These structural components often involve bank stabilization (plantings, rock placements), additions of large wood and boulders (Roni et al. 2015), channel constriction to create chutes, off-channel refuge, and other features (see Figure 15.5). There are good guides on how to allow the natural hydraulics of streams to recreate a diversity of structures (Newbury and Gaboury 1993). In Chapter 7, we covered the importance of natural flow regimes that contribute to the creation and maintenance of natural stream structures. There are good resources on the U.S. Department of Agriculture's Natural Resources Conservation Service website and from other sources.

Figure 15.5   Two examples of physical structures modified to enhance the complexity of in-stream habitats. (A) Wood placement along the edge of the Entiat River in the state of Washington intended to create complexity of structure and flow conditions along the margins of the river. Note the accumulations of finer sediments and wood downstream of the structure where flow is diminished. Photo courtesy Dr. Phil Roni, Cramer Fish Sciences. (B) Small stream with boulder placements to create heterogeneity of flow and habitat. Similar to experimental study reach in Negishi and Richardson (2003). Photo courtesy Sarah Clement.

One form of restoration has been to re-meander streams that have been channelized (see Chapter 7, Figures 7.8A and B and Box 7.1). Streams are often straightened and channelized to facilitate drainage, but end up causing negative impacts, as we discussed earlier. Adding meanders back to streams can restore, to some degree, the natural dynamics of channel evolution that provide for a variety of habitat types. One such example is the Kissimmee River Restoration Project in Florida (see Box 15.1). One of the primary aspects of restoration was the rebuilding of the meandering form of a long section of the Kissimmee River (Anderson 2014). In similar projects, along with reducing the amount of channelization, one can reconnect floodplains, which can contribute to flood control, connections to off-channel wetlands that are critical to many species, and enable the flood-pulse dynamics that are critical to many rivers (Junk et al. 1989; Branton and Richardson 2014). Reconnecting off-channel wetlands and oxbows provides habitat for many freshwater species and can be hot spots for ecosystem processes, such as nutrient cycling and decomposition.

In many instances, removing barriers, especially low-head dams, can provide access for some species to a broader watershed (Hart et al. 2002). Some good success with this has been observed in the northeastern United States where small dams associated with water-powered mills remained as barriers late into the 20th century but were eventually removed. In some places, there have been removals of larger dams, which have provided access to large areas of habitat. For instance, the Elwha Dam and Glines Canyon Dam in the state of Washington were removed and areas upstream that were once inaccessible were rapidly recolonized by fish, including several anadromous salmon species (Pess et al. 2023).

Adding large wood and boulders to streams has become a common engineering activity, but the evidence indicates a range of effects from negative to positive (see Box 15.2). There are studies on the benefits of wood augmentation in streams for some species and age classes, particularly research on population size and growth rates of salmonids (Roni et al. 2015; Hallbert and Keeley 2023). It is necessary to determine whether a physical structure is what is limiting to the specific populations that require aid since certain additions (such as large wood, boulders, and gravels) may result in no change to those target populations (which are often not even measured). Physical structure may not be the most limiting obstacle to ecosystem processes or population dynamics in all cases.

In Chapter 7 we learned about replacing gravel for fish spawning and other purposes downstream of dams. This form of restoration is really a replacement since this intervention is required continually and will likely never return the site to its original, functional ecosystem without ongoing human actions. Nevertheless, this is still considered a restoration activity. Gravel or other sediment additions may also be focused on other target species but is typically a single-benefit option.

As we saw in Chapter 9, many streams have been buried and become part of the storm drain network. In recent years, there has been an interest in *daylighting* or *deculverting* streams—in other words, bringing them back to the surface where space allows. In many ways, this is habitat creation since these watercourses were literally in pipes previously. The work usually involves re-engineering a streambed and allowing the water to flow over the surface. Studies of a few of these projects suggest that many organisms, particularly those with good dispersal stages (such as adult aquatic insects), can recolonize relatively quickly (Wild et al. 2011; Neale and Moffett 2016).

Restoration of Wetlands

Elevated nutrient levels (see Chapter 4) cause many problems in freshwaters, but properly functioning aquatic ecosystems can also contribute to the processing and removal of nutrients. Wetlands in the United States are estimated to remove approximately 5,800 kilotons of N per year (kt N yr−1) (Jordan et al. 2011), and globally, this ecosystem function is enormous. Thus, wetland restoration can result in a large decrease in N, particularly nitrates, from surface waters. If wetland restoration was better spatially aligned with sources of elevated nutrients, it could lead to greater decreases in dissolved N (Cheng et al. 2020).

Wetlands have been heavily impacted or lost by being drained, degraded by livestock access, or damaged by other human activities. However, wetlands with high ecological functionality can be restored or newly created. One of the first actions is often to block pathways that were designed to drain wetlands by way of removing culverts, filling ditches, or removing drains. Another restoration action is finding a way to recreate or return a more natural hydrological pattern to wetland areas. One such approach is reconstructing dikes further away from streams to re-engage floodplain wetlands during floods (see Chapter 7, Figure 7.2C). These projects are implemented in many countries where there is opportunity—that is, away from built infrastructure (Knox et al. 2022). Such projects need to ensure that changes to hydrology do not have unintended impacts upstream or downstream of the wetland, so a landscape context is required. A range of flood-control measures are available to ensure that excess water does not cause damage (see Chapters 7 and 9). In some plans the basin may be excavated to create a deeper wetland, and in other cases, clean sand might be added to provide a new wetland bottom. Planting appropriate vegetation may be needed in order to return the wetland to a functional state, and it may be necessary to remove invasive species of plants. It may also be necessary to fence a restored wetland in order to exclude livestock. One also needs to consider the source area because sediment-laden water from bare fields will not be conducive to a functional wetland, but sediment retention areas (ponds) upstream may provide a solution. *Wetlandkeepers* and other organizations offer training in methods for wetland conservation and restoration.

Beavers were once common and widely distributed across North America and Eurasia (*Castor canadensis* and *C. fiber*, respectively), but their numbers were decimated in the fur trade era. In several regions of North America and Europe, beavers have been reintroduced to serve as ecosystem engineers and as a nature-based solution to restoring wetlands and streams (for example, see Nummi and Holopainen 2019). The reintroduction of beavers provides a sustainable and persistent means for managing wetlands and contributes to the many species that use such lentic areas. Beaver dams generally raise local groundwater levels and reduce erosion, helping to create perennial wetland areas that are used by other species. The reintroduction of beavers in one landscape resulted in a 10-to-40 centimeter aggradation per year of a severely incised streambed, leading to wide wetland areas (Beechie et al. 2010). In the San Pedro River in Arizona, reintroduction of beavers created wetlands and extensive, productive riparian areas, and is now considered a success story for nature-based restoration (Gibson and Olden 2014). In some places beaver reintroduction has not been as successful since some landscapes have few trees or may have just very young (and skinny) trees. Beaver dams that are made of such a small caliber of wood are prone to failing during floods. A novel solution is to create beaver dam analogs by pounding larger diameter wood pieces across the streams at intervals of a couple of meters (Bouwes et al. 2016). These larger pieces of wood can capture wood pieces in transport, but they also encourage beavers to establish their dams there—and the larger wood stabilizes their dams. In a watershed-scale, experimental installation of beaver dam analogs in the state of Oregon, the increase in dams that were cocreated with beavers generated more complex habitats, including wetlands, and improved the production of steelhead (Bouwes et al. 2016). Beavers can also help make wetland and riparian areas more resistant to fires through higher water table levels (Fairfax and Whittle 2020).

Restoration of Lakes

Restoring lakes most often means reducing nutrients and suspended particles (turbidity). In a Europe-wide study, eutrophication through nutrient enrichment had the single biggest impact on lakes (Birk et al. 2020). There has been a huge effort globally to reduce nutrient inputs—phosphorus (P) and N—but the increasing human pressures in all of our watersheds reduces the net effectiveness of these efforts. Nevertheless, no other method will work if nutrient inputs overwhelm other efforts, so reduction of nutrient sources is necessary. Once lakes have received excessive nutrients, a number of methods are typically used (see the following section, *Adding Nutrients*), including aeration, chemical removal of P, biomanipulation, and even draining of hypolimnetic water (Jilbert et al. 2020).

A tool to manage lake nutrient and oxygen concentrations is aeration. Use of this method to get oxygen into the hypolimnion has been used to reduce anoxia and dissolution of P back to the water (internal loading) and to prevent winter fish kills under the ice (Mercier and Gay 1949). The idea is simply that if anoxic conditions are avoided, then P will remain stored in bottom sediments, often bound to reduced iron. However, there is evidence that this might be too simple, and creates its own problems with altering seasonal stratification. Other approaches have used the addition of chemicals that bind to P, such as Phoslock and others.

Biomanipulation by altering community composition, such as the introduction of a top predator, was considered a promising tool for lake restoration when it was suggested by Shapiro et al. (1975). For instance, by creating a food web that promotes large-bodied grazers, like the crustacean *Daphnia* spp., these consumers are capable of eating a wide range of algae, thereby increasing water clarity and removing nutrients. Two ways of increasing *Daphnia* are to remove planktivorous fish or others that consume *Daphnia*, or to add piscivores that can eat and reduce the numbers of planktivores (Carpenter et al. 1985). This method has had some success in reducing algae abundance, especially larger species such as some of the cyanobacteria that are *nuisance* species (although these can also be inedible and escape consumption, leading to blooms). To sustain the effect, it appears that there is also a need for a shift to an alternative stable state where there is a large increase in the number of macrophytes, which sequester a lot of additional nutrients and provide refuge for young *Daphnia* (Moss 1990). Some studies suggest that at least 30% of a lake bottom should have macrophytes to ensure the effectiveness of biomanipulation. Furthermore, it seems that *Daphnia* is one of the few genera that can suppress algae, given that it is a relatively large-bodied grazer that eats a wide variety of algae, and in some cases, this species has been introduced to locations where it is not native. Longer-term studies have presented evidence of a shift to less edible types of algae (that is, toxic, filamentous, or large) that can reduce the effectiveness of algivory to sustain water clarity and suppress algae (Triest et al. 2016). If managed well, these nature-based methods at biomanipulation could be effective and publicly supported, particularly if they can be done with native species (see

Chapter 11 on the dangers of nonnative species). However, the uncertainty of outcomes and the potential for adverse effects render these methods risky, at best.

Adding Nutrients

In contrast to the large issue of excess nutrients, there are instances where the nutrients that sustained the productivity of freshwaters have been reduced as numbers of anadromous species, such as Pacific salmon, have declined. This has spawned an industry of practitioners trying to replace those nutrients with the goal of *kick-starting* the recovery of productivity. One early attempt included adding inorganic nutrients (agricultural fertilizers) directly (Stockner and MacIsaac 1996), but these nutrients were often washed away rapidly or contributed to nontarget parts of food webs. In general, the addition of inorganic nutrients (such as nitrate and phosphate) to oligotrophic lakes increased productivity and supported higher growth of sockeye salmon smolts. This required continued inputs of nutrients to carry on, and there was concern about downstream water quality impacts. However, it was also apparent that the pathway through which dead salmon contributed to these food webs was not through dissolved nutrients, but a complete package of protein and fatty acids that could be directly incorporated into higher trophic levels, above the level of bacteria and phytoplankton. Since early attempts at adding whole salmon carcasses (Compton et al. 2006; Wipfli and Baxter 2010), which was a very expensive effort, more development of the use of salmon waste parts from fish processing, such as guts, fins, and skin have resulted in a form of *briquette* that shows similar benefits to whole carcass additions. However, current evidence suggests that these ways of increasing freshwater production of salmonids are short-lived and do not persist without ongoing artificial enrichment (Benjamin et al. 2020).

Replacement, Conservation Offsets, and Mitigation Banking

Freshwater habitats are sometimes created as a replacement for damaged or lost environments. This is part of a larger role for conservation actions that is sometimes called mitigation banking or conservation offsets, to provide enhanced or restored (or even preserved) habitats in compensation for damage to similar habitats elsewhere. Constructed wetlands are sometimes created in industrial or agricultural areas as compensation for losses of freshwater habitats from development projects that are often far removed and even in different habitat types (Burgin 2010). In oil sands areas (Alberta, Canada), wetlands are lost as the earth's surface is removed to excavate oil-bearing sand. In response, new wetlands are dug out and planted as replacement or compensation. However, in one study, the constructed wetlands had not attained the biodiversity or functioning of nearby reference wetlands, even 20 years after construction (Kovalenko et al. 2013). In the area of a diamond mine in northern Canada where a stream had been lost, a replacement stream was created and then monitored for 17 years (Jones et al. 2017). Many jurisdictions have principles of *no net loss* of habitats, and offsets or *compensation* projects are allowed to replace areas that are given over to degradation for development; however, studies suggest most of these projects are only marginally successful or not successful at all in replacing lost ecosystem functions (Quigley and Harper 2006; Burgin 2010). A meta-analysis of 577 offset projects for freshwaters concluded that these are not very effective, but that larger projects were generally more effective than small projects (Theis et al. 2020, 2022). These kinds of offset measures should be considered as a last resort since it is not actually possible to replace fully functional, freshwater habitats—but work is ongoing to try to improve outcomes. Salmon, trout, sturgeon, crayfish, and mussel hatcheries are another tool that can be used to sustain species that are otherwise disappearing from our freshwaters or to offset species that are lost during development.

Some aspects of restoration may require active intervention (as opposed to passive restoration) in order to effectively rebuild the biological community. In some cases, dispersal to newly *restored* locations might not happen quickly, if at all. In many projects, one consideration may be to bring populations of colonists (especially invertebrates, algae, and microbes) from other source locations to inoculate a project location. However, as we discussed in Chapter 12, moving species around may also inadvertently introduce pathogens to new places. Introductions of individuals from elsewhere might also introduce genotypes to an area that could disrupt the genetic structure of some populations. Awareness of these potential issues is useful before choosing to move organisms around and creating problems that cannot be fixed later.

PERSPECTIVES

The best solution is for stewardship to avoid damage in the first place, thus preventing the need for restoration. In many jurisdictions, there are concepts of mitigation banking, replacement, and no-net-loss. These concepts assume that we can destroy ecosystems that are in the way of some kind of development and that we can replicate these natural ecosystems elsewhere (Moore and Moore 2013). As emphasized by Roni and Beechie (2013) and Allan et al. (2013), the enormous costs to remedy some problems may yield no positive outcomes if all the stressors acting at a site are not addressed. The combined effects of stressors need to be solved in unison. Do not expect instant responses, as colonists take time to arrive and populations may need years to increase, thus one has to be realistic about expectations of improvement in whatever the target response may be.

Finally, many structural changes to freshwaters require the approval of a licensed professional engineer since there are potential liability issues if the structures should move and cause damage, or if other incidents occur.

An intriguing social aspect of the protection and restoration of freshwaters comes from recent actions recognizing streams as deserving of *personhood* (Clark et al. 2019). In New Zealand and Bangladesh, these declarations of personhood come with rights of nature. This perspective is gaining traction in some places and is an interesting addition to how environmental protection is considered.

Box 15.1   Restoration of the Channelized
Kissimmee River in Florida

The Kissimmee River was originally a shallow, low-gradient, wide system with an extensive floodplain. With such a low gradient, seasonal flooding could sometimes expand the floodplain to almost 45 kilometers wide, which was good for wetland plants, birds, and fish. Its headwaters are near Orlando, and the river once meandered its way south through central Florida before emptying into Lake Okeechobee. A period of prolonged flooding, however, resulted in public demand

continued

for flood control, and in 1948, the U.S. Congress authorized alteration of the 166-kilometer-long reach of the Kissimmee River. The South Florida Water Management District was created to oversee projects in the basin (https://www.sfwmd.gov/). In the 1960s, the river was channelized by cutting and dredging a 10-meter deep, straight channel through the river's meanders, creating the C-38 canal. In addition, there was construction of water control structures to facilitate flood drainage and to create potentially arable land.

The creation of the C-38 canal caused extensive alterations to the wetland ecosystems that were connected to, and downstream of, the Kissimmee River—including the Florida Everglades. While the project provided flood protection, it also destroyed much of the floodplain-dependent ecosystem. More than 90% of the waterfowl that once inhabited the wetlands disappeared, and the number of bald eagle nesting territories decreased by 70%. After the river was transformed into a straight, deep canal, it became oxygen-depleted and the fish and invertebrate community that it supported changed dramatically (see Max Chesnes, TC Palm 7/29/2021). This channelization also increased nutrient inputs and transport, especially P (including stored P from past land use), into Lake Okeechobee through the development of drained lands (dairy production and other uses) and loss of floodplain wetlands (less nutrient uptake). Lake Okeechobee would then release these nutrient loads into the Everglades. Consequently, the Kissimmee River Restoration Plan is a key part of the overall Comprehensive Everglades Restoration Plan.

The Kissimmee River Restoration Project was approved in 1992, began in 1999, and was carried out in four phases. This extensive project involved backfilling the linear channel, removing dikes and other control structures, and recreating the meandering pattern of the river. In 2021 the Kissimmee River Restoration Project was successfully completed. It restored more than 103 km^2 of the river floodplain ecosystem, 81 km^2 of wetlands, and 71 km of the historic river channel (See https://www.sfwmd.gov/our-work/Kissimmee-river) (see Figure 15.6).

Importantly, the project had a very specific set of objectives for the outcomes of restoration that were monitored throughout the process. This demonstrated good planning since objectives

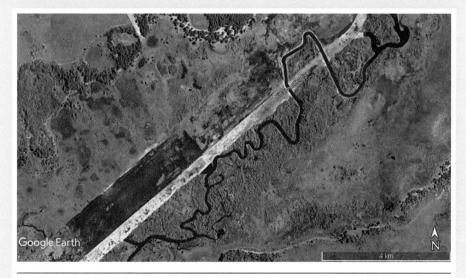

Figure 15.6 The Kissimmee River showing restoration of its former flow path (image taken in 2010). Note the filled in, linear channel that was replaced by recreating the meandering form of the original river. Image: Google Earth, courtesy of USDA Farm Service Agency.

continued

for restoration are often unmeasurable or unspecified. Project objectives included measures of ecological conditions, including hydrology, water quality, and biology (plants, invertebrates, birds, fish). The objectives also had predictions associated with them, including quantitative predictions, in order to judge project success against unrestored conditions (reaches prior to works or reaches that were not restored). Measurements indicate that the hydrological regime of the river has been effectively restored, with higher flows and oxygen concentrations found in the reconnected segments. Sampling of the restored channel segments has been compared in BACI designs, and the evidence indicates that communities of macroinvertebrates and fish are recovering well, as are aspects of the food webs (see Trexler 2006; Koebel and Bousquin 2014).

The Kissimmee River Restoration Project is the largest functioning restoration project of its kind in the world and was a joint effort between the U.S. Army Corps of Engineers and the South Florida Water Management District. (See https://fl.audubon.org/news/kissimmee-river-project-largest-restoration-initiative-its-kind-complete-after-nearly-30-years.)

Box 15.2   Wood Additions

While there is a range of actions for stream restoration, additions of large wood (greater than 10 centimeters in diameter and greater than 1 meter long) have been a frequent part of stream restoration. Some fish and other species use large wood as a security cover and as a refuge from fast currents. Large wood can also influence geomorphology and provide for a diversity of habitats (Roni et al. 2015). Often, land use has removed large wood sources through forest harvest or other land clearing to such an extent that there is little instream wood (decayed, transported, or removed) and not much, if any, near-term supply of trees from the riparian areas. In streams that are large enough for log driving, channels were often cleared of large wood, and the drives of logs down streams removed existing large wood and damaged stream habitat. Many streams historically had enormous amounts of large wood, but much of this was removed in the early industrial era to safeguard boat traffic and to reduce hazards from wood floating into transportation areas or damaging infrastructure downstream (Maser and Sedell 1994). In fact, at one time prior to the 1970s, it was thought that *debris dams* of large wood were an obstruction to the passage of salmonids, and such wood accumulations were actively removed by management agencies.

Projects for adding wood back into streams have been common since the 1990s. Many projects are not monitored beyond noting that the wood has remained in place. However, there are now many studies of the impacts of large wood placements on stream ecosystems, with a wide range of outcomes. In three streams in the state of Michigan, large wood was added (25 pieces/100 meters) to three streams in a BACI design—and in the two years after placements, the productivity of invertebrates increased in one stream, but there was no detectable effect in the other two (Entrekin et al. 2009). In the Basque country in Spain, four streams that received wood placements had a range of outcomes on nutrient and organic matter retention, from no measurable effect to a modest positive effect (Elosegi et al. 2016).

Many wood placements are intended to promote salmonid populations (refer back to Figure 15.5A). The responses of fish populations depend somewhat on the particular fish that is being considered. One of the most detailed studies of fish responses is from Colorado, which included

continued

a BACI design with upstream control reaches and log placement reaches in five streams over two decades (Gowan and Fausch 1996; White et al. 2011). This study showed that initially, older salmonids preferentially moved to log placements, but did not result in overall reach-scale increases in numbers. Over time, there was a net increase in salmonid numbers, but other factors beyond log structures apparently still limited population sizes, although wood placements can lead to increased populations of salmonids in some situations (Roni et al. 2015). Some authors have suggested that the apparent increases are just fish that are aggregating from surrounding parts of the stream reach. Roni (2019) points out that increases in fish numbers are demonstrably linked to increased survival and reproduction. There are many studies that have found improvements to stream populations and function because of wood placements (Roni 2019; Hallbert and Keeley 2023). However, there is a wide range of outcomes and the reasons for the variations are still unclear. Two conclusions are (1) that augmentation of physical habitat can be beneficial, but that it is not always physical structure that is limiting the ecosystem function, and (2) that a landscape perspective is also needed (Beechie et al. 2010; Roni and Beechie 2013).

ACTIVITIES

1. See if any of the Streamkeepers or Wetlandkeepers groups in your region have projects that you could look at or assist with.
2. Ask what targets or objectives they measured their successes against, such as increased fish numbers or higher diversity, and whether they had quantitative targets they were trying to reach (for instance, to double the fish numbers within three years or to return the fish numbers and diversity to preimpact levels).

16

MONITORING

INTRODUCTION

There are many programs for monitoring freshwaters, and the resulting data are critical for evaluating the effectiveness of protection measures, water conservation and allocation, water supply safety, and other management activities. Some obvious measures are discharge rates for lotic systems, stage or lake height for reservoirs and lakes, water chemistry parameters including organic and inorganic pollutants, and pathogens such as *E. coli*. There are also many kinds of programs for monitoring populations of organisms or whole communities, sometimes as a measure of environmental change. One can also measure ecosystem functions, such as primary productivity, nutrient turnover, decomposition rates, and ecosystem respiration. Monitoring sites have been established for such measures in most countries around the world. The specific agencies that are responsible differ by country, such as the U.S. Geological Survey, the U.S. Environmental Protection Agency, Environment and Climate Change Canada, or the United Kingdom's Environment Agency. Many such agencies provide real-time data on their websites, so you can examine data for sites near you. Data are critical to evaluating current conditions against objectives and quantitative targets, and to implement appropriate and timely management responses.

One of the very first requirements for monitoring is to determine exactly why you are monitoring. Often the broad goals of protection (see Figure 16.1) are determined by what the public wants from those who elect them, which typically is water they can drink, swim in, or eat the fish from. Using this information, objectives and priorities are determined. In the case of the Clean Water Act in the United States or the Water Framework Directive (WFD) in the European Union, there are specific targets. Targets may be specific levels of water quality parameters that may not be exceeded, or they may be targets that mandate no change. For instance, states such as Washington have rules that dictate that there should not be a measurable increase (defined as 0.3°C) in water temperature following any action, which is very different from vague terms like "protect the water from temperature increases." A measurable target is much easier to check and report on than general statements that have no specific measurable component. Government agencies are often vague about their targets, which permits them to evade specific standards and the monitoring of them.

Water quality monitoring for nutrients, other solutes, and contaminants is common, but these are often periodic and not continuous. Consequently, some events, such as spills of contaminants, are short-lived and easily missed. That has given rise to biomonitoring efforts (see the upcoming sections on biomonitoring) that use organisms that are living in the ecosystem

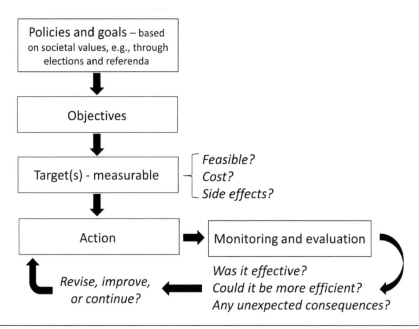

Figure 16.1 Monitoring comes at the end of the decision tree of what needs to be achieved from any management action. Targets should be measurable; for instance, it makes more sense to aim for reducing the Nitrate-N levels to below 10 mg/L, rather than a vague target of "improving water quality." Actions may not be practical or may cost more than some alternatives, which will factor into whether an action occurs. Finally, the outcome should be monitored to determine if the particular action was successful, if it could be made more efficient (and perhaps more cost effective), and if there were any unintended outcomes. Often agencies get locked into continuing to do things that are not effective because they have no quantitative evaluation through monitoring.

as an integrative measure (over time and space) of impacts from alterations to the freshwater environments (Karr 1999). Monitoring may also include assessments of physical conditions, and in the European Union's WFD, one of their measures is hydromorphic (hydrologic and geomorphic) condition. In some locations, continuous, real-time measures of some water quality and quantity are available, and recording equipment makes it possible to have more-or-less continuous traces of discharge, temperature, turbidity, and other measures. However, these stations are costly to install, require regular maintenance to repair and clean, and data management becomes a bigger undertaking. This often limits the number of stations, and availability of reliable, spatially representative data, especially in poorer countries or very large countries.

In addition to monitoring water quality, discharge, sediments, and other physical and chemical measures, we can follow the populations of certain species or even whole communities of organisms through time. What is being measured depends on the objectives. For instance, if the topic of interest is an endangered species, it will be beneficial to know the actual numbers or densities, which requires detailed, structured sampling. If the topic of interest is simply a general idea of where things are more common, then relative abundance measures or indices can be used. These latter kinds of measures for fish are often referred to as catch-per-unit-effort. If the numbers are decreasing, that may be cause for concern, and equally, if numbers are going up, that might suggest overstocking or something that might lead to a density-dependent change in survival or growth.

Another important component of monitoring is what is being used as the *reference* for comparing measures. We avoid the term *control* from experimental studies, as there is also background variation from weather, long-term and short-term climate variations, and even atmospheric concentrations of carbon dioxide and other components. The nature of the appropriate comparison is tied up with what the original reference might have been, perhaps in the preindustrial age. Our waters have been impacted in so many subtle and not-so-subtle ways that we have to be cautious about what we compare to. The concept of a *shifting baseline* was introduced to refer to how the information that we use as a *reference state* will change with personal and recent history and may not refer back to earlier conditions that may no longer exist anywhere (Pauly et al. 1998; Humphries and Winemiller 2009; McIntyre et al. 2016). One example that was mentioned in Chapter 8 is whether we should compare the outcomes of forest harvest protection to old-growth forests (the true reference) or just what all other managed (second-growth) forests are like, the latter of which says we accept a highly modified landscape that may be missing many of the species and functions it once supported—that is, a shifted baseline. Careful consideration of what the appropriate reference condition for comparison is depends on whether we are ready to admit that we have already lost so much from our natural ecosystems that we cannot expect to see those species, structures, and processes again.

Limitations

The European Union's WFD sets targets for what should be considered a *good condition* for a river, which is mostly based on geomorphic features and biota. Some estimates suggest that barely 40% of European rivers are considered to be in good condition (Lemm et al. 2021). Other countries have a range of agencies responsible for the monitoring of freshwaters—both the water and the habitats. In the United States, municipal, state, and federal agencies all have monitoring programs, as do a number of nongovernmental organizations. However, smaller bodies of water are very rarely considered, so we often have no idea of the general condition of small streams (Richardson and Dudgeon 2022) or ponds (Hassall 2014). Monitoring is essential to providing the data needed to manage, protect, and restore our freshwater systems. Knowledge of water supplies (inputs and discharge) and their quality is used for water allocations, and this might be at any level of government, even internationally. Monitoring aspects of water quality affect how governments protect drinking water supplies and water for aquatic life. In the absence of monitoring, we cannot know the effectiveness of management activities, including restoration, to meet the objectives stated by agencies.

Many of the water quality monitoring programs acknowledge difficulty with detecting interactions between stressors. Many monitoring programs are considered early warning systems, and a response would be to more accurately diagnose the cause or causes. The concept of cumulative effects is still challenging for those who are responsible for monitoring and being able to predict interactive effects of stressors rather than just demonstrating their impacts is one of the big challenges for applied, aquatic scientists.

SOLUTIONS

There are many agencies that track species, in terms of their global status, or even just their occurrences. Internationally, the International Union for the Conservation of Nature (IUCN) tracks the status of many species (as was mentioned in Chapter 12), and is a good source of information. The global Freshwater Information Platform (http://www.freshwaterplatform .eu/) is also a good place to find information. In the European Union there is an effort to produce a listing of freshwater biodiversity—called BioFresh—as part of the Freshwater Information Platform. In North America there is NatureServe, a web-based service that is supported by governments. There are national databases, especially for species that are considered at risk. Most states and provinces operate their own Conservation Data Centers, sometimes under different names, but they all track species at a state or province level.

Any monitoring program that is designed to detect the status and changes of freshwater biology requires several design elements in order to provide rigorous, defensible data:

1. The question being addressed by monitoring needs to be clearly articulated and testable.
2. It is important to have control sites or conditions—or more appropriately called *reference* since they are not controls in the sense of a well-controlled experiment.
3. There is a need for replication. For instance, your reference sites should represent the expected range of natural variation so that deviations can be detected from that distribution.
4. There should be a randomization of reference sites, since you do not want to have all the reference sites in one area and all your monitoring site(s) in another area that could be slightly different.
5. It is necessary to consider the types of statistical errors. In most studies we are concerned with falsely saying that something is statistically different when it really is not (type 1 error)—that is, our familiar probability of 5%. In monitoring we may be more concerned with missing something that might be a problem (type 2 error). One way to solve this is to relax our somewhat arbitrary probability level.
6. Finally, we need to consider our statistical power. Most monitoring programs only include a few sites, and with the amount of natural variation between sites, it might not be possible to detect differences, even if they are large. *Statistical power analysis* is a method for determining how many samples are needed in order to detect a significant difference between categories, which depends on how variable the sites are. In many cases we do not have enough sites (replication) to detect even big differences. In Chapter 8 you learned about before-after-control-impact study designs; these plans work well for monitoring if there are data available for at least two sites (one being the project site and at least one reference site) before and after implementation of the project that has the potential to alter a freshwater ecosystem.

Physical and Chemical Measures

Discharge is a fundamental measure for many policy objectives, especially allocations to various users, including natural ecosystems. The ways of measuring discharge were discussed in Chapter 1. The U.S. Geological Survey alone has more than 10,000 sites at which discharge or water height is measured. Temperature is one of the easiest measures to take and there are many models of relatively inexpensive and continuously monitoring data loggers. However, it is more complicated to compare temperatures because the mean, range, maximum, and minimum temperatures have to be considered, all of which have slightly different potential impacts on freshwater ecosystems. One change of concern is the seven-day running average of the daily maximum temperatures, which represents the hottest time of year, and a stressful or lethal temperature limit. Mean temperatures across the year provide some information, but it is probably more useful to break these up by seasons or even months. It may be the daily range of temperatures that is more relevant to some organisms or processes—for instance, it might be important to certain overwintering processes, at least in temperate climates, and warmer winter temperatures could also be a concern. While we often focus on high temperatures, land-use change can also affect how low the water temperatures may go in winter.

At these same measurement stations, data on sediment flux may also be undertaken to determine the transport of sediments (see Chapter 3). It is often efficient to have several measures taken at a common location. Sediment flux and turbidity are useful measures. Turbidity can be measured continuously, provided that instruments are regularly cleaned to avoid biofilm building on the sensors. Total suspended sediment is closely related to turbidity but requires water samples and does capture a bit more of the saltational load (see Chapter 3). In Europe, many of the impacts of fine sediments are from deposited sediment, but management targets are usually based on suspended loads, and these may give different assessments from each other (Jones et al. 2012b).

Water chemistry is still a major tool in monitoring freshwaters. Levels for drinking water and for aquatic life are set for many major contaminants, as we discussed in Chapter 4. There we discussed ecotoxicology as one way to establish standard levels for monitoring. One challenge is how to monitor the large number of chemicals that are present or emerging in our waterways. A lot of organic contaminants require specialized analytic techniques and can be costly. Emerging contaminants can be particularly challenging. However, other groups of chemicals may be correlated, for instance, the concentration of caffeine, artificial sweeteners, hormones, and other personal care products—and these might be good indicators of other chemicals in the wastewater effluents (see Chapter 5).

Biomonitoring for Chemicals

One means of monitoring for contaminants is to expose organisms in the field for known periods of time, and then allow those creatures to accumulate contaminants. This has been done using bivalves and is a standardized method for bioconcentrating contaminants such as heavy metals and organic compounds that may be occurring at low concentrations in water systems. Organisms can bioaccumulate contaminants, and this may be more effective than water quality monitoring for some chemicals. Organisms that are living in water will integrate their exposure over some period of time. Some of the biomonitoring can take advantage of samples to look at the concentration of contaminants. However, this also assumes some details of what the species of interest eats, how much they eat, whether they have moved around (i.e., where does their load of contaminants come from), and how fast they have grown. Thus, one cannot easily determine the rate of accumulation, or if it occurred over a short or long period of time. Bioaccumulation by individual organisms specifically placed in a study site (for instance, clams, such as *Corbicula fluminea*) for a known period of time have been used for this purpose. However, more recent and developing methods allow for the use of a synthetic tool, thereby sparing test organisms.

An alternative method to using living organisms is to use special bioassays, including semipermeable bags filled with some chemical that will bind and accumulate particular contaminants from the environment. Over the last three decades or more, semipermeable membrane devices (SPMDs) and polar organic chemical integrative samplers (POCIS) have been used to sample organic chemicals in freshwaters. These SPMD use a thin-walled polyethylene tube filled with a neutral lipid. Hydrophobic, organic compounds permeate into the device so that the passive accumulation of certain chemicals can be measured (Prest et al. 1992). These devices have also been compared with bioassays, based on live clams that were restrained in cages in the same environments (Prest et al. 1992). The POCIS are effective with some hydrophilic compounds. These approaches have been shown to be useful in estimating the concentrations and rates of bioaccumulation for particular classes of contaminants (Harman et al. 2008).

Biological Measures

We are often interested in numbers or sizes of particular species. In many commercial fisheries (see Chapter 10), the number and sizes (weights) of fish, crustaceans, shellfish, turtles, etc., are recorded. This can provide reasonably accurate, long-term data on trends (assuming minimal poaching and other unreported catches).

Counts of organisms are often used to assess the status of populations or changes in conditions, and as mentioned before, these might be actual counts (or density estimates) or relative abundance (a kind of index). One method that is used for migratory fish (such as salmon or eels) is a fish fence, where individuals moving upstream are captured at a fence where they can be counted. This can also be automated with side-scan radar to count individuals moving past a narrow point, such as in a fish ladder. Also, one might survey anglers in what is often called a creel survey (in some places, it is obligatory) to ask how many fish they caught and of what species and size over a certain period of time. This could lead to a measure of the number of fish caught per angler per day, and these could be compared between sites or between years. Other examples might be simple observations per unit time (such as per hour) of dragonflies, turtles, frogs, etc. Another example is rotary traps used to catch a sample of the emigrating fish (salmon, eels, lamprey). These traps are used in a specific cross-section of a stream, in order to get an estimate of the numbers (as well as species and size) that are moving through a river in a known volume of discharge. If the actual discharge of the stream is known, the total numbers can be estimated. It is also possible to just compare the actual numbers caught per hour over a period of time (days or years).

Sometimes, better estimates of numbers are required—especially for species that may be at risk or hard to assess from simple counts as previously described. There are many methods for marking animals so that we can make *mark-recapture* estimates. There are internal and external marks, such as numbered and/or colored wires (Floy tags), elastomer dye (a fluorescing polymer injected just beneath the skin), passive integrated transponder (PIT) tags, numbers or other patterns painted on the exterior (for instance on turtles or arthropods), bee tags, and also tracking devices (discussed more in the following paragraph). In a mark-recapture study, the animals are first collected and then marked, typically with a distinctive mark or number to distinguish individuals. Individually marking animals can give information on age or sex differences, such as growth rates, movement rates, survival rates, etc. A recapture session is used to determine the number of marked and unmarked individuals, and a rough estimate is based on the ratio of marked:unmarked (sometimes referred to as a ratio method) and the number of marked individuals that were released at the first session. There are several assumptions, such as animals do not lose their marks, that being marked does not alter survival rates (such as being more apparent to predators), marked individuals are equally as likely to be captured as unmarked individuals (assuming no change in behavior), and that between capture sessions, the population is *closed*—that is, little or no immigration, emigration, mortality, or births. At least one recapture session is needed to calculate the population size, and the more recapture sessions, the better the estimate. A very simple version would be:

$$N = [(\text{number marked in first session}) \times (\text{number captured in second session})]$$
$$\div (\text{number of marked individuals in second session})$$

where N is the population size (explore this yourself with some made-up numbers). Some common models for estimating numbers are the Lincoln-Peterson or Cormack-Jolly-Seber estimators, but there are many others. There are several good texts that explain mark-recapture methods in greater detail (for example, Krebs 1999; Amstrup et al. 2005).

We also track animals to determine spawning sites, the timing of migration, and to look at daily or seasonal use of habitats. To do this one often uses radiotelemetry, using small radios that are fastened to animals or inserted into them (usually surgically). This can also be accomplished using PIT tags or acoustic tags with stations that can read the tags as they pass by. These data can be very useful to identify times and locations that are critical to conservation efforts. One other tool that can be used is radar, which is essentially the method used in fish finders but can be refined to estimate fish movements. There is also side-scan radar, which can be used to estimate the depth and sizes of fish (or other organisms) and is especially useful in rivers or fish ladders to count the numbers of individuals migrating past in order to estimate stock sizes.

Biomonitoring Based on Communities or Indicators

Water quality and quantity monitoring are used worldwide as ways to protect water sources and to support the objectives of the Clean Water Act (USA), the WFD (European Union), and others. Most government programs dictate various kinds of monitoring to detect the chemical, physical, and biological status of their freshwaters. More and more often these monitoring programs are based on biological integrity and not simply measures of chemical or physical conditions, and sometimes are used instead of such nonbiological measures.

For a century, those responsible for evaluating the status of our surface waters have developed a range of indices that have names such as the Hilsenhoff Biotic Tolerance Index, the Index of Biotic Integrity (IBI), the reference condition approach (RCA), and others. Many measures of water quality are often spot measures and can miss the responses of ecosystems to punctuated events such as chemical spills or chronic impairment (Karr 1993). The general approach is to predict what a biological community at a site should look like based on assessments of appropriate reference sites. Often these biological communities are constrained—that is, there are indices for fish, macroinvertebrates, algae, or other taxonomically defined groups. Once a reference condition for a particular ecosystem is established, it is possible to contrast the actual structure (species composition or a trait-based index) versus the observed community (Bailey et al. 2004) (see Figure 16.2). A recurring challenge is to find appropriate reference sites, as these may no longer exist in a world of global change, or there may be sufficient geomorphic differences that a chosen site may not be representative (Stoddard et al. 2006).

In general, most of the indices developed are based on knowing the tolerance to pollution and other stressors of a range of organisms and calculating the numerical shift from less

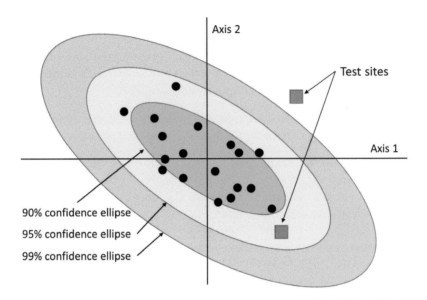

Figure 16.2 Ellipses around reference sites for the RCA. An ordination of communities (with many species present) provides a multidimensional plot, with only two dimensions shown here. The black circles represent the reference community—that is, what should be expected if there is no (or little) environmental degradation, and the ellipses represent different confidence intervals (just as in a 95% confidence interval around a single estimated mean). The squares can represent *test* sites, which may or may not exceed the confidence intervals around the reference sites. Note that it is often difficult to find any reference sites in some landscapes, and that may be one of the challenges. Redrawn after Bailey et al. (2004).

tolerant to more tolerant species. Several early versions of these indices were developed for invertebrates. The Hilsenhoff Biotic Index, or Biotic Tolerance Index, is based on knowing the relative tolerance, mostly to organic pollution, of stream invertebrates. The relative abundance of taxa is scaled by their tolerance to arrive at an index, with one end of the index indicating all intolerant (sensitive) taxa and the other end of the scale would be nothing but tolerant species (Hilsenhoff 1987). These observations of tolerance from field studies can be used or experiments can be carried out in order to define the tolerance levels of particular organisms to certain stressors (Larras et al. 2014). These kinds of indices are fine when the tolerance of taxa is known but are difficult when there are no available data for tolerance or when interactions within food webs begin to play a role—for example, when sensitive prey are replaced by tolerant prey which causes shifts in their predators. Similarly, it is possible to use algae, and thus, similar indices based on the tolerance of diatoms were developed (Patrick 1965).

There are taxonomy-based (species, genus, or family levels) or sensitivity-based (tolerances, traits) metrics and many versions have been developed around the world that are tailored to local needs and ecosystems. Biotic Indices (BIs) can be simple, like species diversity or richness—or complex, as seen in multivariate, whole communities. BIs can also be based on functional metrics including life history traits and feeding mechanisms, which are all relative to reference conditions (Birk et al. 2012). One limitation of such methods is taxonomic resolution, which for some taxa (for instance, benthic macroinvertebrates), might be reasonable to genus or even family level, which is less precise, but often sufficient (Reece et al. 2001). For some BIs, the sensitivities of particular taxa are not well-known, or known only for certain classes of stressors, for example, organic pollution versus sediments. Often, shifts in community structure that are relative to reference sites indicate a response to stressors, but do not clearly identify the nature of the stressor, particularly in the case of multiple stressors (Birk et al. 2012). One biological indicator in the European Union is referred to as SPEAR$_{pesticides}$ (SPEcies At Risk for pesticides), using biological traits of invertebrates reflecting their sensitivities to individual pesticides and the mixtures of those contaminants that are typical of most surface waters (Liess et al. 2021). This measure is directly linked to the European Union's water quality classes.

There are many tools that are now available for monitoring freshwaters, and these have been adopted by government agencies, nongovernmental organizations, Streamkeepers, Wetlandkeepers, and others. The various indices need to be calibrated to a particular region and type of freshwater ecosystem and its size, whether it be a stream, lake, or wetland. Calibration also means having appropriate reference conditions for comparison, but it is often difficult to find relatively pristine freshwaters. In many places, groups have developed Benthic-IBIs for their particular regions. Some agencies have attempted to develop databases based on the RCA for the monitoring of freshwaters (such as RIVPACS in the UK and AUSRIVAS in Australia) and the U.S. Environmental Protection Agency (EPA) has state-specific tools for ecological indicators. Other developments have been to elaborate on a *multimetric index* that combines the use of macroinvertebrates and fish for regions of the United States (Herlihy et al. 2020). There are a wide range of such tools, and each has its advantages and disadvantages (Vander Laan and Hawkins 2014).

Biomonitoring by states and nations utilizes well-developed IBI (such as a Benthic-IBI, or B-IBI), the RCA, observed versus expected metrics, or single, sensitive species (for instance, in the EU's WFD). Many of these have been focused on streams, but there are also lake-based metrics used for environmental assessment that employ similar principles. In the European Union, their WFD requires the use of phytoplankton (composition, abundance, biomass) as a measure of lake status, and these metrics, such as the Brettum Index, are used to determine the *ecological quality ratio* (Bergkemper and Weisse 2018). However, short-term variation may not be adequately captured by this set of measures, as was shown during a short-term heat wave that resulted in a large variation over a short period of time in an Austrian lake (Bergkemper and Weisse 2018).

After recognizing that the particular species in a community may differ regionally, practitioners have developed trait-based ways of assessing impacts. Rather than taxonomic entities, species (and even age classes within a species) are assigned trait values; this gets past the species identity issue and makes freshwaters more comparable across larger geographic scales. These traits might include length of life cycle, presence of dispersal traits (for example, flight), tolerance to low flows, etc. (Bonada et al. 2007; Jeliazkov et al. 2020). This approach makes it possible to compare freshwater ecosystems without the constraint of specific taxonomy, which changes geographically.

Another monitoring consideration is what is known as *proper functioning condition* and this considers ecosystem functions and structures more than species. This has been widely used for the assessment of riparian areas, but its intent is to protect the functions that affect freshwater systems (Swanson et al. 2017). The connection of freshwaters to their riparian areas (and watersheds) is a big contributor to the state of surface waters, therefore, providing properly functioning riparian areas is an important conservation action (Feld et al. 2018).

Genetics and Genomics

Genetics analyses have become relatively inexpensive and easily available, and the use of environmental DNA (eDNA) is now becoming widely used for monitoring. In particular, this can be used to detect the presence of species in specific water bodies, based on their shed cells (and DNA) in the water, in a predator's stomach contents, or on the bottom of a lake or stream. For instance, if someone wants to know if a particular fish species occurs in a large watershed, water samples from all the small tributaries could be collected and amplified to detect DNA of that species. Of course, there can also be many false negatives, and the failure to detect a species based on a water sample is not confirmation that it is not present. For instance, DNA may be too degraded to amplify or too diluted to detect. However, this method can save enormous amounts of time by not having to do detailed capture efforts across large landscapes. There can also be false positives, if, for example, DNA gets from one place to another through a canal that the actual species cannot move through. One interesting example is from the grass carp in the Ohio River system (discussed in Chapter 11) where electric *fences* in the Chicago Sanitary and Ship Canal are used to reduce the probability of grass carp reaching Lake Michigan through the canal. In that case, eDNA is being used to see if the carp have made it into Lake Michigan, although the presence of its DNA in Lake Michigan does not mean the fish is actually there.

With the wider use of genetic tools, new metrics based on metabarcoding are being developed (Pawlowski et al. 2019; see Box 16.2). Metabarcoding in freshwater systems is the barcoding of eDNA (or ribonucleic acid) in a manner that allows for the simultaneous identification of many taxa within the same sample (see for example, Elbrecht and Leese 2017; Gleason et al. 2023). This will allow for faster analysis in biomonitoring programs and reduce reliance on time-consuming and costly analyses based on visual identifications and enumeration of samples of organisms, such as communities of macroinvertebrates or phytoplankton. The advent of high-throughput sequencing now makes it possible to identify communities based on entire specimens or extra-cellular DNA (in the water). As with conventional metrics or indices, similar tools need to be developed for eDNA, and there have been a number of indices developed that go beyond taxonomic structure changes. These methods also expand the range of taxa used in biomonitoring to include bacteria, protists, meiofauna, and others for which there are few or no obvious taxonomy-based methods. Supervised machine learning is being used to develop indices based on training data fit to models (Pawlowski et al. 2019). This method can use metabarcoding data calibrated against particular stressor measures. However, these methods can also produce large measures of diversity given the large number of cryptic species among freshwater organisms or taxa not currently defined in genomics databases (Pawlowski et al. 2019).

DECISION SUPPORT TOOLS

Balancing multiple objectives for the protection of freshwater ecosystems versus other uses, such as drinking water supplies, hydropower production, irrigation, etc., requires tools for evaluating the trade-offs involved. There are a range of decision support tools that allow practitioners to evaluate options. Some methods depend on Bayesian inference and usually some valuation of the options—biologists have a difficult time with valuation (what is a kilometer of stream worth?). This is one realm where ecosystem service valuations become important, even if monetary costs are difficult to specify with any precision. Such decision trees and other methods are a field unto themselves and usually involve a lot of policy considerations beyond protection of the freshwater ecosystem. Nevertheless, it is essential that biologists be included in such deliberations.

PERSPECTIVES

Methods using eDNA show enormous promise, but also, there are sometimes contradictions with standard taxonomy-based approaches, especially in terms of quantifying numbers or biomass based on genomics methods.

Box 16.1   Long-Term Monitoring and Its Benefits

Many agencies collect temperature and flow data on a continuous basis from rivers around the world. These data are typically valuable in the short-term for projecting water supplies and flood risks, and for evaluating the safety of water quality. In the longer term, such data provide benefits for understanding the magnitude and locations of global change impacts on waters, including climate change. These data also provide perspective on variability of the amounts, timing, and quality of freshwaters. For instance, long-term data on lake levels are important to planning for infrastructure that is located near shorelines, as in the Laurentian Great Lakes where many tens of millions of people live, and shipping by boat is critical (see Figure 16.3). However, sustaining long-term collection of data is difficult in terms of budget, and it also requires proper calibration and data management. Calibration can be challenging because methods change over time. One good example of the use of long-term data is for flowing waters in the Pacific Northwest of North America—the NorWeST Interagency Stream Temperature Database—where many hundreds of sites have been amalgamated. Integration of such data is valuable, but it requires a large staff to manage and maintain large databases.

Small watershed studies that are associated with forest harvest (see Chapter 8) have been sustained at the H.J. Andrews Hydrological Laboratory (in the state of Oregon) with flow and water chemistry data since 1948, providing the ability to examine long-term effects of forestry and climate changes (Johnson et al. 2021). Given that forest regrowth after harvesting is a many-decade process, such long-term data allows for the study of patterns that were not evident from shorter-term studies (see Chapter 8, Figures 8.5A and B). For example, Burt et al. (2015) were able to separate out the effects due to long-term adjustments in flows from the forests following harvesting against a background of climate patterns associated with El Niño-Southern Oscillations. Such studies at the H.J. Andrews, Coweeta Hydrological Laboratory, and other long-term ecological research sites, are increasingly showing their value as we gain appreciation for long-term climate cycles and global changes.

continued

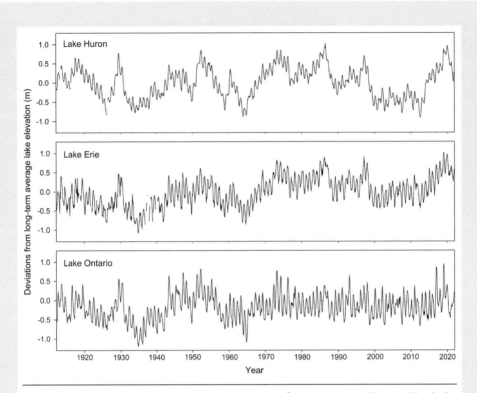

Figure 16.3 Lake elevations of three of the Laurentian Great Lakes showing monthly deviations (in meters) from the long-term (1912 to 2022) averages. Note the similar long-term patterns between lakes in the time series, especially evident in the 1930s. Note also that Lake Ontario's level has been regulated since 1958, whereas the other two lakes have not. Average lake elevations: Lake Ontario (Oswego)—74.947 meters, Lake Erie (Toledo)—174.157 meters, Lake Huron (Mackinaw City)—176.463 meters. Note that the annual cycle in water levels is indicative of spring runoff peaks from the area's winter snowpacks. Data from the U.S. National Oceanographic and Atmospheric Administration.

In Chapter 13 we discussed a large, distributed water quality sampling program that is in the European Union. In that case there are over 52,000 sites that have measurements that are used to determine the status of surface waters (Lemm et al. 2021). Such monitoring programs provide important data for water users, and for agencies to set priorities for controls and remedies for degraded freshwater ecosystems. Another outcome from these programs is the ability to measure the *success* of efforts to protect water systems relative to past measurements. Finally, the ability to detect and define reference conditions can indicate high-priority areas for continued protection and also provide benchmarks against which to compare management and restoration outcomes.

Box 16.2   Metabarcoding to Assess Communities

The increasingly affordable technique of metabarcoding extends the ideas of detection using DNA to another level of becoming a bioassessment tool for freshwaters. The DNA could be from water samples (as in eDNA approaches), from homogenized samples of freshwater communities, or even from the preservatives in which such samples are kept (Blackman et al. 2019). These methods are still being refined, and part of that is calibration to determine how effective and efficient barcoding methods might be for whole communities, such as from samples of benthic invertebrates or plankton. These samples would still require sorting of individuals from the organic materials in such samples, but there are efforts underway so that this step may eventually not be required. Given the amount of effort associated with sorting, and morphological identification and enumeration, genetic techniques could enhance the speed and accuracy and decrease the costs of monitoring programs.

For many taxa, it is not easy to identify specimens to species morphologically—such as immature insect larvae, small species, or cryptic species. This means that metabarcoding eventually could provide better data than traditional sampling. Of course, this also depends on there being appropriate genetic primers for all possible species, some of which are still not well-known. Also, most trials have not been able to demonstrate that the estimates are good at quantifying relative biomass of different taxa, although there is encouraging progress in that too (Blackman et al. 2019). Often, there are more species present than were expected, since we still have incomplete knowledge of groups such as invertebrates and smaller taxa.

Leaf litter entering freshwaters is a key resource for food webs, and much of the decomposition is done by fungi on the leaves. Microbes, such as fungi, are rarely used in biomonitoring but may hold potential. A study from Finland surveyed fungi that were growing on leaf litter from a large number of streams that were experiencing different rates of disturbance by stressors (Jyväsjärvi et al. 2021). Metabarcoding results for fungi from leaves were compared with known values based on macroinvertebrates and diatoms. They found that the fungi taxon richness based on sequencing overestimated the numbers of fungi, but provided a very comparable ability to detect stressor impacts on streams as the macroinvertebrates and diatoms did.

The use of eDNA—and especially the ability to characterize whole communities—is rapidly developing. Early detection of invasive species, assessment of the occurrences of threatened (and therefore usually rare) species, and monitoring of communities are possible with these methods (referred to as metabarcoding). There are also limitations, and false negatives should be expected. For instance, rare species may not shed enough DNA to be detected in a whole watershed, and DNA can degrade rapidly and not be amplified. With cautious application, this method has great potential.

ACTIVITIES

1. What freshwater monitoring programs does your state or province use and for what purpose?
2. Do you think they are effective at providing a good indication of the condition of freshwaters where you live?
3. The US-EPA provides a lot of online data about streams and rivers of the United States at https://mywaterway.epa.gov/ and Canada has something similar at https://wateroffice .ec.gc.ca/mainmenu/real_time_data_index_e.html. Many other countries have similar real-time data sources. Some of these online sources are likely to include streams near you.

REFERENCES

Addicott, J.F., Aho, J.M., Antolin, M.F., Padilla, D.K., Richardson, J.S., and Soluk, D.A. 1987. Ecological neighborhoods: Scaling environmental patterns. *Oikos* 49:340–346.

Alam, S., Gebremichael, M., Li, R., Dozier, J., and Lettenmaier, D.P. 2020. Can Managed Aquifer Recharge mitigate the groundwater overdraft in California's Central Valley? *Water Resources Research* 56: e2020WR027244.

Albert, J.S., Georgia Destouni, G., Duke-Sylvester, S.M., Magurran, A.E., Oberdorff, T., Reis, R.E., Winemiller, K.O., Ripple, W.J. 2021. Scientists' warning to humanity on the freshwater biodiversity crisis. *Ambio* 50:85–94.

Alberts, J.M., Fritz, K.M., and Buffam, I. 2018. Response to basal resources by stream macroinvertebrates is shaped by watershed urbanization, riparian canopy cover, and season. *Freshwater Science* 37:640–652.

Algera, D.A., Rytwinski, T., Taylor, J.T., Bennett, J.R., Smokorowski, K.E., Harrison, P.M., Clarke, K.D., Enders, E.C., Power, M., Bevelhimer, M.S., and Cooke, S.J. 2020. What are the relative risks of mortality and injury for fish during downstream passage at hydroelectric dams in temperate regions? A systematic review. *Environmental Evidence* 9:3. doi.org/10.1186/s13750-020-0184-0.

Allan, J.D., Castillo, M.M., and Capps, K.A. 2021. *Stream Ecology: Structure and function of running waters*. 3rd Edition. Springer International.

Allan, J.D., McIntyre, P.B., Smith, S.D.P., Halpern, B.S., Boyer, G.L., Buchsbaume, A., Burton, G.A., Jr., Campbell, L.M., Chadderton, W.L., Ciborowski, J.J.H., Doran, P.J., Eder, T., Infante, D.M., Johnson, L.B., Joseph, C.A., Marino, A.L., Prusevich, A., Read, J.G., Rose, J.B., Rutherford, E.S., Sowa, S.P., and Steinman, A.D. 2013. Joint analysis of stressors and ecosystem services to enhance restoration effectiveness. *Proceedings of the National Academy of Sciences* 110:372–377. doi/10.1073/pnas.1213841110.

Allen, G.H. and Pavelsky, T.M. 2018. Global extent of rivers and streams. *Science* 361:585–588.

Almeida, L.Z., Sesterhenn, T.M., Rucinski, D.K., and Höök, T.O. 2022. Nutrient loading effects on fish habitat quality: Trade-offs between enhanced production and hypoxia in Lake Erie, North America. *Freshwater Biology* 67:784–800.

Alves, R.N. and Agustí, S. 2020. Effect of ultraviolet radiation (UVR) on the life stages of fish. *Reviews in Fish Biology and Fisheries* 30:335–372.

Aminzadeh, M., Lehmann, P., Or, D. 2018. Evaporation suppression and energy balance of water reservoirs covered with self-assembling floating elements. *Hydrology and Earth System Sciences* 22:4015–4032.

Amstrup, S.C., McDonald, T.L., and Manly, B.F.J. (Eds.). 2005. Handbook of Capture-Recapture Analysis. Princeton University Press.

Amyot, J.-P. and Downing, J. 1997. Seasonal variation in vertical and horizontal movement of the freshwater bivalve *Elliptio complanata* (Mollusca: Unionidae). *Freshwater Biology* 37: 345–354.

Anderson, D.H. 2014. Geomorphic responses to interim hydrology following Phase I of the Kissimmee River Restoration Project, Florida. *Restoration Ecology* 22:367–375.

Anthony, J.L. and Downing, J.A. 2001. Exploitation trajectory of a declining fauna: A century of freshwater mussel fisheries in North America. *Canadian Journal of Fisheries and Aquatic Sciences* 58:2071–2090.

Antonelli, M., Tamea, S., and Yang, H. 2017. Intra-EU agricultural trade, virtual water flows and policy implications. *Science of the Total Environment* 587–588:439–448.

Arlos, M.J., Bragg, L.M., Parker, W.J., and Servos, M.R. 2015. Distribution of selected antiandrogens and pharmaceuticals in a highly impacted watershed. *Water Research* 72:40–50.

Atwood, T.B., Hammill, E., Greig, H.S., Kratina, P., Shurin, J.B., Srivastava, D.S., and Richardson, J.S. 2013. Predator-induced reduction of freshwater carbon dioxide emissions. *Nature Geoscience* 6191–194.

Avlijaš, S., Ricciardi, A. and Mandrak, N.E. 2018. Eurasian tench (*Tinca tinca*): The next Great Lakes invader. *Canadian Journal of Fisheries and Aquatic Sciences* 75:169–179.

Bailey, R.C., Norris, R.H., Reynoldson, T.B. 2004. *Bioassessment of Freshwater Ecosystems Using the Reference Condition Approach*. Springer doi:10.1007/978-1-4419-8885-0.

Baiser, B., Olden, J.D., Record, S., Lockwood, J.L., and McKinney, M.L. 2012. Pattern and process of biotic homogenization in the New Pangaea. *Proceedings of the Royal Society B-Biological Sciences* 279:4772–4777.

Bajer, P.G., Ghosal, R., Maselko, M., Smanski, M.J., Lechelt, J.D., Hansen, G., and Kornis, M.S. 2019. Biological control of invasive fish and aquatic invertebrates: A brief review with case studies. *Management of Biological Invasions* 10:227–254.

Balian, E.V., Segers, H., Lévèque, C., and Martens, K. 2008. The Freshwater Animal Diversity Assessment: An overview of the results. *Hydrobiologia* 595:627–637.

Barbiero, R.P. and Tuchman, M.L. 2004. Changes in the crustacean communities of Lakes Michigan, Huron and Erie following the invasion of the predatory cladoceran *Bythotrephes longimanus*. *Canadian Journal of Fisheries and Aquatic Sciences* 61:2111–2125.

Bärlocher, F. 2012. *The Ecology of Aquatic Hyphomycetes*. Ecological Studies 94, Springer-Verlag.

Barrett, I.C., McIntosh, A.R., Febria, C.M., and Warburton, H.J. 2021. Negative resistance and resilience: Biotic mechanisms underpin delayed biological recovery in stream restoration. *Proceedings of the Royal Society B* 288:20210354.

Bartholomew, A. and Bohnsack, J.A. 2005. A review of catch-and-release angling mortality with implications for no-take reserves. *Reviews in Fish Biology and Fisheries* 15:129–154.

Bauer, B.O., Lorang, M.S., and Sherman, D.J. 2002. Estimating boat-wake-induced levee erosion using sediment suspension measurements. *Journal of Waterway Port Coastal and Ocean Engineering—ASCE* 128: 152–162.

Baumann, R.W. and Kondratieff, B.C. 2010. The stonefly genus *Lednia* in North America (Plecoptera: Nemouridae). *Illiesia*, 6(25):315–327. http://www2.pms-lj.si/illiesia/papers/Illiesi a06-25.pdf.

BC Government. 2018. Manual of British Columbia Hydrometry Standards. https://www2.gov .bc.ca/assets/gov/environment/natural-resource-stewardship/nr-laws-policy/risc/man_bc _hydrometric_stand_v2.pdf.

Beacham, T.D., Wallace, C., Jonse, K., Sutherland, B.J.G., Gummer, C., and Rondeau, E.B. 2021. Estimation of conservation unit and population contribution to Chinook salmon mixed-stock fisheries in British Columbia, Canada, using direct DNA sequencing for single nucleotide polymorphisms. *Canadian Journal of Fisheries and Aquatic Sciences* 78:1422–1434.

Becu, M.H.J., Michalski, T.A., and Richardson, J.S. 2023. Forest harvesting impacts on small, temperate zone lakes: A review. *Environmental Reviews* 31:376–402.

Beechie, T.J., Sear, D.A., Olden, J.D., Pess, G.R., Buffington, J.M., Moir, H., Roni, P., and Pollock, M.M. 2010. Process-based principles for restoring river ecosystems. *BioScience* 60: 209–222.

Beermann, A., Elbrecht, V., Karnatz, S., Ma, L., Matthaei, C.D., Piggott, J.J., and Leese, F. 2018. Multiple-stressor effects on stream macroinvertebrate communities: A mesocosm experiment manipulating salinity, fine sediment and flow velocity. *Science of the Total Environment* 610:961–971.

Behrenfeld, M.J., Halsey, K.H., Boss, E., Karp-Boss, L., Milligan, A.J., and Peers, G. 2021. Thoughts on the evolution and ecological niche of diatoms. *Ecological Monographs* 91: e01457. 10.1002/ecm.1457.

Beketov, M.A., Kefford, B.J., Schäfer, R.B., and Liess, M. 2013. Pesticides reduce regional biodiversity of stream invertebrates. *Proceedings of the National Academy of Science of the United States* 110:11039–11043.

Bellmore, J.R., Pess, G.R., Duda, J.J., O'Connor, J.E., East, A.E., Foley, M.M., Wilcox, A.C., Major. J.J., Shafroth, P.B., Morley, S.A., Magirl, C.S., Anderson, C.W., Evans, J.E., Torgersen, C.E. and Craig, L.S. 2019. Conceptualizing ecological responses to dam removal: If you remove it, what's to come? *BioScience* 69:26–39.

Belsky, A.J., Matzke, A., and Uselman, S. 1999. Survey of livestock influences on stream and riparian ecosystems in the western United States. *Journal of Soil and Water Conservation* 54:419–431.

Benjamin, J.R., Bellmore, J.R., Whitney, E., and Dunham, J.B. 2020. Can nutrient additions facilitate recovery of Pacific salmon? *Canadian Journal of Fisheries and Aquatic Sciences* 77:1601–1611.

Benke, A.C., Cushing, C.E. (Eds.). 2005. *Rivers of North America*. Elsevier, Burlington, MA.

Benke, A.C., Wallace and J.B. 2003. Influence of wood on invertebrate communities in streams and rivers. *American Fisheries Society Symposium* 37:149–177.

Bennett, E.M., Carpenter, S.R., and Caraco, N.F. 2001. Human impact on erodable phosphorus and eutrophication: A global perspective. *BioScience* 51:227–234.

Bergkemper, V. and Weisse, T. 2018. Do current European lake monitoring programmes reliably estimate phytoplankton community changes? *Hydrobiologia* 824:143–162.

Berland, A., Shiflett, S.A., Shuster, W.D., Garmestani, A.S., Goddard, H.C., Herrmann, D.L., and Hopton, M.E. 2017. The role of trees in urban stormwater management. *Landscape and Urban Planning* 162:167–177. doi:10.1016/j.landurbplan.2017.02.017.

Bernhardt, E.S. and Palmer, M.A. 2011. River restoration: The fuzzy logic of repairing reaches to reverse catchment scale degradation. *Ecological Applications* 21:1926–1931.

Bernhardt, E.S., Rosi, E.J., and Gessner, M.O. 2017. Synthetic chemicals as agents of global change. *Frontiers in Ecology and the Environment* 15:84–90.

Bilby, R.E. 1981. Role of organic debris dams in regulating the export of dissolved and particulate matter from a forested watershed. *Ecology* 62:1234–1243.

Birk, S., Bonne, W., Borja, A., Brucet, S., Courrat, A., Poikane, S., Solimini, A., Van De Bund, W., Zampoukas, N., and Hering, D. 2012. Three hundred ways to assess Europe's surface waters:

An almost complete overview of biological methods to implement the water framework directive. *Ecological Indicators* 18:31–41. https://doi.org/10.1016/j.ecolind.2011.10.009.

Birk, S., Chapman, D., Carvalho, L., Spears, B.M., Andersen, H.E., Argillier, C., Auer, S., Baattrup-Pedersen, A., Banin, L., Beklioglu, M., Bondar-Kunze, E., Borja, A., Branco, P., Bucak, T., Buijse, A.D., Cardoso, A.C., Couture, R., Cremona, F., de Zwart, D., Feld, C., Ferreira, M.T., Feuchtmayr, H., Gessner, M., Gieswein, A., Globevnik, L., Graeber, D., Graf, W., Gutierrez-Canovas, C., Hanganu, J., Işkın, U., Järvinen, M., Jeppesen, E., Kotamäki, N., Kuijper, M., Lemm, J.U., Lu, S., Lyche Solheim, A., Mischke, U., Moe, J., Noges, P., Noges, T., Ormerod, S., Panagopoulos, Y., Phillips, G., Posthuma, L., Pouso, S., Prudhomme, C., Rankinen, K., Rasmussen, J.J., Richardson, J., Sagouis, A., Santos, J.M., Schäfer, R.B., Schinegger, R., Schmutz, S., Schneider, S.C., Schülting, L., Segurado, P., Stefanidis, K., Sures, B., Thackeray, S., Turunen, J., Uyarra, M.C., Venohr, M., von der Ohe, P., Willby, N., and Hering, D. 2020. Impacts of multiple stressors on freshwater biota across spatial scales and ecosystems. *Nature Ecology & Evolution* 4:1060–1068.

Bisson, P.A., Gregory, S.V., Nickelson, T.E., and Hall, J.D. 2008. The Alsea Watershed Study: A comparison with other multi-year investigations in the Pacific Northwest. Pp. 259–289 In: Stednick, J.D. (Ed.) *Hydrological and Biological Responses to Forest Practices*. Springer.

Blackman, R.C., Mächler, E., Altermatt, F., et al. 2019. Advancing the use of molecular methods for routine freshwater macroinvertebrate biomonitoring—the need for calibration experiments. *Metabarcoding and Metagenomics* 3:49–57.

Blanchet, S., Loot, G., Bernatchez, L., and Dodson, J.J. 2007. The disruption of dominance hierarchies by a non-native species: An individual-based analysis. *Oecologia* 152:569–581.

Blanchfield, P.J., Kidd, K.A., Docker, M.F., Palace, V.P., Park, B.J., and Postma, L.D. 2015. Recovery of a wild fish population from whole-lake additions of a synthetic estrogen. *Environmental Science & Technology* 49:3136–3144.

Blanchfield, P.J., Rudd, J.W.M., Hrenchuk, L.E., Amyot, M., Babiarz, C.L., Beaty, K.G., Bodaly, R.A.D., Branfireun, B.A., Gilmour, C.C., Graydon, J.A., Hall, B.D., Harris, R.C., Heyes, A., Hintelmann, H., Hurley, J.P., Kelly, C.A., Krabbenhoft, D.P., Lindberg, S.E., Mason, R.P., Paterson, M.J., Podemski, C.L., Sandilands, K.A., Southworth, G.R., St Louis, V.L., Tate, L.S., and Tate, M.T. 2022. Experimental evidence for recovery of mercury-contaminated fish populations. *Nature* 601:74–78.

Blaustein, A.R., Kiesecker, J.M., Chivers, D.P., and Anthony, R.G. 1997. Ambient UV-B radiation causes deformities in amphibian embryos. *Proceedings of the National Academy of Sciences of the United States* 94:13735–13737. https://doi.org/10.1073/pnas.94.25.13735.

Blinn, C.R. and Kilgore, M.A. 2001. Riparian Management Practices: A Summary of State Guidelines. *Journal of Forestry* 99:11–17.

Böhm, M., Dewhurst-Richman, N.I., Seddon, M., Ledger, S.E.H., Albrecht, C., Allen, D., Bogan, A.E. Cordeiro, J., Cummings, K.S., Cuttelod, A., Gustavo Darrigran, G., Darwall, W., Fehér, Z., Gibson, C., Graf, D.L., Köhler, F., Lopes-Lima, M., Pastorino, G., Perez, K.E., Smith, K., van Damme, D., Vinarski, M.V., von Proschwitz, T., von Rintelen, T., Aldridge, D.C., Aravind N.A., Budha, P.B., Clavijo, C., Van Tu, D., Gargominy, O., Ghamizi, M., Haase, M., Hilton-Taylor, C., Johnson, P.D., Kebapçı, U., Lajtner, J., Lange, C.N., Lepitzki, D.A.W., Martínez-Ortí, A., Moorkens, E.A., Eike Neubert, E., Pollock, C.M., Prié, V., Canella Radea, C., Ramirez, R., Ramos, M.A., Santos, S.B., Slapnik, R., Son, M.O., Stensgaard, A., Collen, B. 2020. The conservation status of the world's freshwater molluscs. *Hydrobiologia*. https://doi.org/10.1007/s10750-020-04385-w.

Bonada, N., Doledec, S., and Statzner, B. 2007. Taxonomic and biological trait differences of stream macroinvertebrate communities between Mediterranean and temperate regions: Implications for future climatic scenarios. *Global Change Biology* 13:1658–1671.

Booth, D.B., Roy, A.H., Smith, B., and Capps, K.A. 2016. Global perspectives on the urban stream syndrome. *Freshwater Science* 35:412–420.

Booth, D.B., Scholz, J.G., and Beechie, T.J., et al. 2016. Integrating limiting-factors analysis with process-based restoration to improve recovery of endangered salmonids in the Pacific Northwest, USA. *Water* 8:174.

Boretti, A. and Rosa, L. 2019. Reassessing the projections of the World Water Development Report. *npj Clean Water* 2:15. https://doi.org/10.1038/s41545-019-0039-9.

Bouwes, N., Weber, N., Jordan, C.E., Saunders, W.C., Tattam, I.A., Volk, C., Wheaton, J.M., and Pollock, M.M. 2016. Ecosystem experiment reveals benefits of natural and simulated beaver dams to a threatened population of steelhead (*Oncorhynchus mykiss*). *Scientific Reports* 6:28581.

Bowne, D.R., Cosentino, B.J., Anderson, C.P., et al. 2018. Effects of urbanization on the population structure of freshwater turtles across the United States. *Conservation Biology* 32:1150–1161.

Bracewell, S., Verdonschot, R.C.M., Schäfer, R.B., Bush, A., Lapen, D.R., Paul, J., and Van den Brink, P.J. 2019. Qualifying the effects of single and multiple stressors on the food web structure of Dutch drainage ditches using a literature review and conceptual models. *Science of the Total Environment* 684:727–740.

Bradshaw, A.D. 1984. Ecological principles and land reclamation practice. *Landscape Planning* 11:35–48.

Branco, P., Boavida, I., Santos, J.M., Pinheiro, A., and Ferreira, M.T. 2013. Boulders as building blocks: Improving habitat and river connectivity for stream fish. *Ecohydrology* 6:627–634.

Branton, M.A. and Richardson, J.S. 2014. A test of the umbrella species approach in restored floodplain ponds. *Journal of Applied Ecology* 5:776–785.

Broadmeadow, S.B., Jones, J.G., Langford, T.E.L., Shaw, P.J., and Nisbet, T.R. 2011. The influence of riparian shade on lowland stream water temperatures in southern England and their viability for Brown Trout. *River Research and Applications* 27:226–237.

Bruno, D., Hermoso, V., Sánchez-Montoya, M.M., Belmar O., Gutiérrez-Cánovas, C., and Cañedo-Argüelles, M. 2022. Ecological relevance of non-perennial rivers for the conservation of terrestrial and aquatic communities. *Conservation Biology* 36:e13982.

Bundschuh, M. and McKie, B.G. 2016. An ecological and ecotoxicological perspective on fine particulate organic matter in streams. *Freshwater Biology* 61:2063–2074.

Bunn, S.E. and Arthington, A.H. 2002. Basic principles and ecological consequences of altered flow regimes for aquatic biodiversity. *Environmental Management* 30:492–507.

Burgin, S. 2010. 'Mitigation banks' for wetland conservation: A major success or an unmitigated disaster? *Wetlands Ecology and Management* 18:49–55.

Burnett, N.J., Hinch, S.G., Bett, N.N., Braun, D.C., Casselman, M.T., Cooke, S.J., Gelchu, A., Lingard, S., Middleton, C.T., Minke-Martin, V., and White, C.F.H. 2017. Reducing carryover effects on the migration and spawning success of sockeye salmon through a management experiment of dam flows. *River Research and Applications* 33:3–15.

Burt, T.P., Howden, N.J.K., McDonnell, J.J., Jones, J.A., and Hancock, G.R. 2015. Seeing the climate through the trees: observing climate and forestry impacts on streamflow using a 60-year record. *Hydrological Processes* 29:473–480.

Cantonati, M., Fensham, R.J., Stevens, L.E., Gerecke, R., Glazier, D.S., Goldscheider, N., Knight, R.L., Richardson, J.S., Springer, A.E., and Tockner, K. 2021. An urgent plea for global spring ecosystem protection. *Conservation Biology* 35:378–382.

Cantonati, M., Poikane, S., Pringle, C.M., Stevens, L.E., Turak, E., Heino, J., Richardson, J.S., Bolpagni, R., Borrini, A., Cid, N., Čtvrtlíková, M., Galassi, D.M.P., Hájek, M., Hawes, I., Levkov, Z., Naselli-Flores, L., Saber, A.A., Di Cicco, M., Fiasca, B., Hamilton, P.B., Kubečka, J., Segadelli, S., and Znachor, P. 2020. Characteristics, main impacts, and stewardship of natural and artificial freshwater environments: Consequences for biodiversity conservation. *Water* 12:260.

Cantwell, M.G., Katz, D.R., Sullivan, J.C., Shapley, D., Lipscomb, J., Epstein, J., Juhl, A.R., Knudson, C., and O'Mullan, G.D. 2018. Spatial patterns of pharmaceuticals and wastewater tracers in the Hudson River Estuary. *Water Research* 137:335e343.

Caraco, N.F., Cole, J.J., Raymond, P.A., Strayer, D.L., Pace, M.L., Findlay, S.E.G., and Fischer, D.T. 1997. Zebra mussel invasion in a large, turbid river: Phytoplankton response to increased grazing. *Ecology* 78:588–602.

Carey, D.E. and McNamara, P.J. 2015. The impact of triclosan on the spread of antibiotic resistance in the environment. *Frontiers in Microbiology* 5:780.

Carignan, R., D'Arcy, P., and Lamontagne, S. 2011. Comparative impacts of fire and forest harvesting on water quality in Boreal Shield lakes. *Canadian Journal of Fisheries and Aquatic Sciences* 57(S2):105–117.

Carpenter, K.D., Kuivila, K.M., Hladik, M.L., Haluska, T., and Cole, M.B. 2016. Storm-event-transport of urban-use pesticides to streams likely impairs invertebrate assemblages. *Environmental Monitoring and Assessment* 188:345.

Carpenter, S.R., Kitchell, J.F., and Hodgson, J.R. 1985. Cascading trophic interactions and lake productivity. *BioScience* 35:634–639.

Carvalho, L., Mackay, E.B., Cardoso, A.C., Baattrup-Pedersen, A., Birk, S., Blackstock, K.L., Borics, G., Borja, A., Feld, C.K., Ferreira, M.T., Globevnik, L., Grizzetti, B., Hendry, S., Hering, D., Kelly, M., Langaas, S., Meissner, K., Panagopoulos, Y., Penning, E., Rouillard, J., Sabater, S., Schmedtje, U., Spears, B.S., Venohr, M., van de Bund, W., and Lyche Solheim, A. 2019. Protecting and restoring Europe's waters: An analysis of the future development needs of the Water Framework Directive. *Science of the Total Environment* 658:1228–1238.

Castelle, A.J., Johnson, A.W., and Conolly, C. 1994. Wetland and stream buffer size requirements—a review. *Journal of Environmental Quality* 23:878–882.

Caughley, G. 1994. Directions in conservation biology. *Journal of Animal Ecology* 63:215–244.

Cauvy-Fraunié, S. and Dangles, O. 2019. A global synthesis of biodiversity responses to glacier retreat. *Nature Ecology & Evolution* 3:1675–1685.

Chaffin, B.C., Shuster, W.D., Garmestani, A.S., Furio, B., Albro, S.L., Gardiner, M., Spring, M., and Green, O. 2016. A tale of two rain gardens: Barriers and bridges to adaptive management of urban stormwater in Cleveland, Ohio. *Journal of Environmental Management* 183:431–441.

Chambers, P.A., Lacoul, P., Murphy, K.J., and Thomaz, S.M. 2008. Global diversity of aquatic macrophytes in freshwater. *Hydrobiologia* 595:9–26.

Chapagain, A.K. and Hoekstra, A.Y. 2011. The blue, green and grey water footprint of rice from production and consumption perspectives. *Ecological Economics* 70:749–758.

Chará-Serna, A.M., Epele, L.B., Morrissey, C.A., and Richardson, J.S. 2019. Nutrients and sediment modify the impacts of a neonicotinoid insecticide on freshwater community structure and ecosystem functioning. *Science of the Total Environment* 692:1291–1303.

Cheng, F.Y., Van Meter, K.J., Byrnes, D.K., Basu, N.B. 2020. Maximizing US nitrate removal through wetland protection and restoration. *Nature* 588:625–630.

Christensen, J.R., Macduffee, M., Macdonald, R.W., Whiticar, M., and Ross, P.S. 2005. Persistent organic pollutants in British Columbia grizzly bears: Consequence of divergent diets. *Environmental Science & Technology* 39:6952–6960.

Church, M. 2002. Geomorphic thresholds in riverine landscapes. *Freshwater Biology* 47:541–557.

———. 2015. Channel stability: Morphodynamics and the morphology of rivers. Pp. 281–321 In: Rowinski, P. and Rudecki-Pawlik, A. (Eds.) *Rivers—Physical, Fluvial and Environmental Processes.* Springer International.

Clark, C., Emmanouil, N., Page, J., and Pelizzon, A. 2019. Can You Hear the Rivers Sing? Legal Personhood, Ontology, and the Nitty-Gritty of Governance. *Ecology Law Quarterly* 45:787–844.

Clarke, A., Azulai, D., Dueker, M.E., Vos, M., and Perron, G.G. 2019. Triclosan alters microbial communities in freshwater microcosms. *Water* 11, 961; doi:10.3390/w11050961.

Clarke, G.K.C., Jarosch, A.H., Anslow, F.S., Radić, V., and Menounos, B. 2015. Projected deglaciation of western Canada in the twenty-first century. *Nature GeoScience* 8:372–377.

Clements, W.H., Cadmus, P., Kotalik, C.J., and Wolff, B.A. 2019. Context-dependent responses of aquatic insects to metals and metal mixtures: A quantitative analysis summarizing 24 yr of stream mesocosm experiments. *Environmental Toxicology and Chemistry* 38:2486–2496.

Clements, W.H., Herbst, D.B., Hornberger, M.I., Mebane, C.A., and Short, T.M. 2021. Long-term monitoring reveals convergent patterns of recovery from mining contamination across 4 western US watersheds. *Freshwater Science* 40:407–426.

Clements, W.H., Vieira, N.K.M., and Church, S.E. 2010. Quantifying restoration success and recovery in a metal-polluted stream: A 17-year assessment of physicochemical and biological responses. *Journal of Applied Ecology* 47:899–910.

Colautti, R.I., Ricciardi, A., Grigorovich, I.A., and MacIsaac, H.J. 2004. Is invasion success explained by the enemy release hypothesis? *Ecology Letters* 7:721–733.

Cole, L.J., Brocklehurst, S., Robertson, D., Harrison, W., and McCracken, D.I. 2015. Riparian buffer strips: Their role in the conservation of insect pollinators in intensive grassland systems. *Agriculture, Ecosystems and Environment* 211:207–220.

Collins, A.L., Williams, L.J., Zhang, Y.S., Marius, M., Dungait, J.A.J., Smallman, D.J., Dixon, E.R., Stringfellow, A., Sear, D.A., Jones, J.I., and Naden, P.S. 2013. Catchment source contributions to the sediment-bound organic matter degrading salmonid spawning gravels in a lowland river, southern England. *Science of the Total Environment* 456:181–195.

Compton, J.E., Andersen, C.P., Phillips, D.I., Brooks, J.R., Johnson, M.G., Church, M.R., Hogsett, W.E., Cairns, M.A., Rygiewicz, P.T., McComb, B.C., and Shaff, C.D. 2006. Ecological and water quality consequences of nutrient addition for salmon restoration in the Pacific Northwest. *Frontiers in Ecology and the Environment* 4:18–26.

Compton, T.J., De Winton, M., Leathwick, J.R., and Wadhwa, S. 2012. Predicting spread of invasive macrophytes in New Zealand lakes using indirect measures of human accessibility. *Freshwater Biology* 57:938–948.

Comte, L. and Grenouillet, G. 2013. Do stream fish track climate change? Assessing distribution shifts in recent decades. *Ecography* 36:1236–1246.

Conley, D.J., Paerl, H.W., Howarth, R.W., Boesch, D.F., Seitzinger, S.P., Havens, K.E., Lancelot, C., and Likens G.E. 2009. Controlling eutrophication: Nitrogen and phosphorus. *Science* 323:1014–15.

Connery, K. 2009. Biodiversity and urban design: Seeking an integrated solution. *Journal of Green Building* 4:23–38.

Conroy, E., Turner, J.N., Rymszewicz, A., O'Sullivan, J.J., Bruenc, M., Lawler, D., Lally, H., and Kelly-Quinn, M. 2016. The impact of cattle access on ecological water quality in streams: Examples from agricultural catchments within Ireland. *Science of the Total Environment* 547:17–29.

Cook, K.V., Lennox, R.J., Hinch, S.G., and Cooke, S.J. 2015. Fish out of water: How much air is too much? *Fisheries* 40:452–461.

Cooke, S.J., Vermaire, J.C., Baulch, H.M., Birnie-Gauvin, K., Twardek, W., and Richardson, J.S. 2022. Our failure to protect the stream and its valley: A call to "back off" from riparian development. *Freshwater Science* 41:183–194.

Cooley, S.W., Ryan, J.C., and Smith, L.C. 2021. Human alteration of global surface water storage variability. *Nature* 591:78–81.

Corsi, S.R., Graczyk, D.J., Geis, S.W., Booth, N.L., and Richards, K.D. 2010. A fresh look at road salt: Aquatic toxicity and water-quality impacts on local, regional, and national scales. *Environmental Science and Technology* 44:7376–7382.

Côté, I.M., Darling, E.S., and Brown, C.J. 2016. Interactions among ecosystem stressors and their importance in conservation. *Proceedings of the Royal Society B* 283:20152592.

Coughlan, N.E., Cuthbert, R.N., Potts, S., Cunningham, E.M., Crane, K., Caffrey, J.M., Lucy, F.E., Davis, E., and Dick, J.T.A. 2019. Beds are burning: Eradication and control of invasive Asian clam, *Corbicula fluminea*, with rapid open flame burn treatments. *Management of Biological Invasions* 10:486–499.

Craig, L.S., Olden, J.D., Arthington, A.H., Entrekin, S., Hawkins, C.P., Kelly, J.J., Kennedy, T.A., Maitland, B.M., Rosi, E.J., Roy, A.H., Strayer, D.L., Tank, J.L., West, A.O., and Wooten, M.S. 2017. Meeting the challenge of interacting threats in freshwater ecosystems: A call to scientists and managers. *Elementa—Science of the Anthropocene* 5:72.

Crain, C.M., Kroeker, K., and Halpern, B.S. 2008. Interactive and cumulative effects of multiple human stressors in marine systems. *Ecology Letters* 11:1304–1315.

Cranston, P. and McKie, B. 2006. Aquatic wood—an insect perspective. Pp. 9–14 In: Grove, S.J. and Hanula, J.L. (Eds.) *Insect biodiversity and dead wood: Proceedings of a symposium for the 22nd International Congress of Entomology*. Gen. Tech. Rep. SRS-3. Asheville, NC: U.S. Department of Agriculture Forest Service, Southern Research Station. 106 p.

Cucherousset, J. and Olden, J.D. 2011. Ecological impacts of non-native freshwater fishes. *Fisheries* 36:215–230.

Danehy, R.J., Chan, S.S., Lester, G.T., Langshaw, R.B., and Turner, T.R. 2007. Periphyton and macroinvertebrate assemblage structure in headwaters bordered by mature, thinned and clearcut Douglas-fir stands. *Forest Science* 53:294–307.

Danehy, R.J., Doloff, C.A., and Reeves, G.H. (Eds.). 2022. *Reflections on Forest Management; Can Fish and Fiber Coexist?* American Fisheries Society Symposium 92, Bethesda, Maryland. doi.org/10.47886/9781934874660.

Daniels, R.E. and Allan, J.D. 1981. Life table evaluation of chronic exposure to a pesticide. *Canadian Journal of Fisheries and Aquatic Sciences* 38:485–494.

Daryanto, S., Fu, B., Wang, L., Jacinthe, P.-A., and Zhao, W. 2018. Quantitative synthesis on the ecosystem services of cover crops. *Earth-Science Reviews* 185:357–373.

Datry, T., Bonada, N., and Boulton, A.J. (Eds.). 2017. *Intermittent Rivers and Ephemeral Streams Ecology and Management*. Academic Press.

Davies, B., Biggs, J., Williams, P., Whitfield, M., Nicolet, P., Sear, D., Bray, S., Maund, S. 2008. Comparative biodiversity of aquatic habitats in the European agricultural landscape. *Agriculture, Ecosystems and Environment* 125:1–8.

Davis, J.J., Jackson, P.R., Engel, F.L., LeRoy, J.Z., Neeley, R.N., Finney, S.T., and Murphy, E.A. 2016. Entrainment, retention, and transport of freely swimming fish in junction gaps between commercial barges operating on the Illinois Waterway. *Journal of Great Lakes Research* 42:837–848.

Davis, J.M., Baxter, C.V., Rosi-Marshall, E.J., Pierce, J.L., and Crosby, B.T. 2013. Anticipating stream ecosystem responses to climate change: Toward predictions that incorporate effects via land–water linkages. *Ecosystems* 16:909–922. doi:10.1007/s10021-013-9653-4.

Davy, C.M., Kidd, A.G., and Wilson, C.C. 2015. Development and validation of environmental DNA (eDNA) markers for detection of freshwater turtles. *PLoS ONE* 10(7):e0130965. doi:10.1371/journal.pone.0130965.

Dawson, F.H. 1981. The downstream transport of fine material and the organic-matter balance for a section of a small chalk stream in Southern England. *Journal of Ecology* 69:367–380.

Denoth, M. and Myers, J.H. 2005. Variable success of biological control of *Lythrum salicaria* in British Columbia. *Biological Control* 32:269–279.

De Roos, A.J., Gurian, P.L., Robinson, L.F., Rai, A., Zakeri, I., and Kondo, M. C. 2017. Review of epidemiological studies of drinking-water turbidity in relation to acute gastrointestinal illness. *Environmental Health Perspectives*, 125:086003.

Devlin, S.P., Tappenbeck, S.K., Craft, J.A., Tappenbeck, T.H., Chess, D.W., Whited, D.C., Ellis, B.K., Stanford, J.A. 2017. Spatial and Temporal Dynamics of Invasive Freshwater Shrimp (*Mysis diluviana*): Long-term effects on ecosystem properties in a large oligotrophic lake. *Ecosystems* 20:183–197.

Dextrase, A.J. and Mandrak, N.E. 2006. Impacts of alien invasive species on freshwater fauna at risk in Canada. *Biological Invasions* 8:13–24.

Dickard, M., Gonzalez, M., Elmore, W., Leonard, S., Smith, D., Smith, S., Staats, J., Summers, P., Weixelman, D., and Wyman, S. 2015. Riparian area management: Proper functioning condition assessment for lotic areas. Technical Reference 1737-15. U.S. Department of the Interior, Bureau of Land Management, National Operations Center, Denver, CO.

Dill, W.A. and Cordone, A.J. 1997. History and Status of Introduced Fishes in California, 1871–1996. State of California, The Resources Agency, Department of Fish and Game. *Fish Bulletin* 178. https://escholarship.org/uc/item/5rm0h8qg.

Dillon, P.J. and Rigler, F.H. 1974. A Test of a simple nutrient budget model predicting the Phosphorus concentration in lake water. *Journal of the Fisheries Research Board of Canada* 31:1771–1778.

Dingman, L.W. 2015. *Physical Hydrology*, 3rd Edition. Waveland Press Inc., Long Grove, IL.

Di Veroli, A., Santoro, F., Pallottini, M., Selvaggi, R., Scardazza, F., Cappelletti, D., and Goretti, E. 2014. Deformities of chironomid larvae and heavy metal pollution: From laboratory to field studies. *Chemosphere* 112:9–17.

Dodds, W. and Whiles, M. 2019. *Freshwater Ecology: Concepts and Environmental Applications of Limnology*. 3rd Edition. Elsevier.

Doenz, C.J., Krähenbühl, A.K., Walker, J., Seehausen, O., and Brodersen, J. 2019. Ecological opportunity shapes a large Arctic charr species radiation. *Proceedings of the Royal Society B* 286:20191992.

Donohue, I. and Garcia Molinos, J. 2009. Impacts of increased sediment loads on the ecology of lakes. *Biological Reviews* 84:517–531.

Downing, J.A. 2010. Emerging global role of small lakes and ponds: Little things mean a lot. *Limnetica* 29(1):9–24.

Downing, J.A., Prairie, Y.T., Cole, J.J., Duarte, C.M., Tranvik, L.J., Striegl, R.G., McDowell, W.H., Kortelainen, P., Caraco, N.F., Melack, J.M., and Middelburg, J. 2006. The global abundance and size distribution of lakes, ponds, and impoundments. *Limnology and Oceanography* 51:2388–2397.

Duda, J.J., Anderson, J.H., Beirne, M., Brenkman, S., Crain, P., Mahan, J., McHenry, M., Pess, G., Peters, R., and Winter, B. 2019. Complexities, context, and new information about the Elwha River. *Frontiers in Ecology and the Environment* 17(1):10–11.

Dudgeon, D. 2011. Asian river fishes in the Anthropocene: Threats and conservation challenges in an era of rapid environmental change. *Journal of Fish Biology* 79:1487–1524.

———. 2020. *Freshwater Biodiversity: Status, Threats and Conservation.* Cambridge University Press, Cornwall, UK.

Dudgeon, D., Arthington, A.H., Gessner, M.O., Kawabata, Z.I., and Knowler, D.J. et al. 2006. Freshwater biodiversity: Importance, threats, status and conservation challenges. *Biological Reviews* 81:163–182.

Dudley, T. and Anderson, N.H. 1987. The biology and life cycles of *Lipsothrix* spp. (Diptera: Tipulidae) inhabiting wood in Western Oregon streams. *Freshwater Biology* 17:437–451.

Durance, I. and Ormerod, S.J. 2010. Evidence for the role of climate in the local extinction of a coolwater triclad. *Journal of the North American Benthological Society* 29:1367–1378.

Echols, J.C. 1995. *Review of Fraser River Sturgeon.* Fraser River Action Plan Fishery Management Group, Dept. of Fisheries and Oceans Canada, Vancouver, BC. 33 p.

Eckstein, G.E. 2009. Water scarcity, conflict, and security in a climate change world: Challenges and opportunities for international law and policy. *Wisconsin International Law Journal* 27:409.

Eggert, S.L. and Wallace, J.B. 2007. Wood biofilm as a food resource for stream detritivores. *Limnology and Oceanography* 52:1239–1245.

El Baradei, S. and Al Sadeq, M. 2020. Effect of solar canals on evaporation, water quality, and power production: An optimization study. *Water* 12, 2103. doi:10.3390/w12082103.

Elbrecht, V. and Leese, F. 2017. Validation and development of COI metabarcoding primers for freshwater macroinvertebrate bioassessment. *Frontiers in Environmental Science* 5:11.

Eliason, E.J., Clark, T.D., Hague, M.J., Hanson, L.M., Gallagher, Z.S., Jeffries, K.M., Gale, M.K., Patterson, D.A., Hinch, S.G., and Farrell, A.P. 2011. Differences in thermal tolerance among Sockeye salmon populations. *Science* 332:109–112.

Elmarsafy, M., Tasky, K.L., and Gray, D.K. 2021. Can zooplankton on the North American Great Plains "keep up" with climate-driven salinity change? *Limnology and Oceanography* 66:865–877.

Elmore, A.J. and Kaushal, S.S. 2008. Disappearing headwaters: Patterns of stream burial due to urbanization. *Frontiers in Ecology and the Environment* 6:308–312.

Elosegi, A., Elorriaga, C., Flores, L., Martí, E., and Joserra Díez, J. 2016. Restoration of wood loading has mixed effects on water, nutrient, and leaf retention in Basque mountain streams. *Freshwater Science* 35:41–54.

Emde, S., Rueckert, S., Palm, H.W., and Klimpel, S. 2012. Invasive Ponto-Caspian amphipods and fish increase the distribution range of the Acanthocephalan *Pomphorhynchus tereticollis* in the River Rhine. *PLoS ONE* 7(12):e53218. doi:10.1371/journal.pone.0053218.

Entrekin, S.A., Tank, J.L., Rosi-Marshall, E.J., Hoellein, T.J., and Lamberti, G.A. 2009. Response of secondary production by macroinvertebrates to large wood addition in three Michigan streams. *Freshwater Biology* 54:1741–1758.

Ercoli, F., Lefebvre, F., Delangle, M., Godé, N., Caillon, M., Raimond, R., and Souty-Grosset, C. 2019. Differing trophic niches of three French stygobionts and their implications for conservation of endemic stygofauna. *Aquatic Conservation: Marine and Freshwater Ecosystems* 29:2193–2203.

Estrada, F., Kim, D., and Perron, P. 2021. Spatial variations in the warming trend and the transition to more severe weather in midlatitudes. *Scientific Reports* 11:145. https://doi.org/10.10 38/s41598-020-80701-7.

Fairfax, E. and Whittle, A. 2020. Smokey the Beaver: beaver-dammed riparian corridors stay green during wildfire throughout the western USA. *Ecological Applications* 30:e02225.

FAO. 2020. *The State of World Fisheries and Aquaculture 2020. Sustainability in action.* Rome. https://doi.org/10.4060/ca9229en.

Feely, J.R. and Sorensen, P.W. 2023. Effects of an ensonified bubble curtain and a cyclic sound on blocking 10 species of fishes including 4 invasive carps in a laboratory flume. *Biological Invasions* 25:1973–1989.

Feiner, Z.S., Chong, S.C., Knight, C.T., Lauer, T.E., Thomas, M.V., Tyson, J.T., and Höök, T.O. 2015. Rapidly shifting maturation schedules following reduced commercial harvest in a freshwater fish. *Evolutionary Applications* 8:724–737.

Feld, C.K., Birk, S., Bradley, D.C., Hering, D., Kail, J., and Marzin, A. et al. 2011. From natural to degraded rivers and back again: A test of restoration ecology theory and practice. *Advances in Ecological Research* 44:119–209.

Feld, C.K., Fernandes, M.R., Ferreira, M.T., Hering, D., Ormerod, S.J., Venohr, M., and Gutiérrez-Cánovas, C. 2018. Evaluating riparian solutions to multiple stressor problems in river ecosystems—A conceptual study. *Water Research* 139:381–394.

FEMAT (Forest Ecosystem Management Assessment Team). 1993. Forest ecosystem management: An ecological, economic, and social assessment. U.S. Forest Service, U.S. Department of Commerce, National Oceanic and Atmospheric Administration, National Marine Fisheries Service, United States Bureau of Land Management, Fish and Wildlife Service, Portland, Oregon. (Available from: http://www.blm.gov/or/plans/nwfpnepa/FEMAT-1993 /1993_%20FEMAT-ExecSum.pdf).

Ferreira, V., Koricheva, J., Pozo, J., and Graça, M.A.S. 2016. A meta-analysis on the effects of changes in the composition of native forests on litter decomposition in streams. *Forest Ecology and Management* 364:27–38.

Fischer, A.J., Kerr, L., Sultana, T., and Metcalfe, C.D. 2021. Effects of opioids on reproduction in Japanese Medaka, *Oryzias latipes. Aquatic Toxicology* 236:105873.

Fisheries and Oceans Canada. 2015. *Survey of Recreational Fishing in Canada, 2015.* Ottawa, Canada.

Florido, F.G., Regitano, J.B., Andrade, P.A.M., Andreote, F.D., and Brancalion, P.H.S. 2022. A comprehensive experimental assessment of glyphosate ecological impacts in riparian forest restoration. *Ecological Applications* 32:e02472.

Fluet-Chouinard, E., Funge-Smith, S., and McIntyre, P.B. 2018. Global hidden harvest of freshwater fish revealed by household surveys. *Proceedings of the National Academy of Science of the United States* 115:7623–7628.

Foley, C.J., Feiner, Z.S., Malinich, T.D., and Höök, T.O. 2018. A meta-analysis of the effects of exposure to microplastics on fish and aquatic invertebrates. *Science of the Total Environment* 631–632:550–559.

Foley, J.A., DeFries, R., Asner, G.P., Barford, C., Bonan, G., Carpenter, S.R., Chapin, F.S., Coe, M.T., Daily, G.C., Gibbs, H.K., Helkowski, J.H., Holloway, T., Howard, E.A., Kucharik, C.J., Monfreda, C., Patz, J.A., Prentice, I.C., Ramankutty, N., Snyder, P.K. 2005. Global consequences of land use. *Science* 309:570–574.

Fonseca, D.M. and Hart, D.D. 2001. Colonization history masks habitat preferences in local distributions of stream insects. *Ecology* 82:2897–2910.

Fowler, D., Coyle, M., Skiba, U., Sutton, M.A., Cape, J.N., Reis, S., Sheppard, L.J., Jenkins, A., Grizzetti, B., Galloway, J.N., Vitousek, P., Leach, A., Bouwman, A.F., Butterbach-Bahl, K., Dentener, F., Stevenson, D., Amann, M., and Voss, M. 2013. The global nitrogen cycle in the twenty-first century. *Philosophical Transactions of the Royal Society of London B*, 368: 20130164. https://doi.org/10.1098/rstb.2013.0164.

Francoeur, S.N. and Biggs, B.J.F. 2006. Short-term effects of elevated velocity and sediment abrasion on benthic algal communities. *Hydrobiologia* 561:59–69.

Friedrich, K., Grossman, R.L., Huntington, J., Blanken, P.D., Lenters, J., Holman, K.D., Gochis, D., Livneh, B., Prairie, J., Skeie, E., Healey, N.C., Dahm, K., Pearson, C., Finnessey, T., Hook, S.J., and Kowalski, T. 2018. Reservoir evaporation in the western United States: Current science, challenges, and future needs. *Bulletin of the American Meteorological Society* 99:167–187.

Frissel, C.A., Liss, W.J., Warren, C.E., and Hurley, M.D. 1986. A hierarchical framework for stream habitat classification: Viewing streams in a watershed context. *Environmental Management* 10:199–214.

Frissel, C.A. and Nawa, R.K. 1992. Incidence and causes of physical failure of artificial fish habitat structures in streams of western Oregon and western Washington. *North American Journal of Fisheries Management* 12:182–197.

Fujino, T., Kobori, S., Nomoto, T., Sakai, M., and Gomi, T. 2018. Radioactive cesium contamination and its biological half-life in larvae of *Stenopsyche marmorata* (Trichoptera: Stenopsychidae). *Landscape and Ecological Engineering* 14:37–43.

Gascho Landis, A.M. and Stoeckel, J.A. 2016. Multi-stage disruption of freshwater mussel reproduction by high suspended solids in short- and long-term brooders. *Freshwater Biology* 61:229–238.

Gaston, K.J., Warren, P.H., Thompson, K., and Smith, R.M. 2005. Urban domestic gardens (IV): the extent of the resource and its associated features. *Biodiversity and Conservation* 14:3327–3349.

Gavrilescu, M., Demnerová, K., Aamand, J., Agathos, S., and Fava, F. 2015. Emerging pollutants in the environment: Present and future challenges in biomonitoring, ecological risks and bioremediation. *New Biotechnology* 32:147–156.

Geist, J. 2010. Strategies for the conservation of endangered freshwater pearl mussels (*Margaritifera margaritifera* L.): A synthesis of Conservation Genetics and Ecology. *Hydrobiologia* 644:69–88.

Gessner, M.O. and Tlili, A. 2016. Fostering integration of freshwater ecology with ecotoxicology. *Freshwater Biology* 61:1991–2001.

Gibb, J.P. 2001. Wetland loss and biodiversity conservation. *Conservation Biology* 14:314–317.

Gibson, P.P. and Olden, J.D. 2014. Ecology, management, and conservation implications of North American beaver (*Castor canadensis*) in dryland streams. *Aquatic Conservation: Marine and Freshwater Ecosystems* 24:391–409.

Gilbert, J., Danielopol, D.L., and Stanford, J.A. (Eds.). 1994. *Groundwater Ecology*. Academic Press, Toronto.

Gilpin, M.E. and Soulé, M.E. 1986. Minimum viable populations: Processes of species extinction. Pp 19–34 In M. E. Soulé (Ed.) *Conservation biology: The science of scarcity and diversity*. Sinauer Associates, Sunderland, Massachusetts, USA.

Gleason, J.E., Hanner, R.H., and Cottenie, K. 2023. Hidden diversity: DNA metabarcoding reveals hyper-diverse benthic invertebrate communities. *BMC Ecology and Evolution* 23:19.

Gleick, P.H. 2003. Water Use. *Annual Review of Environment and Resources* 28:275–314.

Gomi, T., Moore, R.D., and Dhakal, A.S. 2006. Headwater stream temperature response to clear-cut harvesting with different riparian treatments, coastal British Columbia, Canada. *Water Resources Research* 42, W08437, doi:10.1029/2005WR004162.

Gomi, T., Sidle, R.C., Bryant, M.D., and Woodsmith, R.D. 2001. The characteristics of woody debris and sediment distribution in headwater streams, southeastern Alaska. *Canadian Journal of Forest Research* 31:1386–1399.

Gopal, K., Tripathy, S.S., Bersillon, J.L., and Dubey, S.P. 2007. Chlorination byproducts, their toxicodynamics and removal from drinking water. *Journal of Hazardous Materials* 140:1–6.

Gottesfeld, A.S., Hassan, M.A., and Tunnicliffe, J.F. 2008. Salmon bioturbation and stream process. *American Fisheries Society Symposium* 65:175–193.

Gounand I., Harvey, E., Little, C.J., and Altermatt, F. 2018. Meta-Ecosystems 2.0: Rooting the theory into the field. *Trends in Ecology and Evolution* 33:36–46.

Govindarajulu, P., Altwegg, R., and Anholt, B.R. 2005. Matrix model investigation of invasive species control: Bullfrogs on Vancouver Island. *Ecological Applications* 15:2161–2170.

Gowan, C. and Fausch, K.D. 1996. Long-term demographic responses of trout populations to habitat manipulation in six Colorado streams. *Ecological Applications* 6:931–946.

Graf, N., Battes, K.P., Cimpean, M., Dittrich, P., Entling, M.H., Link, M., . . . and Schäfer, R.B. 2019. Do agricultural pesticides in streams influence riparian spiders? *Science of The Total Environment* 660:126–135.

Green, R.H. 1979. *Sampling design and statistical methods for environmental biologists*. Wiley, New York City, New York, USA.

Greig, H.S., Kratina, P., Thompson, P.L., Palen, W.J., Richardson, J.S., and Shurin, J.B. 2012. Warming, eutrophication, and predator loss amplify subsidies between aquatic and terrestrial ecosystems. *Global Change Biology* 18:504–514.

Grill, G., Lehner, B., Thieme, M., Geenen, B., Tickner, D., and Antonelli, et al. 2019. Mapping the world's free-flowing rivers. *Nature* 569:215–221.

Gronewold, A.D., Do, H.X., Mei, Y., and Stow, C.A. 2021. A tug-of-war within the hydrologic cycle of a continental freshwater basin. *Geophysical Research Letters* 48:e2020GL090374.

Guasch, H., Ricart, M., Lopez-Doval, J., Bonnineau, C., Proia, L., Morin, S., Muñoz, I., Romaní, A.M., and Sabater, S. 2016. Influence of grazing on triclosan toxicity to stream periphyton. *Freshwater Biology* 61:2002–2012.

Haegerbaeumer, A., Mueller, M.-T., Fueser, H., and Traunspurger, W. 2019. Impacts of micro- and nano-sized plastic particles on benthic invertebrates: A literature Review and gap analysis. *Frontiers in Environmental Science* 7:17. doi:10.3389/fenvs.2019.00017.

Haghighi, E., Madani, K., and Hoekstra, A.Y. 2018. The water footprint of water conservation using shade balls in California. *Nature Sustainability* 1:358–360. doi.org/10.1038/s41893 -018-0092-2.

Hall, J.D., Brown, G.W., and Lantz, R.L. 1987. The Alsea Watershed Study: A retrospective, pp. 399–416. In: E.O. Salo and T.W. Cundy, Eds. Streamside Management: Forestry and Fishery Interactions. University of Washington Institute of Forest Resources, Seattle, WA.

Hallbert, T.B. and Keeley, E.R. 2023. Instream complexity increases habitat quality and growth for cutthroat trout in headwater streams. *Canadian Journal of Fisheries and Aquatic Sciences* 80:992–1005.

Halvorson, H.M., Wyatt, K.H., and Kuehn, K.A. 2020. Ecological significance of autotroph–heterotroph microbial interactions in freshwaters. *Freshwater Biology* 65:1183–1188.

Hanrahan, B.R., Tank, J.L., and Dee, M.M. et al. 2018. Restored floodplains enhance denitrification compared to naturalized floodplains in agricultural streams. *Biogeochemistry* 141 Special Issue: 419–437.

Harding, G., Griffiths, R.A., and Pavajeau, L. 2016. Developments in amphibian captive breeding and reintroduction programs. *Conservation Biology* 30:340–349.

Haro, R.J., Bailey, S.W., Northwick, R.M., Rolf-Hus, K.R., Sandheinrich, M.B. and Wiener, J.G. 2013. Burrowing dragonfly larvae as biosentinels of methylmercury in freshwater food webs. *Environmental Science & Technology*, 47:8148–8156.

Harrison, L.R., Legleiter, C.J., Wydzga, M.A., and Dunne, T. 2011. Channel dynamics and habitat development in a meandering, gravel bed river. Water Resources Research 47, W04513, doi:10.1029/2009WR008926.

Hart, B., Walker, G., Katupitiya, A., and Doolan, J. 2020. Salinity management in the Murray–Darling basin, Australia. *Water* 12:1829. doi:10.3390/w12061829.

Hart, D.D. and Merz, R.A. 1998. Predator-prey interactions in a benthic stream community: A field test of flow-mediated refuges. *Oecologia* 114:263–273.

Hart, D.D., Johnson, T.E., Bushaw-Newton, K.L., Horwitz, R.J., Bednarek, A.T., Charles, D.F., Kreeger, D.A., and Velinsk, D.J. 2002. Dam removal: Challenges and opportunities for ecological research and river restoration. *BioScience* 52:669–681.

Hassall, C. 2014. The ecology of urban ponds. *WIREs Water* 1:187–206.

Hassan, M.A. Gottesfeld, A.S., Montgomery, D.R., Tunnicliffe, J.F., Clarke, G.K.C., Wynn, G., Jones-Cox, H., Poirier, R., MacIsaac, E., Herunter, H., and Macdonald, S.J. 2008. Salmon-driven bed load transport and bed morphology in mountain streams. *Geophysical Research Letters* 35: article L04405.

Hassan, M.A., Roberge, L., Church, M., More, M., Donner, S.D., Leach, J., and Ali, K.F. 2017. What are the contemporary sources of sediment in the Mississippi River? *Geophysical Research Letters* 44:8919–8924. doi:10.1002/2017GL074046.

Hassett, B.A., Sudduth, E.B., and Somers, K.A. et al. 2018. Pulling apart the urbanization axis: patterns of physiochemical degradation and biological response across stream ecosystems. *Freshwater Science* 37:653–672.

Hausner, M.B., Wilson, K.P., Gaines, D.B., Suárez, F., Scoppettone, G.G., and Tyler, S.W. 2016. Projecting the effects of climate change and water management on Devils Hole pupfish (*Cyprinodon diabolis*) survival. *Ecohydrology* 9:560–573.

Hawkins, C.P., Kershner, J.L., Bisson, P.A., Bryant, M.D., Decker, L.M., Gregory, S.V., McCullough, D.A., Overton, C.K., Reeves, G.H., Steedman, R.J., and Young, M.K. 1993. A hierarchical approach to classifying stream habitat features. *Fisheries* 18(6):3–12.

Haxton, T.J. and Cano, T.M. 2016. A global perspective of fragmentation on a declining taxon—the sturgeon (Acipenseriformes). *Endangered Species Research* 31:203–210.

He, F., Zarfl, C., Bremerich, V., Henshaw, A., Darwall, W., Tockner, K., and Jähnig, S.C. 2017. Disappearing giants: A review of threats to freshwater megafauna. *WIREs Water* 2017: 4:e1208. doi: 10.1002/wat2.1208.

Heard, S.B. 1994. Pitcher-plant midges and mosquitoes: A processing chain commensalism. *Ecology* 75:1647–1660.

Hecky, R.E., Mugidde, R., Ramlal, P.S., Talbot, M.R., Kling, G.W. 2010. Multiple stressors cause rapid ecosystem change in Lake Victoria. *Freshwater Biology* 55 (Suppl. 1): 19–42.

Heino, M., Díaz Pauli, B., and Dieckmann, U. 2015. Fisheries-induced evolution. *Annual Review of Ecology, Evolution and Systematics* 46:461–480.

Herb, W.R., Janke, B., Mohseni, O., and Stefan, H.G. 2008. Thermal pollution of streams by runoff from paved surfaces. *Hydrological Processes* 22:987–999.

Herbst, D.B., Bogan, M.T., Roll, S.K., and Safford, H.D. 2012. Effects of livestock exclusion on in-stream habitat and benthic invertebrate assemblages in montane streams. *Freshwater Biology* 57:204–217.

Herlihy, A.T., Sifneos, J.C., Hughes, R.M., Peck, D.V., and Mitchell, R.M. 2020. The relation of lotic fish and benthic macroinvertebrate condition indices to environmental factors across the conterminous USA. *Ecological Indicators* 112:105958.

Hester, E.T. and Doyle, M.W. 2011. Human impacts to river temperature and their effects on biological processes: A quantitative synthesis. *Journal of the American Water Resources Association* 47:571–587.

Hevrøy, T.H., Golz, A.-L., Hansen, E.L., Xie, L., and Bradshaw, C. 2019. Radiation effects and ecological processes in a freshwater microcosm. *Journal of Environmental Radioactivity* 203:71–83.

Hicks, K.A., Loomer, H.A., Fuzzen, M.L.M., Kleywegt, S., Tetreault, G.R., McMaster, M.E., and Servos, M.R. 2017. $\delta^{15}N$ tracks changes in the assimilation of sewage-derived nutrients into a riverine food web before and after major process alterations at two municipal wastewater treatment plants. *Ecological Indicators* 72:747–758.

Hildrew, A. and Giller, P. 2023. *The Biology and Ecology of Streams and Rivers*, 2nd Ed. Oxford University Press.

Hildrew, A.G. and Townsend, C.R. 1976. Distribution of 2 predators and their prey in an iron rich stream. *Journal of Animal Ecology* 45:41–57.

Hill, M.J., Ryves, D.V., White, J.C., and Wood, P.J. 2016. Macroinvertebrate diversity in urban and rural ponds: Implications for freshwater biodiversity conservation. *Biological Conservation* 201:50–59.

Hillebrand, H., Donohue, I., Harpole, W.S., Hodapp, D., Kucera, M., Lewandowska, A.M., Merder, J., Montoya, J.M., and and Freund, J.A. 2020. Thresholds for ecological responses to global change do not emerge from empirical data. *Nature Ecology & Evolution* 4:1502–1509.

Hilsenhoff, W.L. 1987. An improved biotic index of organic stream pollution. *The Great Lakes Entomologist* 20:31–39.

Hintz, W.D., Fay, L., and Relyea, R.A. 2022. Road salts, human safety, and the rising salinity of our fresh waters. *Frontiers in Ecology and the Environment* 20(1):22–30.

Hodgson, E.E., Wilson, S.M., and Moore, J.W. 2020. Changing estuaries and impacts on juvenile salmon: A systematic review. *Global Change Biology* 26:1986–2001. doi:10.1111/gcb.14997.

Hoekstra, A.Y. and Mekonnen, M.M. 2012. The water footprint of humanity. *Proceedings of the National Academy of Sciences, USA* 109:3232–3237.

Holbrook, C.M., Bergstedt, R.A., Barber, J., Bravener, G.A., Jones, M.L., and Krueger, C.C. 2016. Evaluating harvest-based control of invasive fish with telemetry: performance of sea lamprey traps in the Great Lakes. *Ecological Applications* 26:1595–1609.

Hood, G.A., Manaloor, V., and Dzioba, B. 2018. Mitigating infrastructure loss from beaver flooding: A cost–benefit analysis. *Human Dimensions of Wildlife* 23:146–159.

Hortle, K.G. 2007. Consumption and the yield of fish and other aquatic animals from the Lower Mekong Basin (Mekong River Commission, Vientiane, Laos), MRC Technical Paper No. 16.

Humphries, P. and Winemiller, K.O. 2009. Historical impacts on river fauna, shifting baselines, and challenges for restoration. *Bioscience* 59(8):673–684.

Huner, J. (Ed.) 1994. *Freshwater Crayfish Aquaculture in North America, Europe, and Australia Families Astacidae, Cambaridae, and Parastacidae.* CRC Press.

Huss, H., Schwyn, U., Bauder, A., and Farinotti, D. 2021. Quantifying the overall effect of artificial glacier melt reduction in Switzerland, 2005–2019. *Cold Regions Science and Technology* 184:103237.

Hutchinson, G.E. 1983. *A Treatise on Limnology* (4 volumes). Wiley-Interscience.

Hynes, H.B.N. 1970. *The Ecology of Running Waters.* University of Toronto Press.

The Institute for Fisheries Resources. 1996. *The Cost of Doing Nothing: The Economic Burden of Salmon Declines in the Columbia River Basin.* The Institute for Fisheries Resources, Eugene, OR.

Irvine, R.L., Oussoren, T., Baxter, J.S., and Schmidt, D.C. 2009. The effects of flow reduction rates on fish stranding in British Columbia, Canada. *River Research and Applications* 25:405–415.

Islam, M.A., Jacob, M.V., and Antunes, E. 2021. A critical review on silver nanoparticles: From synthesis and applications to its mitigation through low-cost adsorption by biochar. *Journal of Environmental Management* 281:111918.

Izagirre, O., Serr, A., Guasch, H., and Elosegi, A. 2009. Effects of sediment deposition on periphytic biomass, photosynthetic activity and algal community structure. *Science of the Total Environment* 407:5694–5700.

Jackson, C.R., Sturm, C.A., and Ward, J.M. 2001. Timber harvest impacts on small headwater stream channels in the coast ranges of Washington. *Journal of the American Water Resources Association* 37:1533–1549.

Jackson, J.R., VanDeValk, A.J., Brooking, T.E., vanKeeken, O.A., and Rudstam, L.G. 2002. Growth and feeding dynamics of lake sturgeon, *Acipenser fulvescens*, in Oneida Lake, New York: Results from the first five years of a restoration program. *Journal of Applied Ichthyology* 18:439–443.

Jackson, M.C., Loewen, C.J., Vinebrooke, R.D., Chimimba, C.T. 2016. Net effects of multiple stressors in freshwater ecosystems: A meta-analysis. *Global Change Biology* 22:180–189.

Jähnig, S.C., Lorenz, A.W., Hering, D., Antons, C., Sundermann, A., Jedicke, E., and Haase, P. 2011. River restoration success: a question of perception. *Ecological Applications* 21:2007–2015.

Jankowski, T., Livingstone, D.M., Bührer, H., Forster, R., and Niederhauser, P. 2006. Consequences of the 2003 European heat wave for lake temperature profiles, thermal stability, and hypolimnetic oxygen depletion: Implications for a warmer world. *Limnology and Oceanography* 51:815–19.

Januchowski-Hartley, S.R., McIntyre, P.B., Diebel, M., Doran, P.J., Infante, D.M., Joseph, C., and Allan, J.D. 2013. Restoring aquatic ecosystem connectivity requires expanding inventories of both dams and road crossings. *Frontiers in Ecology and the Environment* 11:211–217.

Jaramillo, F. and Destouni, G. 2015. Local flow regulation and irrigation raise global human water consumption and footprint. *Science* 350:1248–1251.

Jardim de Queiroz, L., Doenz, C.J., Altermatt, F., Alther, R., Borko, S., Brodersen, J., Gossner, M.M., Graham, C., Matthews, B., McFadden, I.R., Pellissier, L., Schmitt, T., Selz, O.M., Villalba, S., Rüber, L., Zimmermann, N.E., and Seehausen, O. 2022. Climate, immigration and speciation shape terrestrial and aquatic biodiversity in the European Alps. *Proceedings of the Royal Society B* 289:20221020.

Jardine, T.D., Kidd, K.A., Cunjak, R.A., and Arp, P.A. 2009. Factors affecting water strider (Hemiptera: Gerridae) mercury concentrations in lotic systems. *Environmental Toxicity and Chemistry* 28:1480–1492.

Jeliazkov, A. et al. 2020. A global database for metacommunity ecology, integrating species, traits, environment and space. *Scientific Data* 7: article 6.

Jeziorski, A. and Smol, J.P. 2017. The ecological impacts of lakewater calcium decline on softwater boreal ecosystems. *Environmental Reviews* 25:245–253.

Jeuland, M. 2020. The economics of dams. *Oxford Review of Economic Policy* 36:45–68.

Jeziorski, A., Yan, N.D., Paterson, A.M., DeSellas, A.M., Turner, M.A., Jeffries, D.S., Keller, B., Weeber, R.C., McNicol, D.K., Palmer, M.E., McIver, K., Arseneau, K., Ginn, B.K., Cumming, B.F., and Smol, J.P. 2008. The widespread threat of calcium decline in fresh waters. *Science* 322:1374–1377.

Jilbert, T., Couture, R.-M., Huser, B.J., and Salonen, K. 2020. Preface: Restoration of eutrophic lakes: Current practices and future challenges. *Hydrobiologia* 847:4343–4357. doi.org/10 .1007/s10750-020-04457-x.

Johannessen, C., Helm, P., and Metcalfe, C.D. 2021. Detection of selected tire wear compounds in urban receiving waters. *Environmental Pollution* 287:117659.

Johnson, M.F., Thorne, C.R., Castro, J.M., Kondolf, G.M., Mazzacano, C.S., Rood, S.B., and Westbrook, C. 2020. Biomic river restoration: A new focus for river management. *River Research and Application* 36:3–12.

Johnson, S.L., Henshaw, D., Downing, G., Wondzell, S., Schulze, M., Kennedy, A., Cohn, G., Schmidt, S.A., and Jones, J.A. 2021. Long-term hydrology and aquatic biogeochemistry data from H. J. Andrews Experimental Forest, Cascade Mountains, Oregon. *Hydrological Processes* 35:e14187.

Jones, I. and Smol, J. (Eds.) 2023. *Wetzel's Limnology: Lake and River Ecosystems*, 4th Edition. Academic Press ISBN: 9780128227015.

Jones, J.I., Collins, A.L., Naden, P.S., and Sear, D.A. 2012a. The relationship between fine sediment and macrophytes in rivers. *River Research and Applications* 28:1006–1018.

Jones, J.I., Murphy, J.F., Collins, A.L., Sear, D.A., Naden, P.S., and Armitage, P.D. 2012b. The impact of fine sediment on macro-invertebrates. *River Research and Applications* 28:1055–1071.

Jones, N.E., Scrimgeour, G.J., and Tonn, W. 2017. Lessons learned from an industry, government and university collaboration to restore stream habitats and mitigate effects. *Environmental Management* 59:1–9.

Jones, S.E. and Lennon, J.T. 2015. A test of the subsidy–stability hypothesis: The effects of terrestrial carbon in aquatic ecosystems. *Ecology* 96:1550–1560.

Jordan, S.J., Stoffer, J., and Nestlerode, J.A. 2011. Wetlands as sinks for reactive nitrogen at continental and global scales: A meta-analysis. *Ecosystems* 14:144–155.

Juliano, S.A., Westby, K.M, and Ower, G.D. 2019. Know Your Enemy: Effects of a predator on native and invasive container mosquitoes. *Journal of Medical Entomology* 56:320–328.

Junghans, K., Springer, A.E., Stevens, L.E., and Ledbetter, J.D. 2016. Springs ecosystem distribution and density for improving stewardship. *Freshwater Science* 35:1330–1339.

Junk, W.J., Bayley, P.B., and Sparks, R.E. 1989. The flood-pulse concept in river-floodplain systems. Pp. 110–127 In: Dodge, D.P. (Ed.) Proceedings of the International Large River Symposium. *Canadian Special Publications in Fisheries and Aquatic Sciences* 106.

Jyväsjärvi, J., Lehosmaa, K., Aroviita, J., Turunen, J., Rajakallio, M., Marttila, H., Tolkkinen, M., Mykrä, H., and Muotka, T. 2021. Fungal assemblages in predictive stream bioassessment: A cross-taxon comparison along multiple stressor gradients *Ecological Indicators* 121:106986.

Kampf, S.K., Dwire, K.A., Fairchild, M.P., Dunham, J., Snyder, C.D., Jaeger, K.L., Luce, C.H., and Hammond, J.C. et al. 2021. Managing nonperennial headwater streams in temperate forests of the United States. *Forest Ecology and Management* 497:119523.

Kanakaraju, D., Glass, B.D., and Oelgemöller, M. 2014. Titanium dioxide photocatalysis for pharmaceutical wastewater treatment. *Environmental Chemistry Letters* 12:27–47.

Kang, C.D. and Cervero, R. 2009. From elevated freeway to urban greenway: Land value impacts of the CGC project in Seoul, Korea. *Urban Studies* 46:2771–2794.

Karr, J.R. 1993. Defining and assessing ecological integrity—beyond water-quality. *Environmental Toxicology and Chemistry* 12:1521–1531.

———. 1999. Defining and measuring river health. *Freshwater Biology* 41:221–234.

Kasprzak, P., Shatwell, T., Gessner, M.O., Gonsiorczyk, T., Kirillin, G., Selmeczy, G., Padisák, J., and Engelhardt, C. 2017. Extreme weather event triggers cascade towards extreme turbidity in a clear-water lake. *Ecosystems* 20:1407–1420.

Kaufman, L. 1992. Catastrophic change in species-rich fresh-water ecosystems. *BioScience* 42:846–858.

Kay, A.S. 2002. *Mysis relicta* and kokanee salmon (*Oncorhynchus nerka*) in Okanagan Lake, British Columbia: From 1970 and into the future. M.Sc. Thesis, University of British Columbia.

Kaylor, M.J., Warren, D.R., and Kiffney, P.M. 2017. Long-term effects of riparian forest harvest on light in Pacific Northwest (USA) streams. *Freshwater Science* 36:1–13.

Keddy, P.A. 2010. *Wetland Ecology: Principles and Conservation*. 2nd Ed. Cambridge University Press, Cambridge, UK.

Keefer, M.L., Caudill, C.C., Peery, C.A., and Lee, S.R. 2008. Transporting juvenile salmonids around dams impairs adult migration. *Ecological Applications* 18:1888–1900.

Keeling, P.J. 2004. Diversity and evolutionary history of plastids and their hosts. *American Journal of Botany* 91:1481–1493.

Kelly, V.R., Findlay, S.E.G., and Weathers, K.C. 2019. *Road Salt: The Problem, The Solution, and How to Get There*. Cary Institute of Ecosystem Studies, New York, USA.

Kemp, P.S., Sear, D.A., Collins, A.L., Naden, P., and Jones, J.I. 2011. The impacts of fine sediment on riverine fish. *Hydrological Processes* 25:1800–1821. doi:10.1002/hyp.7940.

Kennen, J.G., Chang, M., Tracy, B.H. 2005. Effects of landscape change on fish assemblage structure in a rapidly growing metropolitan area in North Carolina, USA. *American Fisheries Society Symposium* 47:39–52.

Kidd, K.A., Blanchfield, P.J., Mills, K.H., Palace, V.P., Evans, R.E., Lazorchak, J.M., and Flick, R.W. 2007. Collapse of a fish population after exposure to a synthetic estrogen. *Proceedings of the National Academy of Sciences* 104:8897–8901.

Kidd, K.A., Paterson, M.J., Rennie, M.D., Podemski, C.L., Findlay, D.L., Blanchfield, P.J., Liber, K. 2014. Direct and indirect responses of a freshwater food web to a potent synthetic oestrogen. *Philosophical Transactions of the Royal Society B* 369:20130578. http://dx.doi.org/10.1098/rstb.2013.0578.

Kiffney, P.M., Greene, C.M., Hall, J.E., Davies, and J.R. 2006. Tributary streams create spatial discontinuities in habitat, biological productivity, and diversity in mainstem rivers. *Canadian Journal of Fisheries and Aquatic Sciences* 63:2518–2530.

Kiffney, P.M., Richardson, J.S., and Bull, J.P. 2003. Responses of periphyton and insects to experimental manipulation of riparian buffer width along forest streams. *Journal of Applied Ecology* 40:1060–1076.

Kindervater, E. and Steinman, A.D. 2019. Two-stage agricultural ditch sediments act as Phosphorus sinks in West Michigan. *Journal of the American Water Resources Association* 55:1183–1195.

King, R.B., Stanford, K.M., and Jones, P.C. 2018. Sunning themselves in heaps, knots, and snarls: The extraordinary abundance and demography of island watersnakes. *Ecology and Evolution* 8:7500–7521.

Kirk, J.L., Muir, D.C.M., Antoniades, D., Douglas, M.V., Evans, M., Jackson, T., Kling, H., Lamoureux, S., Lim, D.S., Pienitz, R., Smol, J.P., Stewart, K., Wang, Z., and Yang, F. 2011. Climate change and mercury accumulation in Canadian high and subarctic lakes. *Environmental Science & Technology* 45:964–970.

Kitching, R.L. 2001. Food webs in phytotelmata: "Bottom-up" and "top-down" explanations for community structure. *Annual Review of Entomology* 46:729–760.

Knäbel, A., Meyer, K., Rapp, J., and Schulz, R. 2014. Fungicide field concentrations exceed FOCUS surface water predictions: urgent need of model improvement. *Environmental Science & Technology* 48:455–463.

Knapp, R.A., Hawkins, C.P., Ladau, J., and McClory, J.G. 2005. Fauna of Yosemite National Park lakes has low resistance but high resilience to fish introductions. *Ecological Applications* 15:835–847.

Knapp, R.A., Matthews, K.R., and Sarnelle, O. 2001. Resistance and resilience of alpine lake fauna to fish introductions. *Ecological Monographs* 71:401–421.

Knauer, J., Zaehle, S., Reichstein, M., Medlyn, B.E., Forkel, M., Hagemann, S., and Werner, C. 2017. The response of ecosystem water-use efficiency to rising atmospheric CO_2 concentrations: Sensitivity and large-scale biogeochemical implications. *New Phytologist* 213:1654–1666.

Knox, J.C. 2006. Floodplain sedimentation in the Upper Mississippi Valley: Natural versus human accelerated. *Geomorphology* 79:286–310.

Knox, R.L., Wohl, E.E., and Morrison, R.R. 2022. Levees don't protect, they disconnect: A critical review of how artificial levees impact floodplain functions. *Science of the Total Environment* 837:155773.

Koebel, J.W., Jr. and Bousquin, S.G. 2014. The Kissimmee River Restoration Project and Evaluation Program, Florida, U.S.A. *Restoration Ecology* 22:345–352.

Kolpin, D. W., Furlong, E.T., Meyer, M.T., Thurman, E.M., Zaugg, S.D., Barber, L.B., and Buxton, H.T. 2002. Pharmaceuticals, hormones, and other organic wastewater contaminants in U.S. streams, 1999–2000: A national reconnaissance. *Environmental Science and Technology* 36:1202–1211.

Kominoski, J.S., Moore, P.A., Wetzel, R.G., and Tuchman, N.C. 2007. Elevated CO_2 alters leaf-litter-derived dissolved organic carbon: Effects on stream periphyton and crayfish feeding preference. *Journal of the North American Benthological Society* 26:662–671. http://dx.doi.org/10.1899/07-002.1.

Konar, M. and Marston, L. 2020. The water footprint of the United States. *Water* 12:3286. doi:10. 3390/w12113286.

Kondolf, G.M., Angermeier, P.L., Cummins, K., Dunne, T., Healey, M., Kimmerer, W., Moyle, P.B., Murphy, D., Patten, D., Railsback, S., Reed, D.J., Spies, R., and Twiss, R. 2008. Projecting cumulative benefits of multiple river restoration projects: An example from the Sacramento-San Joaquin River System in California. *Environmental Management* 42:933–945.

Korman, J., Deemer, B.R., Yackulic, C.B., Kennedy, T.A., and Giardina, M. 2023. Drought-related changes in water quality surpass effects of experimental flows on trout growth downstream of Lake Powell reservoir. *Canadian Journal of Fisheries and Aquatic Sciences* 80:424–438.

Kovács-Hostyánszki, A., Espíndola, A., Vanbergen, A.J., Settele, J., Kremen, C., and Dicks, L.V. 2017. Ecological intensification to mitigate impacts of conventional intensive land use on pollinators and pollination. *Ecology Letters* 20:673–689.

Kovalenko, K.E., Ciborowski, J.J.H., Daly, C., Dixon, D.G., Farwell, A.J., Foote, A.L., Frederick, K.R., Gardner Costa, J.M., Kennedy, K., Liber, K., Roy, M.C., Slama, C.A., and Smits, J.E.G. 2013. Food web structure in oil sands reclaimed wetlands. *Ecological Applications* 23:1048–1060.

Kramer, D. 2020. The Great Lakes are filled to their brims, with no signs of receding. *Physics Today* 73:26–29. doi:10.1063/PT.3.4589.

Krebs, C.J. 1999. *Ecological Methodology*, 2nd ed. Addison-Wesley Educational Publishers, Inc.

Kreutzweiser, D.P., Sibley, P.K., Richardson, J.S., and Gordon, A.M. 2012. Introduction and a theoretical basis for using disturbance by forest management activities to sustain aquatic ecosystems. *Freshwater Science* 31:224–231.

Kritzberg, E.S., Hasselquist, E.M., Škerlep, M., Löfgren, S., Olsson, O., Stadmark, J., Valinia, S., Hansson, L.-A., and Laudon, H. 2020. Browning of freshwaters: Consequences to ecosystem services, underlying drivers, and potential mitigation measures. *Ambio* 49:375–390.

Krumm, F., Schuck, A., and Rigling, A. (Eds.). 2020. *How to balance forestry and biodiversity conservation. A view across Europe.* European Forest Institute (EFI), Swiss Federal Institute for Forest, Snow and Landscape Research (WSL), Birmensdorf. 640 p. doi:10.16904/envidat.196.

Kuglerová, L., Ågren, A., Jansson, R., and Laudon, H. 2014. Towards optimizing riparian buffer zones: Ecological and biogeochemical implications for forest management. *Forest Ecology and Management* 334:74–84.

Kuglerová, L., Hasselquist, E.M., Richardson, J.S., Sponseller, R.A., Kreutzweiser, D.P., and Laudon, H. 2017. Management perspectives on *Aqua incognita*: Connectivity and cumulative effects of small natural and artificial streams in boreal forests. *Hydrological Processes* 31:4238–4244.

Kuglerová, L., Jyväsjärvi, J., Ruffing, C., Muotka, T., Jonsson, A., Andersson, E., and Richardson, J.S. 2020. Cutting edge: A comparison of contemporary practices of riparian buffer retention around small streams in Canada, Finland and Sweden. *Water Resources Research* 56(9):e2019WR026381 doi:10.1029/2019WR026381.

Kumar, M. and Kumar, A. 2019. Experimental validation of performance and degradation study of canal-top photovoltaic system. *Applied Energy* 243:102–118. doi.org/10.1016/j.apenergy.2019.03.168.

Kupilas, B., McKie, B.G., Januschke, K., Friberg, N., and Hering, D. 2020. Stable isotope analysis indicates positive effects of river restoration on aquatic-terrestrial linkages. *Ecological Indicators* 113:106242.

Kurtak, D.C. 1978. Efficiency of filter feeding of black fly larvae (Diptera Simuliidae). *Canadian Journal of Zoology* 56:1608–1623.

Langford, T. 1990. *Ecological Effects of Thermal Discharges*. Springer Science & Business Media.

Langford, T.E.L., Langford, J., and Hawkins, S.J. 2012. Conflicting effects of woody debris on stream fish populations: Implications for management. *Freshwater Biology* 57:1096–1111.

Larras, F., Keck, F., Montuelle, B., Rimet, F., and Bouchez, A., 2014. Linking diatom sensitivity to herbicides to phylogeny: A step forward for biomonitoring? *Environmental Science & Technology* 48:1921–1930. doi.org/10.1021/es4045105.

Larsen, S., Muehlbauer, J.D., and Marti, E. 2016. Resource subsidies between stream and terrestrial ecosystems under global change. *Global Change Biology* 22:2489–2504. doi:10.1111/gcb.13182.

Larsen, S., Pace, G., and Ormerod, S.J. 2011. Experimental effects of sediment deposition on the structure and function of macroinvertebrate assemblages in temperate streams. *River Research and Applications* 27:257–267. doi:10.1002/rra.1361.

Larson, K.L., Hoffman, J., and Ripplinger, J. 2017. Legacy effects and landscape choices in a desert city. *Landscape and Urban Planning* 165:22–29.

Le, T.D.H., Schreiner, V.C., Kattwinkel, M., and Schäfer, R.B. 2021. Invertebrate turnover along gradients of anthropogenic salinisation in rivers of two German regions. *Science of the Total Environment* 753:141986.

Leach, J.A., Kelleher, C., Kurylyk, B.L., Moore, R.D., and Neilson, B.T. 2023. A primer on stream temperature processes. *WIREs Water* 10:e1643.

Leach, J.A., Moore, R.D., Hinch, S.G., and Gomi, T. 2012. Estimation of forest harvesting-induced stream temperature changes and bioenergetics consequences for cutthroat trout in a coastal stream in British Columbia, Canada. *Aquatic Sciences* 74:427–441.

LeChevallier, M.W. and Norton, W.D. 1992. Examining relationships between particle counts and Giardia, Cryptosporidium, and turbidity. *Journal of the American Water Works Association* 84:54–60.

Lee, D.-Y., Lee, H., Trevors, J.T., Weir, S.C., Thomas, J.L., and Habash, M. 2014. Characterization of sources and loadings of fecal pollutants using microbial source tracking assays in urban and rural areas of the Grand River Watershed, Southwestern Ontario. *Water Research* 53:123–131.

Lee, P., Smyth, C., and Boutin, S. 2004. Quantitative review of riparian buffer width guidelines from Canada and the United States. *Journal of Environmental Management* 70:165–180.

Lefcort, H., Freedman, Z., House, S., and Pendleton, M. 2008. Hormetic effects of heavy metals in aquatic snails: Is a little bit of pollution good? *Ecohealth* 5:10–17.

Le Gall, M., Palt, M., Kail, J., Hering, D., and Piffady, J. 2022. Woody riparian buffers have indirect effects on macroinvertebrate assemblages of French rivers, but land use effects are much stronger. *Journal of Applied Ecology* 59:526–536.

Lemm, J.U., Venohr, M., Globevnik, L., Stefanidis, K., Panagopoulos, Y., van Gils, J., Posthuma, L., Kristensen, P., Feld, C.K., Mahnkopf, J., Hering, D., and Birk, S. 2021. Multiple stressors determine river ecological status at the European scale: Towards an integrated understanding of river status deterioration. *Global Change Biology* 27:1962–1975.

Leopold, L.B., Wolman, M.G., and Miller, J.P. 1964. *Fluvial Processes in Geomorphology*. W.H. Freeman, San Francisco.

Lepori, F., Gaul, D., Palm, D., and Malmqvist, B. 2006. Food-web responses to restoration of channel heterogeneity in boreal streams. *Canadian Journal of Fisheries and Aquatic Sciences* 63:2478–2486.

Lévêque, C., Oberdorff, T., Paugy, D., Stiassny, M.L.J., and Tedesco, P.A. 2007. Global diversity of fish (Pisces) in freshwater. In: Balian, E.V., Lévêque, C., Segers, H., and Martens, K. (Eds.) Freshwater Animal Diversity Assessment. Developments in Hydrobiology, vol. 198. Springer, Dordrecht. doi.org/10.1007/978-1-4020-8259-7_53.

Levy, D.A. 1991. Acoustic analysis of diel vertical migration behavior of *Mysis relicta* and Kokanee (*Oncorhynchus nerka*) within Okanagan Lake, British Columbia. *Canadian Journal of Fisheries and Aquatic Sciences* 48:67–72.

Lienert, J., Koller, M., Konrad, J., McArdell, C.S., and Schuwirth, N. 2011. Multiple-criteria decision analysis reveals high stakeholder preference to remove pharmaceuticals from hospital wastewater. *Environmental Science and Technology* 45:3848–3857.

Liess, M., Foit, K., Knillmann, S., Schäfer, R.B., and Liess, H.-D. 2016. Predicting the synergy of multiple stress effects. *Scientific Reports* 6:32965.

Liess, M., Liebmann, L., Vormeier, P., Weisner, O., Altenburger, R., Borchardt, D., Brack, W., Chatzinotas, A., Escher, B., Foit, K., Gunold, R., Henz, S., Hitzfeld, K.L., Schmitt-Jansen, M., Kamjunke, N., Kaske, O., Knillmann, S., Krauss, M., Küster, E., Link, M., Lück, M., Möder, M., Müller, A., Paschke, A., Schäfer, R.B., Schneeweiss, A., Schreiner, V.C., Schulze, T., Schüürmann, G., von Tümpling, W., Weitere, M., Wogram, J., and Reemtsma, T. 2021. Pesticides are the dominant stressors for vulnerable insects in lowland streams. *Water Research* 201: 117262.

Likens, G.E., Driscoll, C.T., and Buso, D.C. 1996. Long-term effects of acid rain: Response and recovery of a forest ecosystem. *Science* 272:244–246.

Lindberg, T.T., Berhardt, E.S., Bier, R., Helton, A.M., Merola, R.B., Vengosh, A., and Di Giulio, R.T., 2011. Cumulative impacts of mountaintop mining on an Appalachian watershed. *Proceedings of the National Academy of Science of the United States* 108:20929–20934.

Lopez, J.G., Tor-ngern, P., Oren, R., Kozii, N., Laudon, H., and Hasselquist, N.J. 2021. How tree species, tree size, and topographical location influenced tree transpiration in northern boreal forests during the historic 2018 drought. *Global Change Biology* 27:3066–3078.

Lorenz, A.W., Jähnig, S.C., and Hering, D. 2009. Re-meandering German lowland streams: Qualitative and quantitative effects of restoration measures on hydromorphology and macroinvertebrates. *Environmental Management* 44:745–754.

Louch, J., Tatum, V., Allen, G., Hale, V.C., McDonnell, J., Danehy, R.J., and Ice, G. 2016. Potential risks to freshwater aquatic organisms following a silvicultural application of herbicides in Oregon's Coast Range. *Integrated Environmental Assessment and Management* 13:396–409.

Louette, G., Devisscher, S., and Adriaens, T. 2013. Control of invasive American bullfrog *Lithobates catesbeianus* in small shallow water bodies. *European Journal of Wildlife Research* 59:105–114. doi:10.1007/s10344-012-0655-x.

Louhi, P., Muotka, T., and Richardson, J.S. 2017. Sediment addition reduces the importance of predation on ecosystem functions in experimental stream channels. *Canadian Journal of Fisheries and Aquatic Sciences* 74:32–40.

Louhi, P., Mykrä, H., Paavola, R., Huusko, A., Vehanen, T., Mäki-Petäys, A., and Muotka, T. 2011. Twenty years of stream restoration in Finland: Little response by benthic macroinvertebrate communities. *Ecological Applications* 21:1950–1961.

Louhi, P., Vehanen, T., Huusko, A., Maki-Petays, A., and Muotka, T. 2016. Long-term monitoring reveals the success of salmonid habitat restoration. *Canadian Journal of Fisheries and Aquatic Sciences* 73:1733–1741.

Lowe, S., Browne, M., Boudjelas, S., and De Poorter, M. 2000. *100 of the World's Worst Invasive Alien Species: A selection from the Global Invasive Species Database*. Published by The Invasive Species Specialist Group (ISSG) a specialist group of the Species Survival Commission (SSC) of the World Conservation Union (IUCN), 12 pp.

Lydeard, C., Cowie, R.H., Ponder, W.F., Bogan, A.E., Bouchet, P., Clark, S.A., Cummings, S., Frest, T.J., Gargominy, O., Herbert, D.G., Hershler, R., Perez, K.E., Roth, B., Seddon, M., Strong, E.E., and Thompson, F.G. 2004. The global decline of nonmarine mollusks. *BioScience* 54:321–330.

Lytle, D.A. and Smith, R.L. 2004. Exaptation and flash flood escape in the giant water bugs. *Journal of Insect Behavior* 17:169–178.

Macaulay, S.J., Hageman, K.J., Piggott, J.J., Juvigny-Khenafou, N.P.D., and Matthaei, C.D. 2021. Warming and imidacloprid pulses determine macroinvertebrate community dynamics in experimental streams. *Global Change Biology* 27:5469–5490.

MacIsaac, E.A. 2010. Salmonids and the hydrologic and geomorphic features of their spawning streams in British Columbia. Pp. 461–478. In: Pike, R.G., Redding, T.E., Moore, R.D., Winkler, R.D., and Bladon, K.D. (Eds.). B.C. Ministry of Forests and Range Research Branch, Victoria, BC, Canada. URL: http://www.for.gov.bc.ca/hfd/pubs/Docs/Lmh/Lmh66/Lmh 66_ch14.pdf.

MacLennan, M.M. and Vinebrooke, R.D. 2021. Exposure order effects of consecutive stressors on communities: The role of co-tolerance. *Oikos* doi:10.1111/oik.08884.

Maire, A., Thierry, E., Viechtbauer, W., and Daufresne, M. 2019. Poleward shift in large-river fish communities detected with a novel meta-analysis framework. *Freshwater Biology* 64:1143–1156.

Malmqvist, B.R. and Rundle, S., 2002. Threats to the running water ecosystems of the world. *Environmental Conservation* 29:134–153.

Marczak, L.B., Hoover, T.M., and Richardson, J.S. 2007. Trophic interception: How a boundary-foraging organism influences cross-ecosystem fluxes. *Oikos* 116:1651–1662.

Marshall, B.E. 2018. Guilty as charged: Nile perch was the cause of the haplochromine decline in Lake Victoria. *Canadian Journal of Fisheries and Aquatic Sciences* 75:1542–1559.

Marshall, J.C., Acuña, V., Allen, D.C., Bonada, N., Boulton, A.J., Carlson, S.M., Dahm, C.N., Datry, T., Leigh, C., Negus, P., Richardson, J.S., Sabater, S., Stevenson, R.J., Steward, A.L., Stubbington, R., Tockner, K., and Vander Vorste, R. 2018. Protecting US river health by maintaining the legal status of their temporary waterways. *Science* 361:856–857.

Marti, E., Variatza, E., and Balcazar, J.L. 2014. The role of aquatic ecosystems as reservoirs of antibiotic resistance. *Trends in Microbiology* 22:36–41.

Martínez, A., Larrañaga, A., Pérez, J., Descals, E., Basaguren, A., and Pozo, J. 2013. Effects of pine plantations on structural and functional attributes of forested streams. *Forest Ecology and Management* 310:147–155.

Maser, C. and Sedell, J.R. 1994. *From the Forest to the Sea: The Ecology of Wood in Streams, Rivers, Estuaries and Oceans*. St. Lucie Press, FL, USA.

McCaleb, M.M. and McLaughlin, R.A. 2008. Sediment trapping by five different sediment detention devices on construction sites. *Transactions of the ASABE* 51:1613–1621.

McColl-Gausden, E.F., Weeks, A.R., Coleman, R.A., Robinson, K.L., Song, S., Raadik, T.A., and Tingley, R. 2021. Multispecies models reveal that eDNA metabarcoding is more sensitive than backpack electrofishing for conducting fish surveys in freshwater streams. *Molecular Ecology.* 30:3111–3126.

McIntyre, A.P., Hayes, M.P., Ehinger, W.J., Estrella, S.M., Schuett-Hames, D.E., Timothy Quinn, T. 2018. *Effectiveness of Experimental Riparian Buffers on Perennial Non-fish-bearing Streams on Competent Lithologies in Western Washington*. Washington State Forest Practices Board Forest Practices Adaptive Management Program, Washington State Department of Natural Resources.

McIntyre, P.B., Liermann, C.A.R., and Revenga, C. 2016. Linking freshwater fishery management to global food security and biodiversity conservation. *Proceedings of the National Academy of Sciences of the United States of America* 113:12880–12885.

McLaughlin, R.L., Hallett, A., Pratt, T.C., O'Connor, L.M., and McDonald, D.G. 2007. Research to guide use of barriers, traps, and fishways to control sea lamprey. *Journal of Great Lakes Research* 33, Special Issue 2:7–19.

Mehdi, H., Laub, S.C., Synyshyn, C., Salena, M.G., MaCallum, E.S., Muzzatti, M.N., Bowman, J.E., Mataya, K., Bragge, L.M., Servos, M.R., Kidd, K.A., Scott, G.R., and Balshine, S. 2021. Municipal wastewater as an ecological trap: Effects on fish communities across seasons. *Science of the Total Environment* 759:143430.

Meier, J., Marques, D.A., Mwaiko, S., Wagner, C., Excoffier, L., and Seehausen, O. 2017. Ancient hybridization fuels rapid cichlid fish adaptive radiations. *Nature Communications* 8:14363.

Meli, P., Rey Benayas, J.M., Balvanera, P., Martínez Ramos, M. 2014. Restoration enhances wetland biodiversity and ecosystem service supply, but results are context-dependent: A meta-analysis. *PLoS One* 9(4):e93507. doi:10.1371/journal.pone.0093507.

Melis, T.S., Walters, C.J., Korman, J. 2015. Surprise and opportunity for learning in Grand Canyon: the Glen Canyon Dam Adaptive Management Program. *Ecology and Society* 20:22. http://dx.doi.org/10.5751/ES-07621-200322

Mellina, E. and Hinch, S.G. 2009. Influences of riparian logging and in-stream large wood removal on pool habitat and salmonid density and biomass: A meta-analysis. *Canadian Journal of Fisheries and Aquatic Sciences* 39:1280–1301.

Meng, Q. 2017. The impacts of fracking on the environment: A total environmental study paradigm. *Science of the Total Environment* 580:953–957.

Menz, F.C. and Seip, H.M. 2004. Acid rain in Europe and the United States: An update. *Environmental Science & Policy* 7:253–265.

Mercier, P. and Gay, S. 1949. Station d'aération au lac de Bret. *Aquatic Sciences* 11:423–429.

Mermillod-Blondin, F. and Rosenberg, R. 2006. Ecosystem engineering: The impact of bioturbation on biogeochemical processes in marine and freshwater benthic habitats. *Aquatic Sciences* 68:434–442.

Merritt, R.W., Cummins, K.W., and Berg, M.B. (Eds.). 2019. *An Introduction to the Aquatic Insects of North America*, 5th Edition. Kendall-Hunt Publishing Company, IA.

Meyer, J. L., Paul, M.J., and Taulbee, W.K. 2005. Stream ecosystem function in urbanizing landscapes. *Journal of the North American Benthological Society* 24:602–612.

Meyer, J.L. and Wallace, J.B. 2000. Lost linkages and lotic ecology: Rediscovering small streams. Pp. 295–317 In: Press, M.C., Huntly, N.J., Levin, S. (Eds.). *Ecology: Achievement and challenge: the 41st Symposium of the British Ecological Society*. Blackwell Science, Oxford, UK.

Meyer-Jacob, C., Michelutti, N., Paterson, A.M., Cumming, B.F., Keller, W., and Smol, J.P. 2019. The browning and re-browning of lakes: Divergent lake-water organic carbon trends linked to acid deposition and climate change. *Scientific Reports* 9:16676.

Miara, A., Vörösmarty, C.J., Macknick, J.E., Tidwell, V.C., Fekete, B., Corsi, F., and Newmark, R. 2018. Thermal pollution impacts on rivers and power supply in the Mississippi River watershed. *Environmental Research Letters* 13:034033.

Michalak, A.M., Anderson, E.J., Beletsky, D., Boland, S, Bosche, N.S., Bridgeman, T.B., Chaffin, J.D., Cho, K., Confesor, R., Daloğlu, I., DePinto, J.V., Evans, M.A., Fahnenstiel, G.L., He, L., Ho, J.C., Jenkins, L., Johengen, T.H., Kuo, K.C., LaPorte, E., Liu, X., McWilliams, M.R., Moore, M.R., Posselt, D.J., Richards, R.P., Scavia, D., Steiner, A.L., Verhamme, E., Wright, D.M., and Zagorski, M.A. 2013. Record-setting algal bloom in Lake Erie caused by agricultural and meteorological trends consistent with expected future conditions. *Proceedings of the National Academy of Science of the USA* 110:6448–6452.

Michie, L.E., Hitchcock, J.N., Thiem, J.D., Boys, C.A., and Mitrovic, S.M. 2020. The effect of varied dam release mechanisms and storage volume on downstream river thermal regimes. *Limnologica* 81:125760.

Mihelcic, J.R. and Zimmerman, J.B. 2014. *Environmental Engineering: Fundamentals, Sustainability, Design*. Second Edition. John Wiley & Sons, Inc., Hoboken, NJ.

Millar, R.G. 2000. Influence of bank vegetation on alluvial channel patterns. *Water Resources Research* 36:1109–1118.

Milliman, J.D. and Farnsworth, K.L. 2011. *River Discharge to the Coastal Ocean: A global Synthesis*. Cambridge University Press.

Milner, A.M., Khamisa, K., Battin, T.J., Brittain, J.E., Barrand, N.E., Füreder, L., Cauvy-Fraunié, S., Gíslason, G.M., Jacobsen, D., Hannah, D.M., Hodson, A.J., Hood, E., Lencioni, V., Ólafsson, J.S., Robinson, C.T., Tranter, M., and Brown, L.E. 2017. Glacier shrinkage driving global changes in downstream systems. *Proceedings of the National Academy of Science* 114:9770–9778.

Mims, M.C. and Olden, J.D. 2013. Fish assemblages respond to altered flow regimes via ecological filtering of life history strategies. *Freshwater Biology* 58:50–62.

Montgomery, D.R. 1999. Process domains and the river continuum. *Journal of the American Water Resources Association* 35:397–410.

Montgomery, D.R. 2007. Soil erosion and agricultural sustainability. *Proceedings of the National Academy of Science* 104:13268–13272.

Moore, K.D. and Moore, J.W. 2013. Ecological restoration and enabling behavior: A new metaphorical lens? *Conservation Letters* 6:1–5.

Moore, R.D., Gronsdahl, S., and McCleary, R. 2020. Effects of forest harvesting on warm-season low flows in the Pacific Northwest: A review. *Confluence* 4(1):1–29. doi:10.22230/jwsm.2020v 4n1a35.

Moore, R.D., Sidle, R.C., Eaton, B., Gomi, T., and Wilford, D. 2016. Chapter 7. Water and watersheds. Pp. 161–188 In: Innes, J.L. and Tikina, A.V. (Eds.) *Sustainable Forest Management: From Concept to Practice*. Taylor & Francis Group. ProQuest.

Moore, R.D., Spittlehouse, D., and Story, A. 2005. Riparian microclimate and stream temperature response to forest harvesting: A review. *Journal of the American Water Resources Association* 41:813–834.

Moore, R.D. and Wondzell, S.M. 2005. Physical hydrology and the effects of forest harvesting in the Pacific Northwest. *Journal of the American Water Resources Association* 41:763–784.

Morden, R., Horne, A., Bond, N.R., Nathan, R., and Olden, J.D. 2022. Small artificial impoundments have big implications for hydrology and freshwater biodiversity. *Frontiers in Ecology and Environment* 20(3):141–146.

Morgan, R.P. and Cushman, S.F. 2005. Urbanization effects on stream fish assemblages in Maryland, USA. *Journal of the North American Benthological Society* 24:643–655.

Morris, O.F., Loewen, C.J.G., Woodward, G., Schäfer, R.B., Piggott, J.J., Vinebrooke, R.D., and Jackson, M.C. 2022. Local stressors mask the effects of warming in freshwater ecosystems. *Ecology Letters* 25:2540–2551.

Moss, B. 1990. Engineering and biological approaches to the restoration from eutrophication of shallow lakes in which aquatic plant communities are important components. Pp. 367–377 In: Gulati, R.D., Lammens, E.H.R.R., Meijer, M.-L., and van Donk, E. (Eds.) *Biomanipulation Tool for Water Management*. Springer, Dordrecht.

Mueller, E.R., Grams, P.E., Hazel, J.E., Jr., and Schmidt, J.C. 2018. Variability in eddy sandbar dynamics during two decades of controlled flooding of the Colorado River in the Grand Canyon. *Sedimentary Geology* 363:181–199.

Muhlfeld, C.C., Cline, T.J., Giersch, J.J., Peitzsch, E., Florentine, C., Jacobsen, D., and Hotaling, S. 2020. Specialized meltwater biodiversity persists despite widespread deglaciation. *Proceedings of the National Academy of Science* 117:12208–12214.

Murphy, M.L., Hawkins, C.P., Anderson, N.H. 1981. Effects of Canopy Modification and Accumulated Sediment on Stream Communities. *Transactions of the American Fisheries Society* 110:469–478.

Muthayya, S., Sugimoto, J.D., Montgomery, S., and Maberly, G.F. 2014. An overview of global rice production, supply, trade, and consumption. *Annals of the New York Academy of Sciences* 1324:7–14.

Naiman, R.J., Bilby, R.E., and Bisson, P.A. 2000. Riparian Ecology and Management in the Pacific Coastal Rain Forest. *BioScience* 50:996–1011. doi.org/10.1641/0006-3568(2000)050[0 996:REAMIT]2.0.CO;2.

Neale, M.W. and Moffett, E.R. 2016. Re-engineering buried urban streams: Daylighting results in rapid changes in stream invertebrate communities. *Ecological Engineering* 87:175–184.

Neave, F.B., Booth, R.M.W., Philipps, R.R., Keffer, D.A., Bravener, G.A., and Coombs, N., 2021. Changes in native lamprey populations in the Great Lakes since the onset of sea lamprey (*Petromyzon marinus*) control. *Journal of Great Lakes Research* 47:S378–S387.

Negishi, J.N., Richardson, and J.S. 2003. Responses of organic matter and macroinvertebrates to placements of boulder clusters in a small stream of southwestern British Columbia, Canada. *Canadian Journal of Fisheries and Aquatic Sciences* 60:247–258.

Nevers, M.B., Byappanahalli, M.N., Morris, C.C., Shively, D., Przybyla-Kelly, K., Spoljaric, A.M., Dickey, J., and Roseman, E.F. 2018. Environmental DNA (eDNA): A tool for quantifying the abundant but elusive round goby (*Neogobius melanostomus*). *PLoS ONE* 13(1):e0191720.

Newbury, R. and Gaboury, M. 1993. Exploration and rehabilitation of hydraulic habitats in streams using principles of fluvial behaviour. *Freshwater Biology* 29:195–210.

Newman, M.C. 2019. *Fundamentals of Ecotoxicology: The Science of Pollution*, Fifth Edition. CRC Press, Boca Raton, FL.

Nieman, C.L. and Gray, S.M. 2019. Visual performance impaired by elevated sedimentary and algal turbidity in walleye *Sander vitreus* and emerald shiner *Notropis atherinoides*. *Journal of Fish Biology* 95:186–199.

Niemeyer, R.J., Bladon, K.D., and Woodsmith, R.D. 2020. Long-term hydrologic recovery after wildfire and post-fire forest management in the interior Pacific Northwest. *Hydrological Processes* 34:1182–1197.

Nilsson, C., Polvi, L.E., Gardeström, J., Hasselquist, E.M., Lind, L., and Sarneel, J.M. 2015. Riparian and in-stream restoration of boreal streams and rivers: Success or failure? *Ecohydrology* 8:753–764.

Nilsson, C., Reidy, C.A., Dynesius, M., and Revenga, C. 2005. Fragmentation and flow regulation of the world's large river systems. *Science* 308:405–408.

Norman, J.D. and Whitledge, G.W. 2015. Recruitment sources of invasive Bighead carp (*Hypopthalmichthys nobilis*) and Silver carp (*H. molitrix*) inhabiting the Illinois River. *Biological Invasions* 17:2999–3014.

Nummi, P. and Holopainen, S. 2019. Restoring wetland biodiversity using research: Whole-community facilitation by beaver as framework. *Aquatic Conservation: Marine and Freshwater Ecosystems* 30:1798–1802.

O'Beirne, M.D., Werne, J.P., Hecky, R.E., Johnson, T.C., Katsev, S., and Reavie, E.D. 2017. Anthropogenic climate change has altered primary productivity in Lake Superior. *Nature Communications* 8:15713.

Odum, W.E. 1988. Comparative ecology of tidal freshwater and salt marshes. *Annual Review of Ecology and Systematics* 19:147–176. doi.org/10.1146/annurev.es.19.110188.001051.

Olden, J.D., Comte, L., and Xingli Giam, X. 2018. The Homogocene: A research prospectus for the study of biotic homogenisation. *NeoBiota* 37:23–36. doi:10.3897/neobiota.37.22552.

Olden, J.D. and Rooney, T.P. 2006. On defining and quantifying biotic homogenization. *Global Ecology and Biogeography* 15:113–120.

Oppenheimer, J., Eaton, A., Badruzzaman, M., Haghani, A.W., and Jacangelo, J.G. 2011. Occurrence and suitability of sucralose as an indicator compound of wastewater loading to surface waters in urbanized regions. *Water Research* 45:4019–4027.

Orr, J.A., Vinebrooke, R.D., Jackson, M.C., Kroeker, K.J., Kordas, R.L., Mantyka-Pringle, C., Van den Brink, P.J., De Laender, F., Stoks, R., Holmstrup, M., Matthaei, C.D., Monk, W.A., Penk, M.R., Leuzinger, S., Schäfer, R.B., and Piggott, J.J. 2020. Towards a unified study of multiple stressors: Divisions and common goals across research disciplines. *Proceedings of the Royal Society B* 287:20200421.

Östlund, L., Laestander, S., Aurell, G., and Hörnberg, G. 2022. The war on deciduous forest: Large-scale herbicide treatment in the Swedish boreal forest 1948 to 1984. *Ambio* 51:1352–1366.

Oudshoorn, F.W., Kristensen, T., and Nadimi, E.S. 2008. Dairy cow defecation and urination frequency and spatial distribution in relation to time-limited grazing. *Livestock Science* 113:62–73.

Overmann, J., Beatty, J.T., Hall, K.J., Pfennig, N., and Northcote, T.G. 1991. Characterization of a dense, purple sulphur bacterial layer in a meromictic salt lake. *Limnology and Oceanography* 36:846–859.

Pacioglu, O., Cornut, J., Gessner, M.O., and Kasprzak, P. 2016. Prevalence of indirect toxicity effects of aluminium flakes on a shredder-fungal-leaf decomposition system. *Freshwater Biology* 61:2013–2025.

Padilla, D.K. and Williams, S.L. 2004. Beyond ballast water: Aquarium and ornamental trades as sources of invasive species in aquatic ecosystems. *Frontiers in Ecology and the Environment* 2:131–138.

Pagnucco, K.S., Maynard, G.A., Fera, S.A., Yan, N.D., Nalepa, T.F., and Ricciardi, A. 2015. The future of species invasions in the Great Lakes-St. Lawrence River basin. *Journal of Great Lakes Research* 41, Suppl. 1:96–107. doi.org/10.1016/j.jglr.2014.11.004.

Palace, V.P., Evans, R.E., Wautier, K.G., Mills, K.H., Blanchfield, P.J., Park, B.J., Baron, C.L., and Kidd, K.A. 2009. Interspecies differences in biochemical, histopathological and population responses in four wild fish species exposed to ethynylestradiol added to a whole lake. *Canadian Journal of Fisheries and Aquatic Sciences* 66:1920–1935.

Palmer, M.A., Bernhardt, E.S., Schlesinger, W.H., Eshleman, K.N., Foufoula-Georgiou, E., Hendryx, M.S., Lemly, A.D., Likens, G.E., Loucks, O.L., Power, M.E., White, P.S., and Wilcock, P.R. 2010. Mountaintop Mining Consequences. *Science* 327:148–149.

Palmer, M.A., Hondula, K.L., and Koch, B.J. 2014. Ecological restoration of streams and rivers: Shifting strategies and shifting goals. *Annual Review of Ecology, Evolution and Systematics* 45:247–269.

Palta, M.M., Grimm, N.B., and Groffman, P.M. 2017. "Accidental" urban wetlands: Ecosystem functions in unexpected places. *Frontiers in Ecology and the Environment* 15:248–256.

Papalexiou, S.M., Montanari, A. 2019. Global and regional increase of precipitation extremes under global warming. *Water Resources Research* 55:4901–4914.

Patrick, R. 1965. Algae as indicators of pollution. In: *Biological Problems in Water Pollution*, U.S. Dept. of Health, Education & Welfare, Cincinnati, Ohio, PHS Publ. 999-WP-25, pp. 225–231.

Paul, M.J. and Meyer, J.L. 2001. Streams in the urban landscape. *Annual Review of Ecology and Systematics* 32:333–365.

Pauly, D., Christensen, V., Dalsgaard, J., Froese, R., and Torres, F. 1998. Fishing down marine food webs. *Science* 279:860–863.

Pawlowski, J., Kelly-Quinn, M., Altermatt, F., Apothéloz-Perret-Gentil, L., Beja, P., Boggero, A., Borja, A., Bouchez, A., Cordier, T., Domaizon, I., Feio, M.J., Filipe, A.F., Fornaroli, R., Graf, W., Herder, J., van der Hoorn, B., Jones, J.I., Sagova-Mareckova, M., Moritz, C., Barquín, J., Piggott, J.J., Pinna, M., Rimet, F., Rinkevich, B., Sousa-Santos, C., Specchia, V., Trobajo, R., Vasselon, V., Vitecek, S., Zimmerman, J., Weigand, A., Leese, F., and Kahlert, M. 2018. The future of biotic indices in the ecogenomic era: Integrating (e)DNA metabarcoding in biological assessment of aquatic ecosystems. *Science of the Total Environment* 637–638:1295–1310.

Pearson, J., Dunham, J., Bellmore, J.R., and Lyons, D. 2019. Modeling control of Common Carp (*Cyprinus carpio*) in a shallow lake-wetland system. *Wetlands Ecology and Management* 27:663–682.

Penaluna, B.E., Railsback, S.F., Dunham, J.B., Johnson, S., Bilby, R.E., and Skaugset, A.E. 2015. The role of the geophysical template and environmental regimes in controlling stream-living trout populations. *Canadian Journal of Fisheries and Aquatic Sciences* 72:893–901.

Perales, K.M., Hansen, G.J.A., Hein, C.L., Mrnak, J.T., Roth, B.M., Walsh, J.R., and Vander Zanden, M.J. 2021. Spatial and temporal patterns in native and invasive crayfishes during a 19-year whole-lake invasive crayfish removal experiment. *Freshwater Biology.* 66:2105–2117.

Perkin, E.K., Hölker, F., Richardson, J.S., Sadler, J.P., Wolter, C., and Tockner, K. 2011. The influence of artificial light on stream and riparian ecosystems: Questions, challenges, and perspectives. *Ecosphere* 2: art. 122.

Perkin, E.K., Hölker, F., and Tockner, K. 2014. The effects of artificial night lighting on adult aquatic and terrestrial insects. *Freshwater Biology* 59:368–377.

Pess, G.R., McHenry, M., Denton, K., Anderson, J.H., Liermann, M.C., Peters, R., McMillan, J., Brenkman, S., Bennett, T., Mahan, J., Duda, J., and Hanson, K. 2023. Initial response of Chinook salmon (*Oncorhynchus tshawytscha*) and steelhead (*Oncorhynchus mykiss*) to removal of two dams on the Elwha River, Washington State, USA *Frontiers in Ecology and Evolution* in press.

Peterjohn, W.T. and Correll, D.L. 1984. Nutrient dynamics in an agricultural watershed: observations on the role of a riparian forest. *Ecology* 65:1466–1475.

Peterson, M.G., Hunt, L., Donley Marineau, E.E., and Resh, V.H. 2017. Long-term studies of seasonal variability enable evaluation of macroinvertebrate response to an acute oil spill in an urban Mediterranean-climate stream. *Hydrobiologia* 797:319–333.

Pfaff, P.J., Hase, K.J., and Gido, K.B. 2023. Community assembly of prairie farm ponds: Build it and they will come, stock it and they won't. *Canadian Journal of Fisheries and Aquatic Sciences* 80:287–297.

Phelps, N.B.D., Bueno, I., Poo-Munoz, D.A., Knowles, S.J., Massarani, S., Rettkowski, R., Shen, L., Rantala, H., Phelps, P.L.F., and Escobar, L.E. 2019. Retrospective and predictive investigation of fish kill events. *Journal of Aquatic Animal Health* 31:61–70.

Pick, F.R. 2016. Blooming algae: A Canadian perspective on the rise of toxic cyanobacteria. *Canadian Journal of Fisheries and Aquatic Sciences* 73:1149–1158.

Piégay, H., Gregory, K.J., and Bondarey, V. et al. 2005. Public perception as a barrier to introducing wood in rivers for restoration purposes. *Environmental Management* 36:665–674.

Pierce, R., Podner, C., and Carim, K. 2013. Response of wild trout to stream restoration over two decades in the Blackfoot River Basin, Montana. *Transactions of the American Fisheries Society* 142:68–81.

Piggott, J.J., Townsend, C.R., and Matthaei, C.D. 2015a. Climate warming and agricultural stressors interact to determine stream macroinvertebrate community dynamics. *Global Change Biology* 21:1887–1906.

———. 2015b. Re-conceptualizing synergism and antagonism among multiple stressors. *Ecology & Evolution* 5:1538–1547.

Pikitch, E.K., Doukakis, P., Lauck, L., Chakrabarty, P., and Erickson, D.L. 2005. Status, trends and management of sturgeon and paddlefish fisheries. *Fish and Fisheries* 6:233–265.

Pimentel, D., Zuniga, R., and Morrison, D. 2005. Update on the environmental and economic costs associated with alien-invasive species in the United States. *Ecological Economics* 52:273–288.

Piovia-Scott, J., Sadro, S., Knapp, R.A., Sickman, J., Pope, K.L., and Chandra, S. 2016. Variation in reciprocal subsidies between lakes and land: Perspectives from the mountains of California. *Canadian Journal of Fisheries and Aquatic Sciences* 73:1691–1701.

Poff, N.L., Allan, J.D., Bain, M.B., Karr, J.R., Prestegaard, K.L., Richter, B.D., Sparks, R.E., and Stromberg, J.C. 1997. The Natural Flow Regime. *BioScience* 47:769–784.

Poff, N.L. and Hart, D.D. 2002. How dams vary and why it matters for the emerging science of dam removal. *Bioscience* 52:659–668.

Poff, N.L., Richter, B.D., Arthington, A.H., Bunn, S.E., Naiman, R.J., Kendy, E., Acreman, M., Apse, C., Bledsoe, B.P., Freeman, M.C., Henriksen, J., Jacobson, R.B., Kennen, J.G., Merritt, D.M., O'Keefe, J.H., Olden, J.D., Rogers, K., Tharme, R.E., and Warner, A. 2010. The ecological limits of hydrologic alteration (ELOHA): A new framework for developing regional environmental flow standards. *Freshwater Biology* 55:147–170.

Poff, N.L.R. and Zimmerman, J.K.H. 2010. Ecological responses to altered flow regimes: A literature review to inform the science and management of environmental flows. *Freshwater Biology* 55:194–205.

Pomeranz, J.P.F., Warburton, H.J., Harding, J.S. 2019. Anthropogenic mining alters macroinvertebrate size spectra in streams. *Freshwater Biology* 64: 81–92.

Pope, K.L., Piovia-Scott, J., Lawler, S.P. 2009. Changes in aquatic insect emergence in response to whole-lake experimental manipulations of introduced trout. *Freshwater Biology* 54:982–993.

Post, J.R. 2013. Resilient recreational fisheries or prone to collapse? A decade of research on the science and management of recreational fisheries. *Fisheries Management and Ecology* 20:99–110.

Postel, S. 1999. *Pillar Of Sand: Can The Irrigation Miracle Last?* W.W. Norton & Co., NY.

Posthuma, L., Zijp, M.C., De Zwart, D., Van de Meent, D., Globevnik, L., Koprivsek, M., Focks, A., Van Gils, J., Sebastian Birk, S. 2020. Chemical pollution imposes limitations to the ecological status of European surface waters. *Scientific Reports* 10:14825.

Poteat, M.D. and Buchwalter, D.B. 2014. Four reasons why traditional metal toxicity testing with aquatic insects is irrelevant. *Environmental Science & Technology* 48:887–888.

Poulton, B.C., Kroboth, P.T., George, A.E., Chapman, D.C., Bailey, J., McMurray, S.E., and Faiman, J.S. 2019. First examination of diet items consumed by wild-caught Black Carp (*Mylopharyngodon piceus*) in the US. *American Midland Naturalist* 182:89–108.

Pratt, T.C., O'Connor, L.M., Stacey, J.A., Stanley, D.R., Mathers, A., Johnson, L.E., Reid, S.M., Verreault, G., and Pearce, J. 2019. Pattern of *Anguillicoloides crassus* infestation in the St. Lawrence River watershed. *Journal of Great Lakes Research* 45:991–997.

Preau, C., Nadeau, I., Sellier, Y., Isselin-Nondedeu, F., Bertrand, R., Collas, M., Capinha, C., and Grandjean, F. 2020. Niche modelling to guide conservation actions in France for the endangered crayfish *Austropotamobius pallipes* in relation to the invasive *Pacifastacus leniusculus*. *Freshwater Biology* 65:304–315.

Prepas, E.E. and Burke, J.M. 1997. Effects of hypolimnetic oxygenation on water quality in Amisk Lake, Alberta, a deep, eutrophic lake with high internal phosphorus loading rates. *Canadian Journal of Fisheries and Aquatic Sciences* 54:2111–2120.

Prest, J.W.M., Burns, S.A., Weismuller, T., Martin, M., and Huckins, J.N. 1992. Passive water sampling via semipermeable membrane devices (SPMDs) in concert with bivalves in the Sacramento/San Joaquin River delta. *Chemosphere* 25:1811–1823.

Preston, S.J., Keys, A., and Roberts, D. 2007. Culturing freshwater pearl mussel *Margaritifera margaritifera*: A breakthrough in the conservation of an endangered species. *Aquatic Conservation: Marine and Freshwater Ecosystems* 17:539–549.

Quigley, J.T. and Harper, D.J. 2006. Effectiveness of fish habitat compensation in Canada in achieving no net loss. *Environmental Management* 37:351–366.

Rabinowitz, D. 1981. Seven forms of rarity. Pp 205–217, In: Synge, H. (Ed.) *The Biological Aspects of Rare Plant Conservation*. Wiley, NY.

Rahel, F.J. and Olden, J.D. 2008. Assessing the effects of climate change on aquatic invasive species. *Conservation Biology* 22:521–533.

Railsback, S.F. 2016. Why it is time to put PHABSIM out to pasture. *Fisheries* 41:720–725.

Railsback, S.F., Gard, M., Harvey, B.C., White, J.L., and Zimmerman, J.K.H. 2013. Contrast of degraded and restored stream habitat using an individual-based salmon model. *North American Journal of Fisheries Management* 33:384–399.

Rajakallio, M., Jyväsjärvi, J., Muotka, T., and Aroviita, J. 2021. Blue consequences of the green bioeconomy: Clear-cutting intensifies the harmful impacts of land drainage on stream invertebrate biodiversity. *Journal of Applied Ecology* 58:1523–1532.

Ramey, T. and Richardson, J.S. 2017. Terrestrial invertebrates in the riparian zone: Mechanisms underlying their unique diversity. *BioScience* 67:808–819.

Raptis, C.E., van Vliet, M.T.H., and Pfister, S. 2016. Global thermal pollution of rivers from thermoelectric power plants. *Environmental Research Letters* 11:104011.

Redondo-Hasselerharm, P.E., Falahudin, D., Peeters, E.T.H.M., and Koelmans, A.A. 2018. Microplastic effect thresholds for freshwater benthic macroinvertebrates. *Environmental Science and Technology* 52:2278–2286.

Reece, P.F., Reynoldson, T.B., Richardson, J.S., and Rosenberg, D.M. 2001. Implications of seasonal variation for biomonitoring with predictive models in the Fraser River catchment, British Columbia. *Canadian Journal of Fisheries and Aquatic Sciences* 58:1411–1418.

Reiber, L., Knillmann, S., Kaske, O., Atencio, L.C., Bittner, L., Albrecht, J.E., Götz, A., Fahl, A.-K., Beckers, L.-M., Krauss, M., Henkelmann, B., Schramm, K.-W., Inostroza, P.A., Schinkel, L., Brauns, M., Weitere, M., Brack, W., and Matthias Liess, M. 2021. Long-term effects of a catastrophic insecticide spill on stream invertebrates. *Science of the Total Environment* 768:144456.

Reid, A.J., Carlson, A.K., Creed, I.F., Eliason, E.J., Gell, P.A., Johnson, P.T.J., Kidd, K.A., MacCormack, T.J., Olden, J.D., Ormerod, S.J., Smol, J.P., Taylor, W.W., Tockner, K., Vermaire, J.C., Dudgeon, D., Cooke, S.J. 2019. Emerging threats and persistent conservation challenges for freshwater biodiversity. *Biological Reviews* 94: 849–873.

Reid, D.A., Hassan, M.A., and Floyd, W. 2016. Reach scale contributions of road surface sediment to the Honna River, Haida Gwaii, BC. *Hydrological Processes* 30:3450–3465.

Reiser, D.W. and Hilgert, P.J. 2018. A practioner's perspective on the continuing technical merits of PHABSIM. *Fisheries* 43:278–283.

Rempel, L.L., Richardson, J.S., and Healey, M.C. 2000. Macroinvertebrate community structure along gradients of hydraulic and sedimentary conditions in a large, gravel-bed river. *Freshwater Biology* 45:57–73.

Reshadi, M.A.M., Hasani, S.S., Nazaripour, M., McKay, G., and Bazargan, A. 2021. The evolving trends of landfill leachate treatment research over the past 45 years. *Environmental Science and Pollution Research* doi.org/10.1007/s11356-021-14274-x.

Ricart, M., Guasch, H., Alberch, M., Barceló, D., Bonnineau, C., Geiszinger, A. Farré, M.I., Ferrer, J., Ricciardi, F., Romaní, A.M., Morin, S., Proia, L., Sala, L., Sureda, D., and Sabater, S. 2010. Triclosan persistence through wastewater treatment plants and its potential toxic effects on river biofilms. *Aquatic Toxicology* 100:346–353.

Ricciardi, A. 2001. Facilitative interactions among aquatic invaders: Is an "invasional meltdown" occurring in the Great Lakes? *Canadian Journal of Fisheries and Aquatic Sciences* 58:2513–2525.

———. 2007. Are modern biological invasions an unprecedented form of global change? *Conservation Biology* 21:329–336.

Ricciardi, A., Hoopes, M.F., Marchetti, M.P., and Lockwood, J.L. 2013. Progress toward understanding the ecological impacts of nonnative species. *Ecological Monographs* 83:263–282.

Ricciardi, A. and MacIsaac, H.J. 2000. Recent mass invasion of the North American Great Lakes by Ponto–Caspian species. *Trends in Ecology & Evolution* 15:62–65.

Ricciardi, A., Neves, R.J., and Rasmussen, J.B. 1998. Impending extinctions of North American freshwater mussels (Unionoida) following the zebra mussel (*Dreissena polymorpha*) invasion. *Journal of Animal Ecology* 67:613–619.

Ricciardi. A. and Rasmussen, J.B. 1999. Extinction rates of North American freshwater fauna. *Conservation Biology* 13:1220–1222.

Rice, S.P., Kiffney, P., Greene, C., and Pess, G.R. 2008. The ecological importance of tributaries and confluences. Pp. 209–242 In: Rice, S.P., Roy, A., and Rhoads, B. (Eds.) *River Confluences, Tributaries and the Fluvial Network*. John Wiley & Sons, England.

Richardson, J.S. 2008. Aquatic arthropods and forestry: large-scale land-use effects on aquatic systems in Nearctic temperate regions. *Canadian Entomologist* 140:495–509.

———. 2019. Biological diversity in headwater streams. *Water* 11:366. doi:10.3390/w11020366.

———. 2020. Headwater Streams. In: Goldstein, M.I., and DellaSala, D.A. (Eds.) *Encyclopedia of the World's Biomes*, Vol. 4. Elsevier, Pp. 371–378. https://doi.org/10.1016/B978-0-12-409 548-9.11957-8.

Richardson, J.S. and Dudgeon, D. 2022. Headwater stream ecosystems: An initial evaluation of their threat status. In: DellaSala, D.A. and Goldstein, M.I. (Eds.) *Imperiled: The Encyclopedia of Conservation*, Vol. 2. Elsevier, Pp. 479–484.

Richardson, J.S. and Mackay, R.J. 1991. Lake outlets and the distribution of filter feeders: An assessment of hypotheses. *Oikos* 62:370–380.

Richardson, J.S., Michalski, T., and Becu, B. 2021. Stream inflows to lake deltas: A tributary junction that provides a unique habitat in lakes. *Freshwater Biology* 66:2021–2029.

Richardson, J.S., Moore, R.D., Jackson, C.R., and Kreutzweiser, D.P. 2022. Use and forest practices in the United States: past, present, and future. Pp. 99–142 In Danehy, R.J., Dolloff, C.A., and Reeves, G.H., (Eds.) Reflections on Forest Management: Can fish and fiber coexist? American Fisheries Society, Symposium 92, Bethesda, Maryland.

Richardson, J.S., Naiman, R.J., and Bisson, P.A. 2012. How did fixed-width buffers become standard practice for protecting freshwaters and their riparian areas from forest harvest practices? *Freshwater Science* 31:232–238.

Richardson, J.S., Naiman, R.J., Swanson, F.J., and Hibbs, D.E. 2005. Riparian communities associated with Pacific Northwest headwater streams: Assemblages, processes, and uniqueness. *Journal of the American Water Resources Association* 41:935–947.

Richardson, J.S. and Sato, T. 2015. Resource flows across freshwater-terrestrial boundaries and influence on processes linking adjacent ecosystems. *Ecohydrology* 8:406–415.

Richardson, J.S., Taylor, E., Schluter, D., Pearson, M., and Hatfield, T. 2010. Do riparian zones qualify as critical habitat for endangered freshwater fishes? *Canadian Journal of Fisheries and Aquatic Sciences* 67:1197–1204.

Richardson, J.S. and Thompson, R.M. 2009. Setting conservation targets for freshwater ecosystems in forested catchments. Pp. 244–263 In: Villard, M.-A. and Jonsson, B.-G. (Eds.) *Setting Conservation Targets for Managed Forest Landscapes.* Cambridge University Press.

Richmond, E.K., Rosi-Marshall, E.J., Lee, S.S., Thompson, R.M., and Grace, M. 2016. Antidepressants in stream ecosystems: Influence of selective serotonin reuptake inhibitors (SSRIs) on algal production and insect emergence. *Freshwater Science* 35:845–855. doi:10.1086/687841.

Richter, B. and Thomas, G. 2007. Restoring environmental flows by modifying dam operations—*Ecology and Society* 12(1):12. http://www.ecologyandsociety.org/vol12/iss1/art12/.

Richter, H.E., Gungle, B., Lacher, L.J., Turner, D.S., and Bushman, B.M. 2014. Development of a shared vision for groundwater management to protect and sustain baseflows of the Upper San Pedro River, Arizona, USA. *Water* 6:2519–2538. doi:10.3390/w6082519.

Ries, P., De Jager, N.R., Zigler, S.J., and Newton, T.J. 2016. Spatial patterns of native freshwater mussels in the Upper Mississippi River. *Freshwater Science* 35:934–947.

Riley, W.D., Potter, E.C.E., Biggs, J., Collins, A.L., Jarvie, H.P., Jones, J.I., Kelly-Quinn, M., Ormerod, S.J., Sear, D.A., Wilby, R.L., Broadmeadow, S., Brown, C.D., Chanin, P., Copp, G.H., Cowx, I.G., Grogan, A., Hornby, D.D., Huggett, D., Kelly, M.G., Naura, M., Newman, J.R., and Siriwardena, G.M. 2018. Small water bodies in Great Britain and Ireland: Ecosystem function, human-generated degradation, and options for restorative action. *Science of the Total Environment* 645:1598–1616.

Ritchie, A.C., Warrick, J.A., East, A.E., Magirl, C.S., Stevens, A.W., Bountry, J.A., Randle, T.J., Curran, C.A., Hilldale, R.C., Duda, J.J., Gelfenbaum, G.R., Miller, I.M., Pess, G.R., Foley, M.M., McCoy, R., and Ogston, A.S. 2018. Morphodynamic evolution following sediment release from the world's largest dam removal. *Scientific Reports* 8:13279.

Ritchie, J.C. 1972. Sediment, fish and fish habitat. *Journal of Soil and Water Conservation* 27:124–125.

Rochman, C.M., Brookson, C., Bikker, J., Djuric, N., Earn, A., Bucci, K., Athey, S., Huntington, A., McIlwraith, H., Munno, K., De Fron, H., Kolomijeca, A., Erdle, L., Grbic, J., Bayoumi, M., Borrelle, S.B., Wu, T., Santoro, S., Werbowski, L.M., Zhu, X., Giles, R.K., Hamilton, B.M., Thaysen, C., Kaura, A., Klasios, N., Ead, L., Kim, J., Sherlock, C., Ho, A., and Hunga, C. 2019. Rethinking microplastics as a diverse contaminant suite. *Environmental Toxicology and Chemistry* 38:703–711.

Romanov, A.M., Hardy, J., Zeug, S.C., and Cardinale, B.J. 2012. Abundance, size structure, and growth rates of Sacramento pikeminnow (*Ptychocheilus grandis*) following a large-scale stream channel restoration in California. *Journal of Freshwater Ecology* 27:495–505. doi:10.1080/02705060.2012.674684.

Roni, P. 2019. Does river restoration increase fish abundance and survival or concentrate fish? The effects of project scale, location, and fish life history. *Fisheries* 44:7–19.

Roni, P. and Beechie, T. (Eds.). 2013. *Stream and Watershed Restoration: A guide to restoring riverine processes and habitats.* Wiley-Blackwell, Chichester, UK.

Roni, P., Beechie, T., Pess, G., and Hanson, K. 2015. Wood placement in river restoration: fact, fiction, and future direction. *Canadian Journal of Fisheries and Aquatic Sciences* 72:466–478.

Roscoe, D.W. and Hinch, S.G. 2010. Effectiveness monitoring of fish passage facilities: historical trends, geographic patterns and future directions. *Fish and Fisheries* 11:12–33.

Rosenfeld, J.S. and Naman, S.M. 2021. Identifying and mitigating systematic biases in fish habitat simulation modeling: Implications for estimating minimum instream flows. *River Research and Applications* 37:869–879.

Rosi-Marshall, E.J., Snow, D., Bartelt-Hunt, S.L., Paspalof, A., and Tank, J.L. 2015. A review of ecological effects and environmental fate of illicit drugs in aquatic ecosystems. *Journal of Hazardous Materials* 282:18–25. doi:10.1016/j.jhazmat.2014.06.062.

Ruess, P.J. and Konar, M. 2019. Grain and virtual water storage capacity in the United States. *Water Resources Research* 55:3960–3975.

Rytwinski, T., Kelly, L.A., Donaldson, L.A., Taylor, J.J., Smith, A., Drake, D.A.R., Martel, A.L., Geist, J., Morris, T.J., George, A.L., Dextrase, A.J., Bennett, J.R., and Cooke, S.J. 2021. What evidence exists for evaluating the effectiveness of conservation-oriented captive breeding and release programs for imperilled freshwater fishes and mussels? *Canadian Journal of Fisheries and Aquatic Sciences* 78:1332–1346.

Rytwinski, T., Taylor, J.J., Donaldson, L.A., Britton, J.R., Browne, D.R., Gresswell, R.E., Lintermans, M., Prior, K.A., Pellatt, M.G., Vis, C., and Cooke, S.J. 2019. The effectiveness of non-native fish removal techniques in freshwater ecosystems: A systematic review. *Environmental Reviews* 27:71–94.

Sabater, S., Ludwig, R., and Elosegi, A. (Eds.). 2018. *Multiple stress in river ecosystems. Status, impacts and prospects for the future.* Elsevier, Cambridge, MA. 404 Pp.

Sakai, M., Gomi, T., Naito, R.S., Negishi, J.N., Sasaki, M., Toda, H., Nunokawa, M., and Murase, K. 2015. Radiocesium leaching from contaminated litter in forest streams. *Journal of Environmental Radioactivity* 144:15–20.

Sakai, M., Gomi, T., Nunokawa, M., Wakahara, T., and Onda, Y. 2014. Soil removal as a decontamination practice and radiocesium accumulation in tadpoles in rice paddies at Fukushima. *Environmental Pollution* 187:112–115.

Scavia, D., Allan, J.D., Arend, K. K., Bartell, S., Beletsky, D., Bosch, N.S., . . . , and Zhou, Y. 2014. Assessing and addressing the re-eutrophication of Lake Erie: Central basin hypoxia. *Journal of Great Lakes Research* 40:226–246.

Schäfer, R.B., Jackson, M., Juvigny-Khenafou, N, Osakpolor, S.E., Posthuma, L., Schneeweiss, A., Spaak, J., and Vinebrooke, R. 2023. Chemical mixtures and multiple stressors: Same but different? *Environmental Toxicology and Chemistry* 2023: in press.

Schäfer, R.B., Kuhn, B., Malaj, E., Konig, A., and Gergs, R. 2016. Contribution of organic toxicants to multiple stress in river ecosystems. *Freshwater Biology* 61:2116–2128.

Schäfer, R.B. and Piggott, J.J. 2018. Advancing understanding and prediction in multiple stressor research through a mechanistic basis for null models. *Global Change Biology* 24:1817–1826.

Schäfer, R.B., von der Ohe, P.C., Rasmussen, J., Kefford, B.J., Beketov, M.A., Schulz, R., and Liess, M. 2012. Thresholds for the effects of pesticides on invertebrate communities and leaf breakdown in stream ecosystems. *Environmental Science & Technology* 46:5134–5142.

Scheder, C., Lerchegger, B., Jung, M., Csar, D., and Gumpinger, C. 2014. Practical experience in the rearing of freshwater pearl mussels (*Margaritifera margaritifera*): Advantages of a work-saving infection approach, survival, and growth of early life stages. *Hydrobiologia* 735:203–212.

Schindler, D.E., Carter, J.L., Francis, T.B., Lisi, P.J., Askey, P.J., and Sebastian, D.C. 2012. *Mysis* in the Okanagan Lake food web: A time-series analysis of interaction strengths in an invaded plankton community. *Aquatic Ecology* 46:215–227.

Schindler, D.W. 1974. Eutrophication and recovery in experimental lakes: Implications for lake management. *Science* 184:897–899. doi:10.1126/science.184.4139.897.

———. 2001. The cumulative effects of climate warming and other human stresses on Canadian freshwaters in the new millennium. *Canadian Journal of Fisheries and Aquatic Sciences* 58:18–29.

Schindler, D.W. and Donahue, W.F. 2006. An impending water crisis in Canada's western prairie provinces. *Proceedings of the National Academy of Sciences of the United States* 103:7210–7216.

Schindler, D.W., Mills, K.H., Malley, D.F., Findlay, D.L., Shearer, J.A., Davies, I.J., Turner, M.A., Linsey, G.A., and Cruikshank, D.R. 1985. Long term ecosystem stress: The effect of years of acidification on a small lake. *Science* 228:1395–1401.

Schlesinger, W.H. and Jasechko, S. 2014. Transpiration in the global water cycle. *Agricultural and Forest Meteorology* 189:115–117.

Schluter, D. and McPhail, J.D. 1992. Ecological character displacement and speciation in sticklebacks. *American Naturalist* 140:85–108.

Schreiner, V.C., Bakanov, N., Kattwinkel, M., Könemann, S., Kunz, S., Vermeirssen, E.L.M., and Schäfer, R.B. 2020. Sampling rates for passive samplers exposed to a field-relevant peak of 42 organic pesticides. *Science of the Total Environment* 740:140376.

Scrimgeour, G.J. and Kendall, S., 2003. Effects of livestock grazing on benthic invertebrates from a native grassland ecosystem. *Freshwater Biology* 48:347–362.

Seehausen, O., Terai, Y., Magalhaes, I.S., Carleton, K.L., Mrosso, H.D.J., Miyagi, R., van der Sluijs, I., Schneider, M.V., Maan, M.E., Tachida, H., Imai, H., and Okada, N. 2008. Speciation through sensory drive in cichlid fish. *Nature* 455:620–626.

Segadelli, S., Grazzini, F., Adorni, M., De Nardo, M.T., Fornasiero, A., Chelli, A., and Cantonati, M. 2020. Predicting extreme-precipitation effects on the geomorphology of small mountain catchments: Towards an improved understanding of the consequences for freshwater biodiversity and ecosystems. *Water* 12:79. doi:10.3390/w12010079.

Seitz, N.E., Westbrook, C.J., and Noble, B.F. 2011. Bringing science into river systems cumulative effects assessment practice. *Environmental Impact Assessment Review* 31:172–179.

Semlitsch, R.D., Todd, B.D., Blomquist, S.M., Calhoun, A.J.K., Gibbons, J.W., Gibbs, J.P., Graeter, G.J., Harper, E.B., Hocking, D.J., Hunter, M.L., Jr., Patrick, D.A., Rittenhouse, T.A.G., and Rothermel, B.B. 2009. Effects of timber harvest on amphibian populations: Understanding mechanisms from forest experiments. *BioScience* 59:853–862.

Shapiro, J., Lamarra, V., and Lynch, M. 1975. Biomanipulation: An ecosystem approach to lake restoration. Pp. 85–96 In: Brezonik, P.L. and Fox, J.L. (Eds.) *Proceedings of the Symposium on Water Quality Management Through Biological Control*. University of Florida.

Sharma, S., Richardson, D.C., Woolway, R.I., Imrit, M.A., Bouffard, D., Blagrave, K., Daly, J., Filazzola, A., Granin, N., Korhonen, J., Magnuson, J., Marszelewski, W., Matsuzaki, S.S., Perry, W., Robertson, D.M., Rudstam, L.G., Weyhenmeyer, G.A., and Yao, H. 2021. Loss of ice cover, shifting phenology, and more extreme events in Northern Hemisphere lakes. *Journal of Geophysical Research: Biogeosciences* 126:e2021JG006348.

Shaw, E.A. and Richardson, J.S. 2001. Effects of fine inorganic sediment on stream invertebrate assemblages and rainbow trout (*Oncorhynchus mykiss*) growth and survival: Implications of exposure duration. *Canadian Journal of Fisheries and Aquatic Sciences* 58:2213–2221.

Shumilova, O., Tockner, K., Thieme, M., Koska, A., and Zarfl, C. 2018. Global water transfer megaprojects: A potential solution for the water-food-energy nexus? *Frontiers in Environmental Science* 6:150.

Shurin, J.B., Gruner, D.S., and Hillebrand, H. 2006. All wet or dried up? Real differences between aquatic and terrestrial food webs. *Proceedings of the Royal Society B, Biological Sciences* 273:1–9.

Sidle, R.C., Tsuboyama, Y., Noguchi, S., Hosoda, I., Fujieda, M., and Shimizu, T. 2000. Stormflow generation in steep forested headwaters: A linked hydrogeomorphic paradigm. *Hydrological Processes* 14:369–385.

Sievers, M., Parris, K.M., Swearer, S.E., and Hale, R. 2018. Stormwater wetlands can function as ecological traps for urban frogs. *Ecological Applications* 28:1106–1115.

Skonberg, E.R. 2014. Pipeline Design for Installation by Horizontal Directional Drilling—ASCE Manuals and Reports on Engineering Practice (MOP) No. 108. American Society of Civil Engineers ISBN: 0-7844-1350-9, 978-0-7844-1350-0.

Smith, P., House, J.I., Bustamante, M., Sobocká, J., Harper, R., Genxing, P., West, P.C., Clark, J.M., Adhya, T., Rumpel, C., Paustian, K., Kuikman, P., Cotrufo, M.F., Elliott, J.A., McDowell, R., Griffiths, R.I., Asakawa, S., Bondeau, A., Jain, A.K., Meersmans, J., and Pugh, T.A.M. 2016. Global change pressures on soils from land use and management. *Global Change Biology* 22:1008–1028.

Smith, R.D., Sidle, R.C., and Porter, P.E. 1993. Effects on bedload transport of experimental removal of woody debris from a forest gravel-bed stream. *Earth Surface Processes and Landforms* 18:455–468.

Solomon, C.T., Jones, S.E., Weidel, B.C., Buffam, I., Fork, M.L., Karlsson, J., Larsen, S., Lennon, J.T., Read, J.S., Sadro, S., and Saros, J.E. 2015. Ecosystem consequences of changing inputs of terrestrial dissolved organic matter to lakes: Current knowledge and future challenges. *Ecosystems* 18:376–389.

Somers, K.A., Bernhardt, E.S., Grace, J.B., Hassett, B.A., Sudduth, E.B., Wang, S., and Urban, D.L. 2013. Streams in the urban heat island: Spatial and temporal variability in temperature. *Freshwater Science* 32:309–326.

Spears, B.M., Chapman, D.S., Carvalho, L., Feld, C.K., Gessner, M.O., Piggott, J.J., Banin, L.F., Gutiérrez-Cánovas, C., Solheim, A.L., Richardson, J.A., Schinegger, R., Segurado, P., Thackeray, S.J., and Birk, S. 2021. Making waves. Bridging theory and practice towards multiple stressor management in freshwater ecosystems. *Water Research* 196:116981.

Spencer, C.N., Potter, D.S., Bukantis, R.T., and Stanford, J.A. 1999. Impact of predation by *Mysis relicta* on zooplankton in Flathead Lake, Montana, USA. *Journal of Plankton Research* 21:51–64.

Spoelstra, J., Schiff, S.L., and Brown, S.J. 2013. Artificial sweeteners in a large Canadian river reflect human consumption in the watershed. *PLoS One* 8(12):e82706.

Sprules, W.G. and Barth, L.E. 2015. Surfing the biomass size spectrum: Some remarks on history, theory, and application. *Canadian Journal of Fisheries and Aquatic Sciences* 73:477–495.

Srivastava, D.S. 2006. Habitat structure, trophic structure and ecosystem function: Interactive effects in a bromeliad–insect community. *Oecologia* 149:493–504.

Staentzel, C., Kondolf, G.M., Schmitt, L., Combroux, I., Barillier, A., and Jean-Nicolas Beisel, J.-N. 2020. Restoring fluvial forms and processes by gravel augmentation or bank erosion below dams: A systematic review of ecological responses. *Science of the Total Environment* 706:135743.

Stammler, K.L, Yates, A.G., Bailey, R.C. 2013. Buried streams: Uncovering a potential threat to aquatic ecosystems. *Landscape and Urban Planning* 114:37–41.

Stanford, J.A. and Ward, J.V. 2001. Revisiting the serial discontinuity concept. *River Research and Applications* 17:303–310.

Stanley, E.H. and Doyle, M.W. 2002. A geomorphic perspective on nutrient retention following dam removal. *Bioscience* 52:693–701.

Statzner, B. 2012. Geomorphological implications of engineering bed sediments by lotic animals. *Geomorphology* 157–158:49–65.

Statzner, B. and Beche, L.A. 2010. Can biological invertebrate traits resolve effects of multiple stressors on running water ecosystems? *Freshwater Biology* 55:80–119.

Stednick, J.D. 2008. Effects of timber harvesting on streamflow in the Alsea Watershed Study. Chapter 2. Pp. 19–36 In, J.D. Stednick (Ed.) *Hydrological and Biological Responses to Forest Practices*. Springer, New York.

Stephens, C.M., Lall, U., Johnson, F.M., and Marshall, L.A. 2021. Landscape changes and their hydrologic effects: Interactions and feedbacks across scales. *Earth-Science Reviews* 212:103466. doi.org/10.1016/j.earscirev.2020.103466.

Sterner, R.W. and Elser, J.J. 2002. *Ecological Stoichiometry: The Biology of Elements from Molecules to the Biosphere*. Princeton University Press.

Stevens, C.J. 2016. How long do ecosystems take to recover from atmospheric nitrogen deposition? *Biological Conservation* 200:160–167.

Stevenson, R.J., Bothwell, M.L., and Lowe, R.L. 1996. *Algal Ecology: Freshwater benthic ecosystems*. Academic Press, Elsevier.

Stevenson, R.J., Bothwell, M.L., and Lowe, R.L., Thorp, J.H. 1995. *Algal ecology: Freshwater benthic ecosystem*. Academic Press.

Stewart, G.B., Bayliss, H.R., Showler, D.A., Sutherland, W.J., and Pullin, A.S. 2009. Effectiveness of engineered in-stream structure mitigation measures to increase salmonid abundance: A systematic review. *Ecological Applications* 19:931–941.

Stewart-Oaten, A. and Bence, J.R. 2001. Temporal and spatial variation in environmental impact assessment. *Ecological Monographs* 71:305–339.

Stockner, J.G. and MacIsaac, E.A. 1996. British Columbia lake enrichment programme: Two decades of habitat enhancement for sockeye salmon. *Regulated Rivers: Research & Management* 12:547–561.

Stockner, J.G. and Shortreed, K.S. 1989. Algal picoplankton production and contribution to food-webs in oligotrophic British Columbia lakes. *Hydrobiologia* 173:151–166.

Stoddard, J.L., Larsen, D.P., Hawkins, C.P., Johnson, R.K., and Norris, R.H. 2006. Setting expectations for the ecological condition of streams: The concept of reference condition *Ecological Applications* 16:1267–1276.

Stoler, A.B. and Relyea, R.A. 2011. Living in the litter: The influence of tree litter on wetland communities. *Oikos* 120:862–872.

Strahler, A.N. 1957. Quantitative analysis of watershed geomorphology. *Transactions of the American Geophysical Union* 38:913–920.

Strait, J.T., Eby, L.A., Kovach, R.P., Muhlfeld, C.C., Boyer, M.C., Amish, S.J., Smith, S., Lowe, W.H., and Luikart, G. 2021. Hybridization alters growth and migratory life-history expression of native trout. *Evolutionary Applications* 14:821–833.

Strayer, D.L. 2010. Alien species in fresh waters: Ecological effects, interactions with other stressors, and prospects for the future. *Freshwater Biology* 55 (Suppl. 1):152–174.

Strayer, D.L. and Dudgeon, D. 2010. Freshwater biodiversity conservation: Recent progress and future challenges. *Journal of the North American Benthological Society* 29:344–358.

Strayer, D.L., Eviner, V.T., Jeschke, J.M., and Pace, M.L. 2006. Understanding the long-term effects of species invasions. *Trends in Ecology & Evolution* 21:645–651.

Strayer, D.L., Fischer, D.T., Hamilton, S.K., Malcom, H.M., Pace, M.L., and Solomon, C.T. 2020. Long-term variability and density dependence in Hudson River *Dreissena* populations. *Freshwater Biology* 65:474–489.

Strecker, A.L., Campbell, P.M., and Olden, J.D. 2011. The aquarium trade as an invasion pathway in the Pacific Northwest. *Fisheries* 36(2):74–85.

Streib, L., Juvigny-Khenafou, N., Heer, H., Kattwinkel, M., and Schäfer, R.B. 2022. Spatiotemporal dynamics drive synergism of land use and climatic extreme events in insect metapopulations. *Science of The Total Environment* 814:152602.

Sturtevant, R.A., Mason, D.M., Rutherford, E.S., Elgin, A., Lower, E., and Martinez, F. 2019. Recent history of nonindigenous species in the Laurentian Great Lakes; An update to Mills et al., 1993 (25 years later). *Journal of Great Lakes Research* 45:1011–1035.

Sudduth, E.B., Hassett, B.A., Cada, P., and Bernhardt, E.S. 2011. Testing the field of dreams hypothesis: Functional responses to urbanization and restoration in stream ecosystems. *Ecological Applications* 21:1972–1988.

Suleiman, M., Daugaard, U., Choffat, Y., Zheng, X., and Petchey, O.L. 2022. Predicting the effects of multiple global change drivers on microbial communities remains challenging. *Global Change Biology* 28:5575–5586.

Sundermann, A., Stoll, S., and Haase, P. 2011. River restoration success depends on the species pool of the immediate surroundings. *Ecological Applications* 21:1962–1971.

Surendran, U., Raja, P., Jayakumar, M., and Subramoniam, R. 2021. Use of efficient water saving techniques for production of rice in India under climate change scenario: A critical review. *Journal of Cleaner Production* 309:127272.

Sutherland, A.B. and Meyer, J.L. 2007. Effects of increased suspended sediment on growth rate and gill condition of two southern Appalachian minnows. *Environmental Biology of Fishes* 80:389–403.

Swanson, S., Kozlowski, D., Hall, R., Heggem, D., and Lin, J. 2017. Riparian proper functioning condition assessment to improve watershed management for water quality. *Journal of Soil and Water Conservation* 72:168–182.

Sweeney, B.W. and Newbold, J.D. 2014. Streamside forest buffer width needed to protect stream water quality, habitat, and organisms: A literature review. *Journal of the American Water Resources Association* 50:560–584.

Syvitski, J.P.M. 2003. Supply and flux of sediment along hydrological pathways: Research for the 21st century. *Global and Planetary Change* 39:1–11.

Syvitski, J.P.M. and Kettner, A. 2011. Sediment Flux and the Anthropocene. *Philosophical Transactions of the Royal Society A: Mathematical, Physical and Engineering Sciences* 369:957–975. https://doi.org/10.1098/rsta.2010.0329.

Taabu-Munyaho, A., Marshall, B.E., Tomasson, T., Marteinsdottir, G. 2016. Nile perch and the transformation of Lake Victoria. *African Journal of Aquatic Science* 41:127–142.

Tabor, R.A., Bell, A.T.C., Lantz, D.W., Gregersen, C.N., Berge, H.B., and Hawkins, D.K. 2017. Phototaxic behavior of subyearling salmonids in the nearshore area of two urban lakes in western Washington State. *Transactions of the American Fisheries Society* 146:753–761.

Tachet, H., Richoux, P., Bournaud, M., and Usseglio-Polatera, P. 2010. *Invertébrés d'eau douce : systématique, biologie, écologie*. CNRS Editions, Paris.

Taguchi, V.J., Olsen, T.A., Natarajan, P., Janke, B.D., Gulliver, J.S., Finlay, J.C., and Stefan, H.G. 2020. Internal loading in stormwater ponds as a phosphorus source to downstream waters. *Limnology and Oceanography Letters* doi:10.1002/lol2.10155.

Takao, A., Negishi, J.N., Nunokawa, M., Gomi, T., and Nakahara, O. 2006. Potential influences of a net-spinning caddisfly (Trichoptera : *Stenopsyche marmorata*) on stream substratum stability in heterogeneous field environments. *Journal of the North American Benthological Society* 25:545–555.

Tanentzap, A.J., Kielstra, B.W., Wilkinson, G.M., Berggren, M., Craig, N., del Giorgio, P.A., Grey, J., Gunn, J.M., Jones, S.E., Karlsson, J., Solomon, C.T., and Pace, M.L. 2017. Terrestrial support of lake food webs: Synthesis reveals controls over cross-ecosystem resource use. *Science Advances* 3:e1601765.

Tank, J.L., Rosi-Marshall, E.J., Royer, T.V., Whiles, M.R., Griffiths, N.A., Frauendorf, T.C., and Treering, D.J. 2010. Occurrence of maize detritus and a transgenic insecticidal protein (Cry1Ab) within the stream network of an agricultural landscape. *Proceedings of the National Academy of Science, USA* 107:17645–17650.

Taylor, E. B., Boughman, J.W., Groenenboom,M., Sniatynski, M., Schluter, D., and J. Gow, J.L. 2006. Speciation in reverse: Morphological and genetic evidence of the collapse of a three-spined stickleback (*Gasterosteus aculeatus*) species pair. *Molecular Ecology* 15:343–355.

Taylor, M.E.D. and Paszkowski, C.A. 2018. Postbreeding movement patterns and habitat use of adult Wood Frogs (*Lithobates sylvaticus*) at urban wetlands. *Canadian Journal of Zoology* 96:521–532.

Templeton, N.P., Vivoni, E.R., Wang, Z.-H., and Schreiner-McGraw, A.P. 2018. Quantifying water and energy fluxes over different urban land covers in Phoenix, Arizona. *Journal of Geophysical Research: Atmospheres* 123:2111–2128.

Terrell, E.E., Emery, W.H.P., and Beaty, H.E. 1978. Observations on *Zizania texana* (Texas wildrice), an endangered species. *Bull. Torrey Botanical Club* 105:50–57.

Tetreault, G.R., Kleywegt, S., Marjan, P., Bragg, L., Arlos, M., Fuzzen, M., Smith, B., Moon, T., Massarsky, A., Metcalfe, C., Oakes, K., McMaster, M.E., and Servos, M.R. 2021. Biological responses in fish exposed to municipal wastewater treatment plant effluent *in situ*. *Water Quality Research Journal* 56:83–99.

Theis, S., Koops, M.A., and Poesch, M.S. 2022. A Meta-analysis on the effectiveness of offsetting strategies to address harm to freshwater fishes. *Environmental Management* 70:793–807.

Theis, S., Ruppert, J.L.W., Roberts, K.N., Minns, C.K., Koops, M., and Poesch, M.S. 2020. Compliance with and ecosystem function of biodiversity offsets in North American and European freshwaters. *Conservation Biology* 34:41–53.

Thomas, S.M., Griffiths, S.W., and Ormerod, S.J. 2016. Beyond cool: Adapting upland streams for climate change using riparian woodlands. *Global Change Biology* 22:310–324.

Thorbergsdóttir, I.M., Gíslason, S.R., Ingvason, H.R., and Einarsson, Á. 2004. Benthic oxygen flux in the highly productive subarctic Lake Myvatn, Iceland: In situ benthic flux chamber study. *Aquatic Ecology* 38:177–189.

Thorp, J.H., Rogers, C. (Eds.) 2014. *Thorp and Covich's Freshwater Invertebrates: Ecology and General Biology*, 4th Edition. Academic Press. doi.org/10.1016/C2010-0-65590-8.

Thorp, J.H., Rogers, C., and Covich, A. (Eds.). 2016. *Thorp and Covich's Freshwater Invertebrates: Keys to Nearctic Fauna*, 4th Edition. Academic Press.

Tidwell, V.C., Kobos, P.H., Malczynski, L.A., Klise, G., and Castillo, C.R. 2012. Exploring the water-thermoelectric power nexus. *Journal of Water Resources Planning and Management* 138:491–501.

Till, A., Rypel, A.L., Bray, A., and Fey, S.B. 2019. Fish die-offs are concurrent with thermal extremes in north temperate lakes. *Nature Climate Change* 9:637–642.

Tilman, D. 1999. Global environmental impacts of agricultural expansion: The need for sustainable and efficient practices. *Proceedings of the National Academy of Sciences* 96:5995–6000.

Tockner, K. and Stanford, J.A. 2002. Riverine flood plains: Present state and future trends. *Environmental Conservation* 29:308–330.

Tonkin, J.D., Olden, J.D., Merritt, D.M., Reynolds, L.V., Rogosch, J.S., and Lytle, D.A. 2021. Designing flow regimes to support entire river ecosystems. *Frontiers in Ecology and the Environment* 19:326–333.

Trexler, J.C. 2006. Restoration of the Kissimmee River: A conceptual model of past and present fish communities and its consequences for evaluating restoration success. *Restoration Ecology* 3:195–210.

Triest, L., Stiers, I., and Van Onsem, S. 2016. Biomanipulation as a nature-based solution to reduce cyanobacterial blooms. *Aquatic Ecology* 50:461–483.

Tuchman, N.C., Wetzel, R.G., Rier, S.T., Wahtera, K.A., and Teeri, J.A. 2002. Elevated atmospheric CO_2 lowers leaf litter nutritional quality for stream ecosystem food webs. *Global Change Biology* 8:163–170.

Turschwell, M.P., Connolly, S.R., Schäfer, R.B., De Laender, F., Campbell, M.D., Mantyka-Pringle, C., Jackson, M.C., Kattwinkel, M., Sievers, M., and Ashauer, R., et al. 2022. Interactive effects of multiple stressors vary with consumer interactions, stressor dynamics and magnitude. *Ecology Letters* 25:1483–1496.

Turunen, J., Elbrecht, V., Steinke, D., and Aroviita, J. 2021. Riparian forests can mitigate warming and ecological degradation of agricultural headwater streams. *Freshwater Biology* 66:785–798.

Underwood, A.J. 1992. Beyond BACI: The detection of environmental impacts on populations in the real, but variable, world. *Journal of Experimental Marine Biology and Ecology* 161:145–178.

USDA. https://www.ers.usda.gov/topics/farm-practices-management/crop-livestock-practices/manure-management/ accessed 9 Feb 2021.

U.S. EPA. 2017. National Water Quality Inventory: Report to Congress. EPA 841-R-16-011.

Ussery, E.J., McMaster, M.E., Servos, M.R., Miller, D.H., and Munkittrick, K.R. 2021. A 30-year study of impacts, recovery, and development of critical effect sizes for endocrine disruption in White Sucker (*Catostomus commersonii*) exposed to bleached-kraft pulp mill effluent at Jackfish Bay, Ontario, Canada. *Frontiers in Endocrinology* 12:664157 doi.org/10.3389/fendo.2021.664157.

Vander Laan, J.J. and Hawkins, C.P. 2014. Enhancing the performance and interpretation of freshwater biological indices: An application in arid zone streams. *Ecological Indicators* 36:470–482.

Vannote, R.L., Minshall, G.W., Cummins, K.W., Sedell, J.R., and Cushing, C.E. 1980. The river continuum concept. *Canadian Journal of Fisheries and Aquatic Sciences* 37:130–137.

Van Woerkom, T., van Beek, R., Middelkoop, H., and Bierkens, M.F.P. 2021. Global sensitivity analysis of groundwater related dike stability under extreme loading conditions. *Water* 13:3041.

Vaughn, C.C. 2010. Biodiversity losses and ecosystem function in freshwaters: Emerging conclusions and research directions *BioScience* 60:25–35.

Venkiteswaran, J.J., Schiff, S.L., and Taylor, W.D. 2015. Linking aquatic metabolism, gas exchange, and hypoxia to impacts along the 300-km Grand River, Canada. *Freshwater Science* 34:1216–1232.

Venn, A.A., Loram, J.E., and Douglas, A.E. 2008. Photosynthetic symbioses in animals. *Journal of Experimental Botany* 59:1069–1080. doi.org/10.1093/jxb/erm328.

Venter, H.J. and Bøhn, T. 2016. Interactions between Bt crops and aquatic ecosystems: a review. *Environmental Toxicology and Chemistry* 35:2891–2902.

Verberk, W.C.E.P., van Noordwijk, C.G.E., and Hildrew, A.G. 2013. Delivering on a promise: Integrating species traits to transform descriptive community ecology into a predictive science. *Freshwater Science* 32:531–547.

Verdonschot, R.C.M., Keizer-Vlek, H.E., Verdonschot, P.F.M. 2011. Biodiversity value of agricultural drainage ditches; a comparative analysis of the aquatic invertebrate fauna of ditches and small lakes. *Aquatic Conservation: Marine and Freshwater Ecosystems* 21:715–727.

Verlicchi, P., Galletti, A., and Aukidy, M.A. 2013. Hospital wastewaters: Quali-quantitative characterization and for strategies for their treatment and disposal. Pp. 225–252 In: Sharma, S.K., Sanghi, R., Eds. *Wastewater Reuse and Management*. Dordrecht, Germany: Springer.

Vertessy, R.A., Watson, F.G.R., and O'Sullivan, S.K. 2001. Factors determining relations between stand age and catchment water balance in mountain ash forests. *Forest Ecology and Management* 143:13–26.

Vidon, P.G., Welsh, M.K., and Hassanzadeh, Y.T. 2019. Twenty years of riparian zone research (1997–2017): Where to next? *Journal of Environmental Quality* doi:10.2134/jeq2018.01.0009.

Vinebrooke, R.D., Cottingham, K.L., Norberg, J., Scheffer, M., Dodson, S.I., Maberly, S.C., and Sommer, U. 2004. Impacts of multiple stressors on biodiversity and ecosystem functioning: The role of species co-tolerance. *Oikos* 104:451–457.

Violin, C.R., Cada, P., Sudduth, E.B., Hassett, B.A., Penrose, D.L., and Bernhardt, E.S. 2011. Effects of urbanization and urban stream restoration on the physical and biological structure of stream ecosystems. *Ecological Applications* 21:1932–1949.

Vogel, S. 1996. *Life in Moving Fluids: The Physical Biology of Flow*, Second Edition. Princeton University Press.

Vörösmarty, C.J., McIntyre, P.B., Gessner, M.O., Dudgeon, D., Prusevich, A., Green, P., Glidden, S., Bunn, S.E., Sullivan, C.A., Reidy Liermann, C., and Davies, P.M. 2010. Global threats to human water security and river biodiversity. *Nature* 467:555–561.

Vörösmarty, C.J., Meybeck, M., Fekete, B., and Sharma, K., 1997. The potential impact of neo-Castorization on sediment transport by the global network of rivers. In: Walling, D.E. and Probst, J.L. (Eds.) Human Impact on Erosion and Sedimentation. (Proc. Rabat Symposium, April 1997), IAHS Publication No. 245. IAHS Press, Wallingford, UK, pp. 261–273.

Vos, M., Hering, D., Gessner, M.O., Leese, F., Schäfer, R.B., Tollrian, R., Boenigk, J., Haase, P., Meckenstock, R., and Baikova, D., et al. 2023. The Asymmetric Response Concept explains ecological consequences of multiple stressor exposure and release. *Science of The Total Environment* 872:162196.

Vymazal, J. 2007. Removal of nutrients in various types of constructed wetlands. *Science of the Total Environment* 380:48–65.

Wada, Y., van Beek, L.P., van Kempen, C.M., Reckman, J.W., Vasak, S., and Bierkens, M.F. 2010. Global depletion of groundwater resources. *Geophysical Research Letters* 37:L20402.

Wagner, A.M., Larson, D.L., DalSoglio, J.A., Harris, J.A., Labus, P., Rosi-Marshall, E.J., and Skrabis, K.E. 2015. A framework for establishing restoration goals for contaminated ecosystems. *Integrated Environmental Assessment and Management* 12:264–272.

Wagner, T., Congleton, J.L., Marsh, D.M., Walker, C.H., Sibly, R.M., Hopkin, S.P., and Peakall, D.B. 2014. *Principles of Ecotoxicology*, 4th Edition. CRC Press, Boca Raton, FL.

Waite, I.R., Van Metre, P.C., Moran, P.W., Konrad, C.P., Nowell, L.H., Meador, M.R., Munn, M.D., Schmidt, T.S., Gellis, A.C., and Carlisle, D.M., et al. 2021. Multiple in-stream stressors degrade biological assemblages in five U.S. regions. *Science of The Total Environment* 800:149350.

Walker, C. 2014. *Ecotoxicology: Effects of Pollutants on the Natural Environment*. Routledge, Taylor & Francis Group.

Walker, R.H., Girard, C.E., Alford, S.L., and Walters, A.W. 2020. Anthropogenic land use change intensifies the effect of low flows on stream fishes. *Journal of Applied Ecology* 57:149–159.

Wallace, J.B. and Benke, A.C. 1984. Quantification of wood habitat in sub-tropical coastal-plain streams. *Canadian Journal of Fisheries and Aquatic Sciences* 41:1643–1652.

Wallace, J.B., Eggert, S.L., Meyer, J.L., and Webster, J.R. 1999. Effects of resource limitation on a detrital-based ecosystem. *Ecological Monographs* 69:409–442.

Wallace, J.B. and O'Hop, J. 1985. Life on a fast pad: Waterlily Leaf Beetle impact on water lilies. *Ecology* 66:1534–1544.

Walling, D.E. and Fang, D. 2003. Recent trends in the suspended sediment loads of the world's rivers. *Global and Planetary Change* 39:111–126.

Walsh, C.J., Roy, A.H., Feminella, J.W., Cottingham, P.D., Groffman, P.M., and Morgan, R.P. 2005. The urban stream syndrome: Current knowledge and the search for a cure. *Journal of the North American Benthological Society* 24:706–723.

Walsh, J.R., Carpenter, S.R., and Vander Zanden, M.J. 2016. Invasive species triggers a massive loss of ecosystem services through a trophic cascade. *Proceedings of the National Academy of Sciences of the United States* 113(15):4081–4085.

Walters, D.M., Fritz, K.M., and Otter, R.R. 2008. The dark side of subsidies: Adult stream insects export organic contaminants to riparian predators. *Ecological Applications* 18:1835–1841.

Wang, C. and Jiang, H.-L. 2016. Chemicals used for *in situ* immobilization to reduce the internal phosphorus loading from lake sediments for eutrophication control. *Critical Reviews in Environmental Science and Technology* 46:947–997.

Wang, W. and Wang, J. 2018. Different partition of polycyclic aromatic hydrocarbon on environmental particulates in freshwater: Microplastics in comparison to natural sediment. *Ecotoxicology and Environmental Safety* 147:648–655.

Warren, M.L., Jr. and Burr, B.M. 1994. Status of freshwater fishes of the United States: Overview of an imperilled fauna. *Fisheries* 19(1):6–18.

Warrick, J.A., Bountry, J.A., East, A.E., Magirl, C.S., Randle, T.J., Gelfenbaum, G., Ritchie, A.C., Pess, G.R., Leung, V., and Duda, J.J. 2015. Large-scale dam removal on the Elwha River, Washington, USA: Source-to-sink sediment budget and synthesis. *Geomorphology* 246:729–750.

Waters, T.F. 1995. Sediment in streams. Sources, biological effects, and control. American Fisheries Society Monograph, American Fisheries Society, Bethesda, MD. 251 pp.

Watson, J.M., Coghlan, S.M., Zydlewski, J., Hayes, D.B., and Kiraly, I.A. 2018. Dam removal and fish passage improvement influence fish assemblages in the Penobscot River, Maine. *Transactions of the American Fisheries Society* 147:525–540.

Weber, A., Scherer, C., Brennholt, N., Reifferscheid, G., and Wagner, M. 2018. PET microplastics do not negatively affect the survival, development, metabolism and feeding activity of the freshwater invertebrate *Gammarus pulex*. *Environmental Pollution* 234:181e189.

Wehr, J.D., Sheath, R.G., and Kociolek, J.P. (Eds.). 2015. Freshwater Algae of North America: Ecology and Classification. 2nd Edition. Academic Press.

Weidlich, E.W.A., Flórido, F.G., Sorrini, T.B., and Brancalion, P.H.S. 2020. Controlling invasive plant species in ecological restoration: A global review. *Journal of Applied Ecology* 57:1806–1817.

Weisner, O., Frische, T., Liebmann, L., Reemtsma, T., Roß-Nickoll, M., Schäfer, R.B., Schäffer, A., Scholz-Starke, B., Vormeier, P., Knillmann, S., Liess, M. 2021. Risk from pesticide mixtures—The gap between risk assessment and reality. *Science of the Total Environment* 796:149017.

Welsh, H.H., Jr. and Ollivier, L.M. 1998. Stream amphibians as indicators of ecosystem stress: A case study from California's Redwoods. *Ecological Applications* 8:1118–1132.

Werner, E.E. and Gilliam, J.F. 1984. The ontogenetic niche and species interactions in size-structured populations. *Annual Review of Ecology and Systematics* 15:393–425.

Wetmore, S.H., Mackay, R.J., and Newbury, R.W. 1990. Characterization of the hydraulic habitat of Brachycentrus occidentalis, a filter-feeding caddisfly. *Journal of the North American Benthological Society* 9:157–169.

Wetzel, R.G. 1983. *Limnology*. Saunders, Philadelphia.

Weyhenmeyer, G.A., Hartmann, J., and Hessen, D.O. et al. 2019. Widespread diminishing anthropogenic effects on calcium in freshwaters. *Scientific Reports* 9:10450.

White, S.L., Gowan, C., Fausch, K.D., Harris, J.G., and Saunders, W.C. 2011. Response of trout populations in five Colorado streams two decades after habitat manipulation. *Canadian Journal of Fisheries and Aquatic Sciences* 68:2057–2063.

Wiggins, G.B., Mackay, R.J., and Smith, I.M. 1980. Evolutionary and ecological strategies of animals in annual temporary pools. *Archiv für Hydrobiologie* 58:97–206.

Wilbur, H.M. 1980. Complex Life Cycles. *Annual Review of Ecology and Systematics* 11:67–93. doi.org/10.1146/annurev.es.11.110180.000435.

Wild, T.C., Bernet, J.F., Westling, E.L., and Lerner, D.N. 2011. Deculverting: Reviewing the evidence on the 'daylighting' and restoration of culverted rivers. *Water and Environment Journal* 25:412–421.

Wilkinson, J.L., Boxall, A.B.A., and Kolpin, D.W. et al. 2022. Pharmaceutical pollution of the world's rivers. *Proceedings of the National Academy of Sciences* 119:e2113947119.

Williams, D.D. 2005. *The Biology of Temporary Waters*. Oxford University Press.

Windsor, F.M., Tilley, R.M., Tyler, C.R., and Ormerod, S.J. 2019. Microplastic ingestion by riverine macroinvertebrates. *Science of the Total Environment* 646:68–74.

Wipfli, M.S. and Baxter, C.V. 2010. Linking ecosystems, food webs, and fish production: Subsidies in salmonid watersheds. *Fisheries* 35:373–387.

Wipfli, M.S. and Gregovich, D.P. 2002. Export of invertebrates and detritus from fishless headwater streams in southeastern Alaska: implications for downstream salmonid production. *Freshwater Biology* 47:957–969.

Wood, R. 2016. Acute animal and human poisonings from cyanotoxin exposure—a review of the literature. *Environment International* 91:276–282.

Wood, S.L.R. and Richardson, J.S. 2009. Impact of sediment and nutrient inputs on growth and survival of tadpoles of the Western Toad. *Freshwater Biology* 54:1120–1134.

Woodward, G., Perkins, D.M., and Brown, L.E. 2010. Climate change and freshwater ecosystems: Impacts across multiple levels of organization. *Philosophical Transactions of the Royal Society B* 365: https://doi.org/10.1098/rstb.2010.0055.

Wotton, R.S., Malmqvist, B., Muotka, T., and Larsson, K. 1998. Fecal pellets from a dense aggregation of suspension-feeders in a stream: An example of ecosystem engineering. *Limnology and Oceanography* 43:719–725.

Wotton, R.S. and Preston, T.M. 2005. Surface films: Areas of water bodies that are often over-looked. *BioScience* 55:137–145.

WWF. 2022. Living Planet Report 2022—Building a nature-positive society. Almond, R.E.A., Grooten, M., Juffe Bignoli, D., and Petersen, T. (Eds.). WWF, Gland, Switzerland.

Yao, W., Rutschmann, P., Sudeep. 2015. Three high flow experiment releases from Glen Canyon Dam on rainbow trout and flannelmouth sucker habitat in Colorado River. *Ecological Engineering* 75:278–290.

Yao, Z., Zheng, X., Liu, C., Lin, S., Zuo, Q., and Butterbach-Bahl, K. 2017. Improving rice production sustainability by reducing water demand and greenhouse gas emissions with bio-degradable films. *Scientific Reports* 7:39855. doi:10.1038/srep39855.

Young, B., Allaire, B.J., Smith, S. 2021. Achieving Sea Lamprey control in Lake Champlain. *Fishes* 6: art. 2.

Young, R.A. and Loomis, J.B. 2014. *Determining the Economic Value of Water: Concepts and Methods*. Routledge, New York, USA.

Zaimes, G.N., Tufekcioglu, M., and Schultz, R.C. 2019. Riparian land-use impacts on stream bank and gully erosion in agricultural watersheds: What we have learned. *Water* 2019, 11, 1343; doi:10.3390/w11071343.

Zech, W.C., Halverson, J.L., and Clement, T.P. 2008. Intermediate-scale experiments to evaluate silt fence designs to control sediment discharge from highway construction sites. *Journal of Hydrologic Engineering* 13:497–504.

Zhang, Y., Richardson, J.S., and Pinto, X. 2009. Catchment-scale effects of forestry practices on benthic invertebrate communities in Pacific coastal streams. *Journal of Applied Ecology* 46:1292–1303.

Zou, K., Thébault, E., Lacroix, G., and Barot, S. 2016. Interactions between the green and brown food web determine ecosystem functioning. *Functional Ecology* 30:1454–1465.

Zubrod, J.P., Bundschuh, M., Arts, G., Brühl, C.A., Imfeld, G., Knäbel, A., Payraudeau, S., Rasmussen, J.J., Rohr, J., Scharmüller, A., Smalling, K., Stehle, S., Schulz, R., and Schäfer, R.B. 2019. Fungicides: An Overlooked Pesticide Class? *Environmental Science & Technology* 53:3347–3365.

Zuniga-Teran, A.A. and Tortajada, C. 2021. Water policies and their effects on water usage: The case of Tucson, Arizona. *Water Utility Journal* 28:1–17.

INDEX

Page numbers followed by *b*, *f*, and *t* refer to boxes, figures, and tables, respectively.